P9-AOZ-597
WITHDRAWN

Psychoneuroendocrinology

Proceedings of the Workshop Conference of the
International Society for Psychoneuroendocrinology, Mieken, September 3–5, 1973

Psychoneuroendocrinology

Editor: *N. Hatotani,* Tsu, Mieken

118 figures and 58 tables, 1974

S. Karger · Basel · München · Paris · London · New York · Sydney

S. Karger · Basel · München · Paris · London · New York · Sydney
Arnold-Böcklin-Strasse 25, CH–4011 Basel (Switzerland)

Printed in Switzerland by Thür AG Offsetdruck, Pratteln
ISBN 3–8055–1711–4

Contents

Regulation of Hypothalamo-Pituitary Function

Foreword

This volume is based on the Workshop of the International Society for Psychoneuroendocrinology held in September 1973 under the auspice of the Department of Psychiatry, Mie University, Japan.

In this workshop, we had three sessions:

(1) Clinical studies of psychoneuroendocrinology.

(2) Stress and biological rhythm.

(3) Regulation of hypothalamo-pituitary function.

These three topics are not separated, but closely related to each other. From the anatomical and physiological basis of neuroendocrine mechanisms, it is obvious that endocrine functions exert a great influence on mental functions and vice versa through changes in basic psychic activities such as instinctive drives, mood and emotion, spontaneity and vigilance of consciousness.

Clinically, it is common that endocrine diseases are often accompanied by various types of mental disorders and, conversely, mental diseases concomitantly show varied endocrine deviations. In spite of such close interdependence between mental and endocrine functions, their relationship is generally nonspecific in nature. There is no hormone specificity in mental symptoms, and the exact mode of hormone action on brain functions is not clear. On the other hand, most endocrine disorders concomitant with psychoses are also considered to be kinds of nonspecific bodily reactions. However, there are a number of cases in which the impairment of neuroendocrine mechanisms seems to be deeply concerned with the pathogenesis of psychoses, and thus endocrine findings can be a useful indicator of therapeutic measures. For instance, in periodic types of affective psychoses it is most likely that disorder of hypothalamic function is responsible for the periodic recurrence of psychoses, the bipolar manifestation of symptoms, and the impairment of emotion and consciousness. Especially in psychoses which occur in relation to the female gonadal function such as menstrual cycle, gestation, puerperium, and menopause, the neuroendocrine mechanism involved in the hypothalamo-pituitary-gonadal system possibly plays a significant role in the onset of psychoses. Moreover, the abnormal circadian rhythm of hormone secretion and impaired hypothalamo-pituitary response to various loading tests are also reported in mental patients.

From these clinical observations, it is suggested that chronobiological approaches are indispensable for psychoneuroendocrinological research. From the chronobiological point of view, derangement of the temporal structure in organism is essential for mental disorders. In the course of the development of mental disorders, desynchronization of internal biorhythms with each other and with environmental rhythms makes its appearance. Among the desynchronization symptoms impairment of the wake-sleep cycle, diurnal changes of mood, and alteration of neuroendocrine rhythms are most common. Such alterations of biological rhythms are caused by either unusual daily schedules or mental distress; in other words, they result from physical or psychological stress. Thus, the relationship between 'stress and biological rhythm' is one of the most significant problems in psychoneuroendocrinology.

The hypothalamo-pituitary system, which is closely connected to the reticular system and the limbic lobe, plays a central part in neuroendocrine mechanisms coordinating autonomic nervous activities and endocrine functions, and thereby functions as a pacemaker of biological rhythms. Although numerous studies elucidated the neurosecretory control of the posterior lobe and the functional properties of hypothalamic releasing and inhibiting factors, details about the mode of action of hypothalamic control of the hypophysis require further elucidation. The function of the hypothalamo-pituitary system as such, is not only regulated by peripheral endocrine glands via feedback mechanisms, but is also controlled by other parts of the brain which are closely related to the hypothalamus. Therefore, to make clear the regulatory mechanisms of the hypothalamo-pituitary function is crucial, not only for understanding neuroendocrine mechanisms, but also for investigating the pathogenesis of psychosomatic and mental disorders. About two-thirds of the papers in this Workshop deal with this issue from various aspects. For instance, they are concerned with hypothalamic-releasing factors, biogenic amines, cyclic-AMP, and hormonal influences on the developmental and reproductive process, etc.

I believe that this volume has a strong commitment to an integrative approach in the multidisciplinary field of research, which extends from the molecular to the behaviour level of organization, and will promote further steps in this particular field.

Noboru Hatotani
Tsu, Japan

Clinical Studies in Psychoneuroendocrinology

Psychoneuroendocrinology. Workshop Conf. Int. Soc. Psychoneuroendocrinology, Mieken 1973, pp. 4–12 (Karger, Basel 1974)

Circadian Rhythm of Plasma Cortisol in Endogenous Psychoses

N. Suwa, I. Yamashita, T. Moroji, K. Yamazaki, F. Okada, Y. Saito, Y. Asano and T. Fujieda

Department of Psychiatry and Neurology, Hokkaido University School of Medicine, Sapporo

There are two reasons for our serious lack of knowledge about the biological basis of endogenous psychoses; schizophrenia and manic-depressive illness. First: no animal, other than the human being, suffers from them. Accordingly, animal experiments so extensively employed in medical research, are of very limited use for the study of endogenous mental illnesses. Second: postmortem examination is of almost no avail in this field of research. Only a live mental patient can be the object of investigation. In order to develop biological research under this difficult condition, it seems essential to exploit a physiological measure which can trace what is occurring in the brain of a living individual.

Electroencephalography (EEG) and pneumoencephalography (PEG) are undoubtedly such measures. However, EEG failed to obtain such a success in endogenous psychoses in the search for cerebral dysfunction as it did splendidly in epilepsy. Some abnormalities of PEG have been reported in schizophrenic subjects (6), but they are not consistent with the established observation that no gross morphological changes exist in the brains of schizophrenics.

We have been engaged for some 20 years in a series of studies on endocrine and autonomic nervous functions of mental patients with endogenous, psychogenic, and organic disorders (25–28). As a result, we found four characteristic features of bodily functions, particularly in cases of deteriorated schizophrenia: a wider distribution of estimated values, greater day-to-day fluctuations, deranged circadian (daily) rhythm, and reduced responsiveness to stress (27).

This paper will concentrate on the circadian pattern of plasma levels of adrenal cortical hormone, because the determination of this rhythm seems to provide us with a clue to evaluate the functional state of the central nervous system (CNS) in a live subject. There is much evidence that patients with localized hypothalamic, temporal lobe or midbrain lesions quite often demonstrate abnormal circadian patterns of plasma cortisol (2, 5, 7, 8, 17, 23, 24). Obviously, the circadian variation of plasma cortisol is dependent on the same daily fluctuation of ACTH and CRF levels, which is under the control of the modulating function of the hypothalamus and other CNS areas. For this reason, it is of interest to investigate circadian change of plasma cortisol in patients with endogenous psychoses, in comparison with those subjects with definite brain lesions.

Methods

Hospitalized patients with various mental disorders, totalling 128, in the psychiatric clinic of the Hokkaido University Hospital, and other institutions affiliated to the University, were chosen for investigation. Blood samples were drawn at 8 a.m., 2 p.m. and 8 p.m. of the first day, and at 2 a.m. and 8 a.m. of the next. Plasma 17-hydroxycorticosteroid (17-OHCS) was determined photometrically in subjects with schizophrenia or general paresis (20), and plasma cortisol was estimated fluoro-photometrically in those with depression (13) or epilepsy (22). In most cases, determination of diurnal rhythm was repeated about monthly, in accordance with variations in the patients' clinical states.

Results

(a) Individual circadian patterns of plasma 17-OHCS in 70 *schizophrenic subjects,* classified into four groups, are shown in part in figures 1 and 2. The first group of patients (fig. 1), who were in complete remission, showed normal and orderly circadian patterns of plasma 17-OHCS, with the highest level early in the morning and the lowest in the late evening. The second group, consisting of subjects with acute symptoms: anxiety, agitation, hallucinatory experiences, etc., exhibited various deviations from the normal form of circadian pattern. The third group of patients, who passed the acute stage and were relatively free from active symptoms, indicated less frequently such a disturbed pattern of circadian rhythm. The last group was composed of cases of deteriorated schizophrenia, whose symptoms were characterized by emotional blunting, disordered thinking and personality disorganization. They demonstrated a total loss of ordinary circadian rhythmicity of plasma 17-OHCS, as shown in figure 2.

These findings are consistent with the results of the longitudinal, follow-up studies of the same subjects. Patients who had completely recovered from their illness constantly showed a normal pattern on every occasion of successive monthly examinations (fig. 3). Acutely disturbed patients usually indicated deranged patterns of circadian cortisol rhythm at the height of psychic aggravation, and later, a normal daily rhythm as their symptoms subsided. The correlation between emotional distress and disturbed circadian pattern of plasma 17-OHCS can also be observed in the laboratory findings of a subject who was in the intermediate stage of illness. He had a fairly stable course of disease, but occasionally complained of an anxiety state accompanied by various hypo-

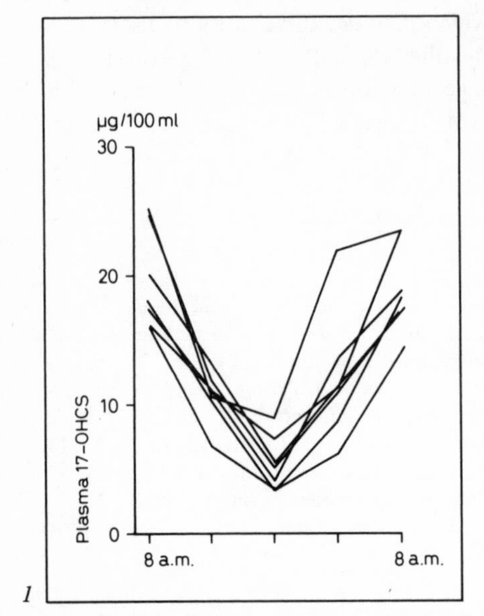

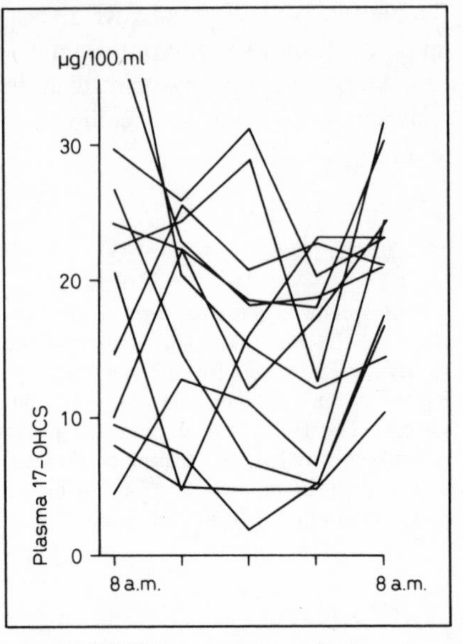

Fig. 1. Individual circadian variations of plasma 17-OHCS in complete remission.

Fig. 2. Individual circadian variations of plasma 17-OHCS in the deteriorated stage of schizophrenia.

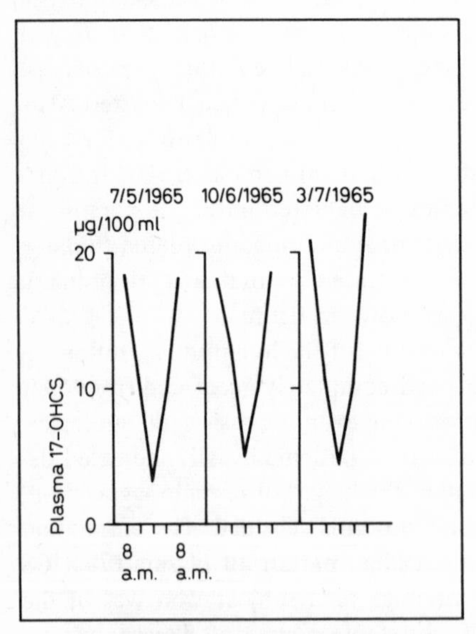

Fig. 3. Serial determinations of circadian pattern of plasma 17-OHCS in a schizophrenic subject in complete remission.

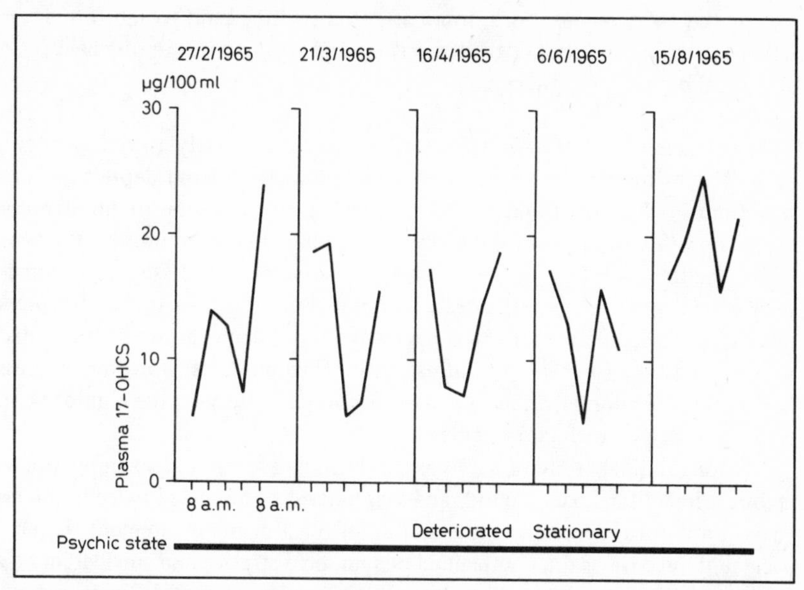

Fig. 4. Serial determinations of circadian pattern of plasma 17-OHCS in a schizophrenic subject in the deteriorated stage of illness.

chondriac discomforts. Circadian variation in plasma 17-OHCS levels was markedly deranged when his psychic state was aggravated, whereas it showed a normal pattern during the periods of time prior to, and after this worsened condition.

On the other hand, circadian pattern of plasma 17-OHCS was continuously irregular and disturbed in cases of deteriorated schizophrenia. Figure 4 illustrates the laboratory results of a representative subject. The case is a 23-year-old male patient who became exceedingly abulic at the age of 13, and, at the time of investigation, was usually dressed in dirty clothes and recumbent all day with his face toward the wall. It is obvious that neither emotional upset nor sleep disorder is responsible for the continuous distortions of circadian 17-OHCS rhythm in the deteriorated cases. They showed no conspicuous changes in their psychic states, and slept regularly according to hospital regulations, during the entire period of laboratory examinations.

To account for this data, two factors should be taken into consideration. First, emotional impact on plasma 17-OHCS levels, and second, any disease processes involving the CNS areas regulating the circadian variation of ACTH secretion. The former seems more substantial for acute cases of schizophrenia, and the latter for deteriorated ones.

In the following sections, more attention will be paid to the former, emotional factor, in our studies of depression; and to the latter, in the investigation of general paresis and epilepsy.

(b) Fifteen cases of *depression* were studied transversely as well as longitudinally. It was disclosed that patients who had suffered from depression for the first time in their lives and easily attained a good remission by antidepressant drugs showed a normal circadian rhythm of plasma cortisol even at the time of hospitalization, while those who presented severely depressive symptoms, or such complicated courses of disease as rapid alteration of manic and depressive phases, or exceedingly protracted depressive state, demonstrated various distortions of circadian cycle. In contrast with schizophrenics, however, depressive patients unexceptionally came to show a normal pattern as the clinical pictures were settled to a satisfactory degree.

A longitudinal study of a 33-year-old man will serve as an example. He has a manifest hereditary background, and was himself hospitalized twice in the last 2 years, once for a depressive state and again for a manic excitement. Lately, he became depressive again, complained of somatic distresses and suicidal ideas, and fell into a stupor-like state. By treatment with antidepressant drugs, he recovered gradually and reached a state of complete remission within a few months. Circadian patterns of plasma cortisol demonstrated a marked disturbance during the early period of hospitalization, then a moderately abnormal variation, and a normal rhythm in the stage of recovery.

These findings seem to indicate that emotional derangements, to a certain degree and of a certain duration, may accompany changes of circadian rhythm of plasma cortisol, as pointed out before in cases of acute schizophrenia.

(c) Twelve of 13 subjects with *general paresis* showed irregular, abnormal patterns of circadian variations. Having been hospitalized for many years, the patients were generally demented to a marked degree, and emotionally shallow and vacant.

It seems reasonable to assume that these changes are not related to emotional turmoil, but due to the extensive involvement of the CNS with the organic process of general paresis itself.

(d) More than half of 29 hospitalized patients with *epilepsy* showed deranged patterns of circadian rhythm. It is noteworthy in this connection that psychic symptoms and organic brain lesions were more frequently and distinctly observed in a group of patients with abnormal circadian patterns than in patients with normal ones. Actually, those epileptic patients who have only slight intellectual regression and/or character deviation, usually indicated a normal circadian cycle throughout the periods of observation. On the other hand, those who have definite brain damage often exhibited disturbed circadian patterns of plasma cortisol.

Two cases will be presented as examples. A 50-year-old man was hit heavily in the right temporal region about 10 years ago. Grand mal and psychomotor seizures soon ensued, and about 5 years later, an episodic attack of auditory hallucinations began to occur. He used to hear the voice of God telling him stories of a religious nature. This condition lasted for a few days and recurred at irregular intervals. His EEG showed a focal discharge at the right temporal region. Circadian variations of plasma cortisol, estimated during, as well as between the episodes of hallucinatory condition, were continuously deranged. Another patient contracted encephalitis at the age of eight. Afterwards, grand mal seizures occurred quite often. When laboratory examinations were performed, he was slow, apathetic, and indifferent to his surroundings. His EEG indicated conspicuous abnormalities, and PEG disclosed a marked dilatation of the whole ventricular system. A complete loss of normal rhythmicity of circadian cortisol pattern was observed in this patient.

Discussion

The circadian variation in adrenal cortical activity in man was first described by *Pincus* (21). In subjects with a normal evening sleep routine, the cyclic plasma cortisol pattern consists of an early morning peak, followed by a downward trend during the day, and a sharp increase at night. This circadian rhythm is stable in a variety of bodily diseases (19). A phase reversal takes place only after a period of approximately 1 week of the reversed day-night living schedule (18).

In regard to the physiological mechanism for the formation of circadian rhythm of plasma cortisol, there were three hypotheses (19): first, different rates of removal of cortisol from plasma at different times of the day; second, variation in the adrenal cortical responsiveness to the same amount of ACTH in plasma; and third, changes in the rate of ACTH secretion regulated by a centrally located mechanism. Recent investigations provided much evidence to preclude the former two suppositions and favor the third explanation. *Ney et al.* (16) and *Cheifetz et al.* (1) reported a circadian variation of ACTH in plasma level as well as pituitary content in the rat. According to *Hiroshige et al.* (4), there is a distinct circadian periodicity in the hypothalamic content of corticotropin releasing factor (CRF) which precedes that of plasma corticosterone, and ACTH, by a constant time interval. *Slusher* (23, 24) also indicated that anterior hypothalamic lesions, or chronic cortisol implant in the median eminence, diminish circadian variation of plasma corticosterone in rats.

In the clinical field, too, reports are accumulating to support the view that circadian cycle of adrenal cortical activity is under the control of the CNS, particularly of the hypothalamus, limbic system and related areas. *Hökfelt and Luft* (5) found in 1959 a loss of diurnal pattern of plasma cortisol in 4 of 8 subjects with suprasellar brain tumors. *Krieger* (7) observed four patients with either focal temporal lobe of pretectal area disease, who showed distinctly abnormal patterns of plasma 17-OHCS. In 1966, *Krieger and Krieger* (8) examined 27 patients with hypothalamic, temporal lobe or pretectal disease, 13 with pituitary disease with minimal evidence of endocrinopathy, and 21 with focal CNS disease outside these areas. It was confirmed that a statistically significant difference from normal controls in respect to circadian pattern was observed in the first group of patients, and to a lesser degree in the second, while the third group showed a substantially normal rhythm of plasma 17-OHCS.

Considering these reports and our own findings, there is good reason to assume, in the authors' opinions, that deteriorated cases of schizophrenia with continuously deranged circadian patterns have certain dysfunctions of the hypothalamus, limbic system and related areas, which are involved in the circadian regulation of CRF and ACTH secretions.

Noteworthy in this respect are recent reports by *Krieger and Krieger* (9); *Krieger et al.* (10); *Krieger and Rizzo* (11); *Ganong* (3), and other investigators, who postulated that synaptically active agents; acetylcholine, serotonin and norepinephrine in the brain, can mediate circadian variations of plasma cortisol by modulating an outpouring of CRF from hypothalamic secretory neurons. The results of our own studies on this subject (15) are rather ambiguous. Postnatal development of circadian rhythms of brain acetylcholine, serotonin and norepinephrine did not precede that of plasma corticosterone in rats. Circadian pattern of hypothalamic norepinephrine content was abolished by the administration of methamphetamine, whereas that of plasma corticosterone was not affected by the same procedure. It remains to be seen what may cause these peculiar dysfunctions.

In summary, our findings on circadian rhythm of plasma cortisol in endogenous psychoses: schizophrenia and manic-depressive illness, provided two presumptions. First, certain disturbances of a neurochemical nature may exist in the CNS in deteriorated schizophrenia, as no gross morphological changes are negated in the brains of schizophrenics. Second, deranged circadian patterns in acute schizophrenia and depression can be explained, for the most part, as concomitant changes of emotional distresses, which have been extensively studied by ourselves (25, 27) and recently reviewed by *Mason* (12).

It is hoped that circadian rhythm of plasma cortisol, which is dependent on the functional state of certain CNS areas, can be an indicator of the CNS function itself, and may serve as a useful means in the biological investigations of endogenous psychoses.

Summary

Circadian (daily) rhythm of plasma cortisol levels was determined in 128 patients with schizophrenia, manic-depressive psychosis, general paresis, or epilepsy. In most cases, the test was repeated about monthly for several months.

Circadian pattern constantly takes a normal form in subjects who have completely recovered from schizophrenic illness, patients with a mildly depressive state, and epileptics having only slight impairment of intelligence and/or character.

This rhythm was deranged to a considerable degree, when such active symptoms as anxiety or excitement were manifest in schizophrenic patients, or when emotional distress was extremely severe in depressed subjects. These changes returned to a normal rhythm as clinical symptoms became satisfactorily settled.

There is a total loss of this circadian rhythmicity in cases of deteriorated schizophrenia, general paresis, and those epileptic patients who have definite brain lesions and conspicuous disturbances of intelligence and personality.

It can be concluded that circadian variation of plasma cortisol is disturbed on two occasions: first, in emotional turmoil, and second, by disease processes involving certain areas of the central nervous system. In regard to endogenous psychoses, schizophrenia seems to have the former mechanism in the acute stage and also the latter process, at least in the chronic, deteriorated stage, while depression shows only the former emotion-related variations. The physiological correlation between endocrine activity and synaptic transmitters was still ambiguous.

References

1 *Cheifetz, P.; Gaffud, N., and Dingman, J.F.:* Effects of bilateral adrenalectomy and continuous light on the circadian rhythm of corticotropin in female rats. Endocrinology, Springfield *82:* 1117–1124 (1968).

2 *Eik-Nes, K. and Clark, L.D.:* Diurnal variation of plasma 17-hydroxycorticosteroids in subjects suffering from severe brain damage. J. clin. Endocrin. *18:* 764–768 (1958).

3 *Ganong, W.F.:* Pharmacological aspects of neuroendocrine integration. Progr. Brain Res. *38:* 41–57 (Elsevier, Amsterdam 1972).

4 *Hiroshige, T.; Sakakura, M., and Itoh, S.:* Diurnal variation of corticotropin releasing activity in the rat hypothalamus. Endocrin. jap. *16:* 465–467 (1969).

5 *Hökfelt, B. and Luft, R.:* The effects of suprasellar tumors on the regulation of adrenocortical function. Acta endocrin., Kbh. *32:* 177–186 (1959).

6 *Huber, G.:* Neuroradiologie und Psychiatrie. Psychiatrie der Gegenwart, vol. 1/1B, pp. 253–290 (Springer, Heidelberg 1964).

7 *Krieger, D.T.:* Diurnal pattern of plasma 17-hydroxycorticosteroids in pretectal and temporal lobe disease. J. clin. Endocrin. *21:* 695–698 (1961).

8 *Krieger, D.T. and Krieger, H.P.:* Circadian variation of the plasma 17-hydroxycorticosteroids in the central nervous system disease. J. clin. Endocrin. *26:* 929–940 (1966).

9 *Krieger, H.P. and Krieger, D.T.:* Effect of central nervous system injection of synaptically active agents on ACTH release. Fed. Proc. *27:* 217 (1968).

10 *Krieger, D.T.; Silverberg, A.T.; Rizzo, F., and Krieger, H.P.:* Abolition of circadian periodicity of plasma 17-OHCS levels in the cat. Amer. J. Physiol. *215:* 959–967 (1968).

11 *Krieger, D.T. and Rizzo, F.:* Serotonin mediation of circadian periodicity of plasma 17-hydroxycorticosteroids. Amer. J. Physiol. *217:* 1703–1707 (1969).

12 *Mason, J.W.:* A review of psychoendocrine research on the pituitary-adrenal cortical system. Psychosom. Med. *30:* 576–607 (1968).
13 *Mattingly, D.:* A simple fluorimetric method for the estimation of free 11-hydroxysteroids in human plasma. J. clin. Path. *15:* 374–379 (1962).
14 *Migeon, C.J.; Tyler, F.H.; Mahoney, J.P.; Florentin, A.A.; Castle, H.; Bliss, E.L., and Samuels, L.T.:* The diurnal variation of plasma levels and urinary excretion of 17-hydroxycorticosteroids in normal subjects, night workers and blind subjects. J. clin. Endocrin. *16:* 622–635 (1956).
15 *Moroji, T.; Yamazaki, K.; Okada, F.; Saito, Y.; Asano, A., and Yamashita, I.:* Circadian rhythm of glucocorticoid secretion; in *Yagi and Yoshida* Neuroendocrine control, pp. 57–81 (University of Tokyo Press, Tokyo 1973).
16 *Ney, R.L.; Shimizu, N.; Nicholson, W.E.; Island, D.P., and Liddle, G.W.:* Correlation of plasma ACTH concentration with adrenocortical response in normal human subjects, surgical patients, and patients with Cushing's disease. J. clin. Invest. *42:* 1669–1677 (1963).
17 *Oppenheimer, J.H.; Fisher, L.V., and Jailer, J.W.:* Disturbances of the pituitary-adrenal interrelationship in diseases of the central nervous system. J. clin. Endocrin. *21:* 1023–1036 (1961).
18 *Orth, D.N.; Island, D.P., and Liddle, G.W.:* Experimental alteration of the circadian rhythm in plasma cortisol concentration in man. J. clin. Endocrin. *27:* 549–555 (1967).
19 *Perkoff, G.T.; Eik-Nes, K.; Nugent, C.A.; Fred, H.L.; Nimer, R.A.; Rush, L.; Samuel, L.T., and Tyler, F.H.:* Studies of the diurnal variation of plasma 17-hydroxycorticosteroids in man. J. clin. Endocrin. *19:* 432–443 (1959).
20 *Peterson, R.E.; Kaner, A., and Guena, S.L.:* Evaluation of Silber-Porter procedure for determination of plasma hydrocortisone. Anal. Chem. *29:* 144–149 (1957).
21 *Pincus, G.:* A diurnal rhythm in the excretion of urinary ketosteroids by young men. J. clin. Endocrin. *3:* 195–199 (1943).
22 *Rudd, B.T.; Sampson, P., and Brooke, B.N.:* A new fluorimetric method of plasma cortisol assay with a study of pituitary-adrenal function using methyrapone (SU. 4885). J. Endocrin. *27:* 317–325 (1963).
23 *Slusher, M.A.:* Effects of chronic hypothalamic lesions on diurnal and stress corticosteroid levels. Amer. J. Physiol. *206:* 1161–1164 (1964).
24 *Slusher, M.A.:* Effects of cortisol implants in the brainstem and ventral hippocampus on diurnal corticosteroid levels. Exp. Brain Res. *1:* 184–194 (1966).
25 *Suwa, N.; Yamashita, I.; Owada, H.; Shinohara, S., and Nakazawa, A.:* Psychic state and adrenocortical function. A psychophysiologic study of emotion. J. nerv. ment. Dis. *134:* 268–276 (1962).
26 *Suwa, N.; Yamashita, I.; Ito, K.; Yoshimura, Y., and Moroji, T.:* Psychic state and gonadal function. A psychophysiologic study of emotion. J. nerv. ment. Dis. *143:* 36–46 (1966).
27 *Suwa, N. and Yamashita, I.:* Psychophysiological studies of emotion and mental disorders (Hokkaido Univ. Med. Libr. Series No. 5, Sapporo 1972).
28 *Yamashita, I.; Moroji, T.; Yamazaki, K.; Kato, H.; Sakashita, A.; Onodera, I.; Ito, K.; Okada, F.; Saito, Y.; Tamakoshi, M., and Suwa, N.:* Neuroendocrinological studies in mental disorders and psychotropic drugs. Part I. On the circadian rhythm of plasma adrenocortical hormone in mental patients and methamphetamine- and chlorpromazine-treated animals.

Author's address: Dr. *N. Suwa,* Department of Psychiatry and Neurology, Hokkaido University School of Medicine, *Sapporo* (Japan)

Psychoneuroendocrinology. Workshop Conf. Int. Soc. Psychoneuroendocrinology,
Mieken 1973, pp. 13–21 (Karger, Basel 1974)

Growth Hormone Secretion in Schizophrenia

Francesca Brambilla, A. Guastalla, A. Guerrini, F. Riggi and M. Recchia
Ospedale Psichiatrico Provinciale Paolo Pini, Ospedale Psichiatrico Provinciale Antonini, and Istituto Farmacologico Mario Negri, Milan

The existence of a strict relationship between endocrine dysfunctions and schizophrenia has been supported by recent data, in regard to cognitive, affective and behavioural disorders and hypothalamo-pituitary-adrenal-gonadal functions (*Eiduson,* 1961; *Sachar,* 1963, 1970; *Persky,* 1968; *Wohlberg et al.,* 1970).

The center of interference between the psychosis and the endocrine disorders seems to be located at the level of the hypothalamus or the suprahypothalamic areas, which react to the psychological imbalance with inhibition or stimulation of the peripheral glands. Our personal experience tends to confirm this hypothesis. However, we observed that not all the schizophrenics present the same modality of disease development, the same symptomatological phases, and the same reactiveness to psychophysical stimuli. Two main categories of patients can be selected, displaying evident differences in regard to cognitive, behavioural and biochemical features.

The first category includes paranoid patients, with onset of the disease in adult age, and a mildly deteriorating processuality with active *poussées.* The endocrine function is generally normal, with a tendency towards ACTH and glucocorticoids hypersecretion in relation to arousal of affects.

The second group includes the hebephrenics and hebephreno-catatonics, with onset of the disease at pubertal or immediate postpubertal age, and a progressively and deeply deteriorating processuality of the illness.

The endocrine features show hyporeactivity of the hypothalamo-pituitary-adrenal axis to stimuli and marked hyposecretion of the pituitary-gonadal system.

The reduced hormonal function seems to correlate with the psychopathological symptomatology of the patients, mainly represented by withdrawal from the environment, maladjustment to reality, passive behaviour, reduced or absent responsiveness to external-internal stimuli. In other words, the patients seem to be psychologically and biochemically unreactive.

Obviously there are subjects who represent an intermediate category and show psychopathological and biochemical features which overlap in the two groups previously described. The psychoendocrine interrelation is confirmed by the finding that specific pharmacological treatments induce a psychological improvement together with a revival of the endocrine secretions. The hebephrenic-hebephrenocatatonics are the patients more progressively deteriorating, both physically and psychologically, and less responsive to therapies, being the most severely ill in terms of disease prognosis.

Therefore, we considered it worthwhile to continue our experiments in this group, examining other hypothalamo-pituitary parameters, in order to control our hypothesis of a functional failure of the diencephalic centers, interfering with the characters and processuality of the disease.

We decided to study the GH secretion in response to insulin stimulation.

In this regard the literature offers very few data.

It is well known that states of emotional arousal can stimulate outbursts of GH secretion, both in animals and humans (*Reichlin,* 1968; *Mason et al.*, 1968; *Greene et al.*, 1969).

In other cases, psychological impairments may inhibit the GH production, as it has been noted in children suffering from the maternal deprivation syndrome (*Powell et al.*, 1967).

Sachar et al. (1971) reported a reduced GH secretion after insulin in middle-aged depressed patients, the phenomenon being not related to age or malnutrition.

On the contrary, according to *Suematsu* (1973), depressed patients present normal GH secretion, with a small percent of subjects showing high hormonal levels.

Schimmelbush et al. (1971) observed low basal values of GH in schizophrenics with long-term hospitalization. The insulin stimulation resulted in normal release of the hormone. Few long hospitalized subjects displayed unresponsiveness to the stimulus, regardless of the kind of therapy previously administered. According to the authors, the relative GH deficiency in these cases is due to the insulin resistance typical of the subjects.

The low basal values of GH could be interpreted as a consequence of the emotional deprivation induced by the long lasting hospital stage. *Takahashi et al.* (1967) found a normal GH response to insulin in schizophrenics, independently from previous pharmacological therapies. Insulin coma treatment induced a good GH response, with a tendency to a delay in reaching the peak value in the second week of therapy. This phenomenon seemed to correlate with a poor response to the treatment. When protracted, the insulin coma therapy tended to induce a decrease in magnitude of the GH peak values, suggesting the presence of a fatigued state of the central mechanism responsible for the hormonal secretion.

Electroshock therapy resulted in elevation of the GH values, with peak levels lower than the ones induced by insulin, and no relation with the glycemic values, suggesting, therefore, a direct stimulation of the hypothalamic areas responsible for the GH-RH secretion. We decided to examine a selected group of chronic hebephrenic subjects, deeply deteriorated, anergic, resistant to different kinds of therapeutic approaches, in order to control if the endocrine dysfunctions previously observed in these patients, in regard to hypothalamo-pituitary-adrenal and gonadal secretions, included other hormonal parameters.

We controlled also the effects of a haloperidol treatment. This drug is antagonistic to noradrenaline and dopamine, and its properties seem to originate from the ability to decrease the permeability of the adrenergic receptors for catecholamines, thus blocking uptake and increasing enzymatic inactivation by *o*-methyl-transferase activity (*Jansen,* 1964). As noradrenaline and dopamine seem to stimulate the release of hypothalamic GH-RH (*Muller,* 1973), the haloperidol, blocking the catecholamines, should decrease the GH secretion.

Data of *Kim et al.* (1971), obtained in mentally normal subjects, confirm this hypothesis.

Therefore, we thought it worthwhile to control the effect of such a treatment in schizophrenics.

Moreover, we tried a short-term therapeutic approach with haloperidol associated to chorionic gonadotrophin. In a previous research, this combined treatment seemed to stimulate the hypothalamo-pituitary-adrenal and gonadal axes, and in the meantime to improve the affective-behavioural features of the hebephrenic patients.

Cases

Data have been obtained in 15 chronic hebephrenic schizophrenics, 9 males and 6 females, aged 16–51 years.

The onset of the disease varied between 3 and 28 years before the beginning of our experiments. Patients with previous histories of cerebropathies, cerebral trauma, organic diseases and overt endocrinopathies were not included in this investigation.

The subjects were hospitalized in our Institute during the experiments. They were given the same common hospital diet, and subjected to the same hygienic rules.

It must be mentioned that they had been previously examined under the profile of the hypothalamo-pituitary adrenal reactivity to stress and hypothalamo-pituitary-gonadal function. All of them displayed the common features of a marked deficiency of both systems, therefore suggesting that the hypothalamic centers were in a state of deeply reduced activity. Such a phenomenon was considered as functional and reversible, as demonstrated by the fact that therapies with haloperidol, and haloperidol + HCG, induced a marked improvement of the biochemical data, and in parallel, of the psychopathological features.

None of the patients had had electroconvulsive and insulin coma therapy during the previous 2 years. They were off psychotropic drugs at least for 10 days before the beginning of our experiment.

Methods

The GH secretion was examined using the insulin stimulation test. All the patients fasted and rested in bed at least 12 h before the experiment and up to the end of it.

Crystalline insulin, 0.1 U/kg of body weight, was used as stimulus. Blood samples were drawn before the intravenous injection of insulin and thereafter at 30, 60, 90 and 120 min, using a cannula inserted in a forearm vein and kept patent by a saline solution infusion. Plasma GH was analyzed by a radioimmunoassay procedure, using the double antibody technique of *Shalch and Parker* (1964). Blood glucose was examined by a glucose oxidase method at the same intervals.

The patients were submitted to the insulin stimulation test twice at the beginning of the experiments before starting the therapy, with an interval of 24 h between the two basal exams.

They were then on haloperidol therapy for 30 days, 6 mg i.m.a.d. to a total dose of 180 mg. During this period the GH secretion was checked at 10, 20 and 30 days of therapy.

The male patients only were then subjected to a combined therapy with haloperidol at the same dose, plus chorionic gonadotrophin 5,000 IU i.m. twice a week, to a total dose of 40,000 IU, for the next 30 days. The patients were checked for GH secretion at 15 and 30 days of therapy.

We did not use the combined therapy in females because of the danger of ovarian cysts developing.

The psychological examination was done daily by two psychiatrists and by the ward staff. Moreover, the patients were rated with a Wittenborn scale, checked at the same intervals as the hormonal controls.

Results

The endocrine data are reported in figures 1–3 inclusive. The statistical evaluation of our data was done according to the method of giving, for every mean value, the confidence level of 5 % ($p = 0.05$). The diagrams report the percent of increment after insulin, compared to the basal preinsulin levels.

It must be mentioned first that all the patients presented a normal glycemic response to insulin, with curves which paralleled the ones of healthy subjects. In this regard, our data differ from those presented in the literature, which seem to support the presence of an impaired glycemic response to insulin in schizophrenia.

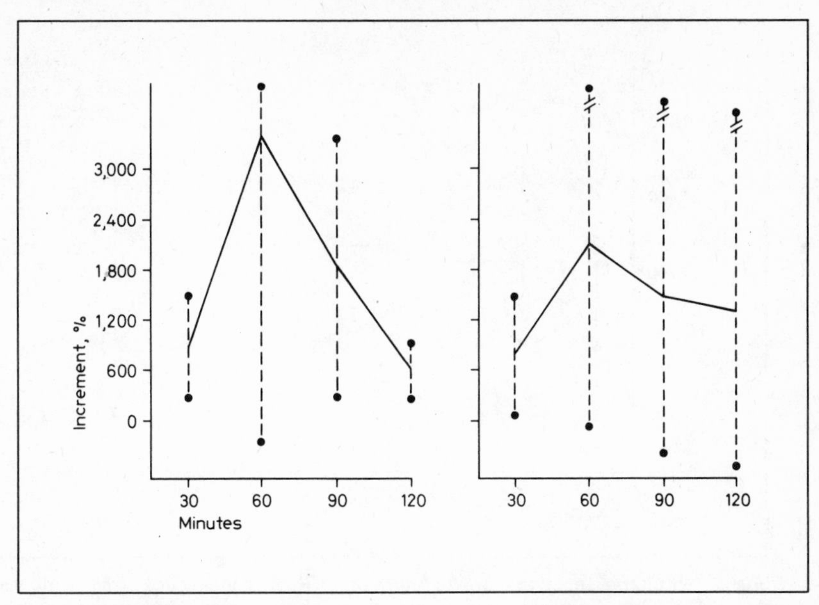

Fig. 1. Pituitary function growth hormone. First and second basal trial.

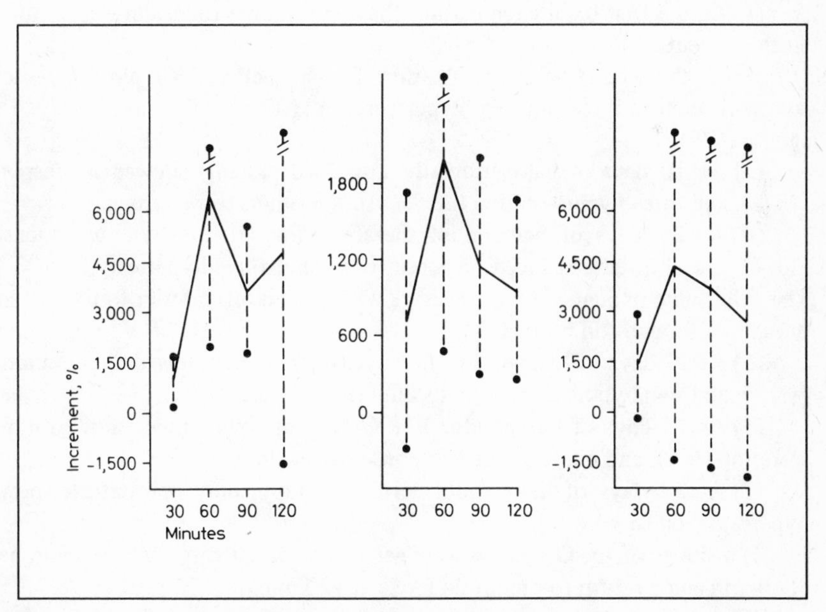

Fig. 2. Pituitary function growth hormone. First, second and third trial with haloperidol.

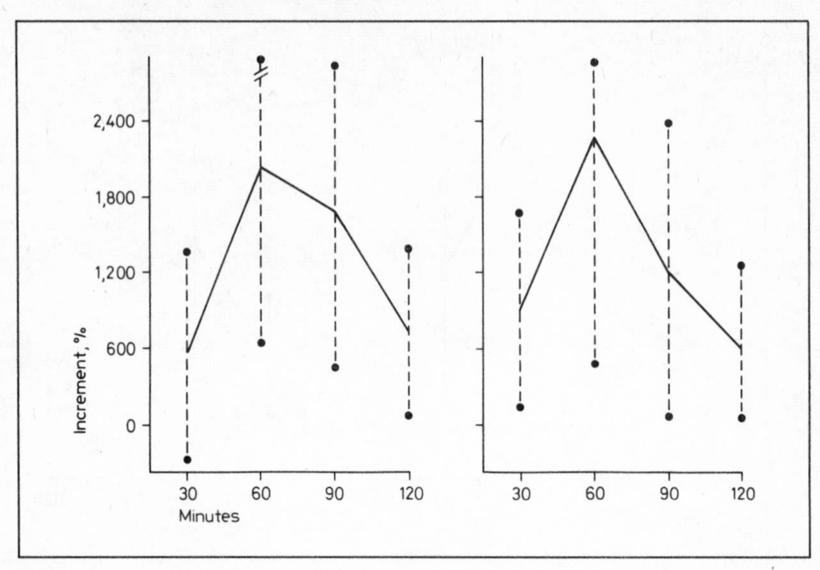

Fig. 3. Pituitary function growth hormone. First and second trial with haloperidol + HCG.

(1) At the first basal examination the GH response to insulin was normal in all the patients.

(2) At the second basal examination, five subjects (33 %), did not respond to the stimulus, three showed hypersecretion (20 %), and one a delayed response (6 %).

(3) At 10 days of haloperidol therapy, two patients showed no response (13 %), and three hypersecretion (20 %). All the others were normal.

(4) At 20 days of haloperidol therapy, two patients were unresponsive (13 %), one showed a reduced response (6 %), and three hypersecretion (20 %). The other nine patients displayed curves with peak levels inferior to the previous ones, even though still normal.

(5) At 30 days of haloperidol therapy, two patients showed hypersecretion (13 %), and two hyporeactivity to insulin (13 %).

(6) At 15 days of haloperidol + HCG therapy, one patient showed hypersecretion (6 %), and one a blunted response to insulin (6 %).

(7) At 30 days of haloperidol + HCG therapy, only one patient showed hypersecretion (6 %).

The slopes of the GH curves were generally normal with the maximum peak at 60 min and a return to preinsulin levels at 120 min.

The psychological examination before the beginning of the treatment revealed: (a) severe thought disorders, mainly represented by coaptive thought, hallucinations, delusions, dissociation and mental deterioration; (b) affectivity disturbances, represented by psychomotor excitement, hostility, anxiety and depression, and (c) behavioural disorders, represented by withdrawal from reality, negativism and apathy.

During the haloperidol treatment we observed a modest improvement of all the symptoms.

During the haloperidol + HCG therapy, the improvement was more evident and statistically significant. A mild increase of psychomotor excitement observed at 15 days of haloperidol + HCG therapy, was totally reduced at the end of the treatment.

Conclusions

Our data seem to offer some preliminary observations. It is evident that the GH response to insulin is normal at the first examination, before the beginning of the therapy.

However, the second insulin stimulation, done at 24-hour intervals, evidenced a blunted response in one-third of the subjects. This phenomenon would suggest an exhaustion in the secretory mechanism either at hypothalamic or pituitary level, and seems to confirm the data of *Takahashi* (1967). The presence in two patients of elevated peaks of secretion excludes that the impairment is a common feature of the hebephrenics.

The results of the three examinations done during the haloperidol therapy, contrast with the data of *Kim et al.* (1971). We observed a normal GH secretion in most of the patients, with a small incidence of hyper- and hyporesponsiveness to insulin. It must be mentioned, however, that the GH response to the stimulus, even though normal, tended to decrease at 20 days of haloperidol treatment.

We observed also that the episodes of hyper- and hyposecretion were not a constant finding in the same patients, but varied from subject to subject and from time to time.

The clinical observation revealed a modest psychological improvement on haloperidol therapy. The association haloperidol plus chorionic gonadotrophin resulted in a frank and statistically significative decrease of most of the symptoms, especially in regard to affectivity and behavioural disorders. In other words, the hormone seemed to potentiate the effect of the psychotropic drug. This phenomenon confirms the data obtained by us in previous researches (*Brambilla et al.*, 1972a, b).

Observing in parallel the biochemical and psychopathological data of every single patient, we noted a strict relationship between extremely elevated peaks of GH secretion in response to insulin, and presence of peculiar affective-behavioural disorders, such as psychomotor excitement and hostility.

Anxiety was seldom accompanied by GH hypersecretion; however, this symptom was observed very infrequently, due to the pharmacological treatment.

Depression seemed not to influence the pattern of hormonal secretion. Among the cognitive disorders, only severe coaptive thought were significantly related to GH hypersecretion. It must be mentioned that this symptom implies acute emotional tension in the patient, probably responsible for the hormonal outburst. On the contrary, negativism, apathy, and withdrawal from reality were generally paralleled by a lack of GH response to insulin.

In other words, arousal of affects seemed to correspond to a massive hormonal secretion, while behavioural-affective anergy paralleled a hypothalamo-pituitary unresponsiveness to stimuli. This finding tends to confirm the results obtained in previous researches in regard to hypothalamo-pituitary-adrenal and gonadal function.

Two findings are worthwhile keeping in mind: the occurrence of a GH secretory exhaustion when the insulin tests are repeated at short intervals, and the observation of a rapid passage from hyper- to hyposecretion in the same patient during the experiments.

These two observations seem to contrast with each other, in that the former points out a deficiency in the regulatory mechanism of secretion, with exhaustion under repeated stimuli, and the latter the possibility of a rapid adaptation of the hypothalamic centers to psychological stresses.

We would tentatively suggest that the impairment of the GH-RH and GH secretion of our patients is due to a disruption of the normal homeostatic mechanism, which in our cases responds excessively to stimuli, and in the meantime, tends to become easily exhausted.

The same explanation could take account of the abnormal response to haloperidol therapy, which is probably related to a decrease threshold of the hypothalamus, which still reacts to reduced amounts of catecholamines. It seems to us that the effect of the combined therapy with haloperidol and HCG, on both the psychological and biochemical parameters, is worthy of more extensive study. As we observed in previous researches, the simultaneous improvement of the affective-behavioural disorders and of the hormonal patterns during the therapy, is strong enough evidence of a strict psychoendocrine relationship.

This therapeutic approach has proved to be effective in anergic, apathetic, deeply deteriorated subjects, totally resistant to the usual schemes of treatment, who represent a group with a very poor prognosis of the disease, in that it induced a more dynamic cognitive, affective and behavioural psychology. The patients, therefore, become more available to both the stimuli of the environment and to psychotherapeutic approaches.

This phenomenon could represent the first step towards a return to a more valid vital circuit.

References

Brambilla, F.; Guerrini, A.; Riggi, F., and Ricciardi, I.: Psychoendocrine correlation in schizophrenia: effects of a psychotropic and hormonal treatment. III. Int. Symp. of Psychoneuroendocrinology, London 1972.

Brambilla, F.; Guerrini, A.; Riggi, F., and Ricciardi, I.: Psychoendocrine investigation in schizophrenia: relationship between pituitary-gonadal function and behaviour (in press, 1973).

Eiduson, S.; Brill, N., and Crumpton, E.: Adrenocortical activity in psychiatric disorders. Arch. gen. Psychiat. *5:* 227 (1961).

Greene, W.; Conron, G., and Schalch, D.: Psychological correlates of growth hormone and adrenal secretory responses in patients undergoing cardiac catheterization. Psychosom. Med. *31:* 450 (1969).

Jansen, P.: A review of the pharmacology of haloperidol and triperidol. Symp. Int. on Haloperidol and Triperidol, Milano 1964.

Kim, S.; Sherman, L.; Kolodni, H.; Benjamin, F., and Singh, A.: Attenuation by haloperidol of human serum growth hormone (HGH) response to insulin. Clin. Res. *19:* 718 (1971).

Mason, J.; Wool, M., and Wherry, F.: Growth hormone response to avoidance sessions in the monkey. Psychosom. Med. *30:* 760 (1968).

Muller, E.: Nervous control of growth hormone secretion. Neuroendocrinology *11:* 338 (1973).

Persky, H.; Zuckerman, M., and Curtis, G.: Endocrine function in emotionally disturbed and normal men. J. nerv. ment. Dis. *146:* 488 (1968).

Powell, G.; Brasel, J.; Raiti, S., and Blizzard, R.: Emotional deprivation and growth retardation simulating idiopathic hypopituitarism. N. Engl. J. Med. *276:* 1279 (1967).

Reichlin, S.: Hypothalamic control of growth hormone secretion and the response to stress. Endocrinology and human behaviour, pp. 256–282 (Oxford Univ. Press, New York 1968).

Sachar, E.; Mason, J.; Kolmer, H., and Artiss, K.: Psychoendocrine aspects of acute schizophrenic reactions. Psychosom. Med. *25:* 510 (1963).

Sachar, E.; Hellman, L., and Fukushima, D.: Cortisol production in depressive illness: a clinical and biochemical clarification. Arch. gen. Psychiat. *23:* 289 (1970).

Sachar, E.; Finkelstein, J., and Hellman, L.: Growth hormone responses in depressive illness. Arch. gen. Psychiat. *25:* 263 (1971).

Schimmelbush, W.; Mueller, P., and Sheps, J.: The positive correlation between insulin resistence and duration of hospitalization in untreated schizophrenia. Brit. J. Psychiat. *118:* 429 (1971).

Schalch, S. and Parker, M.: A sensitive double antibody immunoassay for human growth hormone in plasma. Nature, Lond. *203:* 1141 (1964).

Suematsu, H.: Changes of serum growth hormone in psychosomatic disorders. Int. Congr. Psychosom. Med., Amsterdam 1973.

Takahashi, K.; Takahashi, S.; Honda, Y.; Shizume, K.; Irie, M.; Sakuma, M., and Tsushima, T.: Secretion of human growth hormone during insulin coma and electroshock therapies. Folia psychiat. neurol. jap. *21:* 87 (1967).

Wolhberg, G.; Knapp, P., and Vachon, L.: A longitudinal investigation of adrenocortical function in acute schizophrenia. J. nerv. ment. Dis. *151:* 245 (1970).

Author's address: Prof. *Francesca Brambilla,* Ospedale Psichiatrico Paolo Pini, Via Ippocrate 45, *Milan-Affori* (Italy)

Psychoneuroendocrinology. Workshop Conf. Int. Soc. Psychoneuroendocrinology, Mieken 1973, pp. 22–31 (Karger, Basel 1974)

Endocrine Studies in Depression

M. Endo, J. Endo, M. Nishikubo, T. Yamaguchi and N. Hatotani

Department of Neuropsychiatry and Central Clinical Laboratory, Kyoto University, Kyoto, and Department of Psychiatry, Mie University, Tsu

Introduction

In 1968, one of us *(M.E.)*, working on plasma growth hormone (GH) levels in some depressive patients, found that GH response to insulin hypoglycemia was attenuated during depression in some patients who were repeatedly studied at different stages of their illness (15, 16). Figure 1 illustrates a typical example. Similar results were reported by *Mueller et al.* (24) and *Sachar et al.* (29).

This encouraged us to investigate endocrine function of depressive patients further.

Materials and Methods

Population

Eleven depressive patients, aged 12–60 years (four men and seven women, six bipolar and five monopolar) were studied during illness and after recovery following treatment. All were treated by psychotropic drug medication. Two received electroconvulsive therapy in addition. Four patients were not medicated for 3 or more days before tests at either stage. Equivalent doses were given at both stages in the other seven.

Patients were diagnosed by agreement of more than two doctors. During illness, all patients showed major symptoms such as depressive moods, guilty feelings, suicidal ideas, psychomotor retardation and insomnia, accompanied by minor symptoms like agitation, anxiety, depersonalization, loss of appetite, memory disturbance, fatigability and hypochondria. All these symptoms subsided after several months of treatment.

Procedure

Each patient underwent the same series of tests carried out within 48 h each during both stages of depression and recovery.

On the first day, the insulin tolerance tests (ITT) was carried out by injecting 0.1 U/kg of regular insulin intravenously at 8 a.m. The technique was described in detail elsewhere (17). At 8 p.m. of the same day another sample was drawn, and at midnight 1 mg of dexamethasone was administered orally (dexamethasone test).

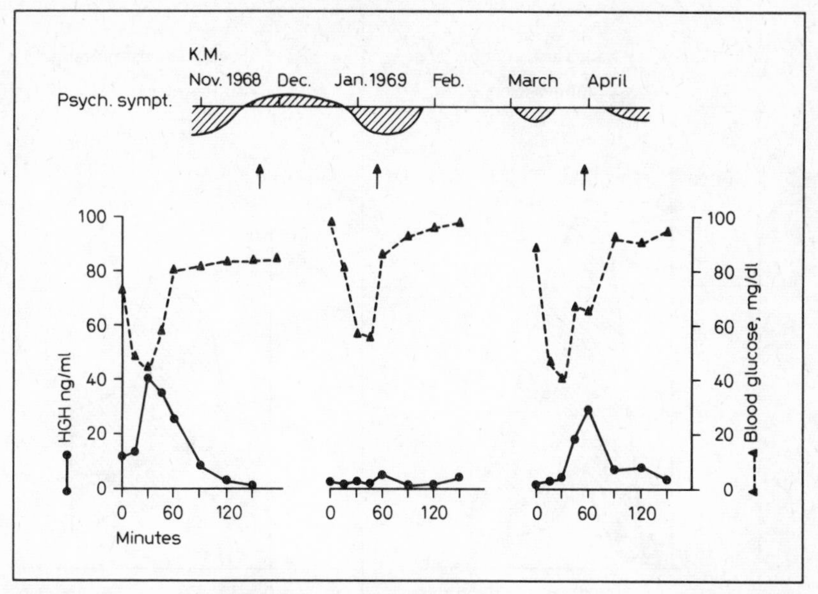

Fig. 1. Blood sugar and plasma GH (HGH) responses to ITT of a 54-year-old female bipolar depressive patient. The psychic state is indicated at the top. The upward shift means hypomanic and downward shift means depressive phases. Time points when the results below were obtained are indicated by arrows. Broken lines indicate blood sugar levels and solid lines indicate plasma GH levels.

On the second day a sample was drawn at 8 a.m. and 0.25 mg of synthetized $^{1-24}$ACTH was injected intravenously (rapid ACTH test). Blood sampling 15–90 min after injection was done following the same procedure as ITT. At midnight, 1 g of metyrapone was administered orally (metyrapone test).

The last sample was drawn at 8 a.m. of the third day.

Levels of blood sugar, plasma GH and cortisol were determined on ITT samples. The first sample of each patient served to evaluate basal morning levels of cortisol and ACTH as well. The sample drawn at 8 p.m. of the first day was used for the evening level of cortisol. Plasma levels of GH and cortisol were also determined on rapid ACTH test samples. The last sample was used to evaluate the plasma ACTH level after metyrapone.

Plasma GH, cortisol and ACTH were measured by radioimmunoassay (17, 26, 36). Blood sugar was measured by the auto-analyzer.

Results

Blood Sugar

Blood sugar levels before and during ITT were almost identical at both clinical stages. The mean ± standard error of mean (SEM) of levels before ITT

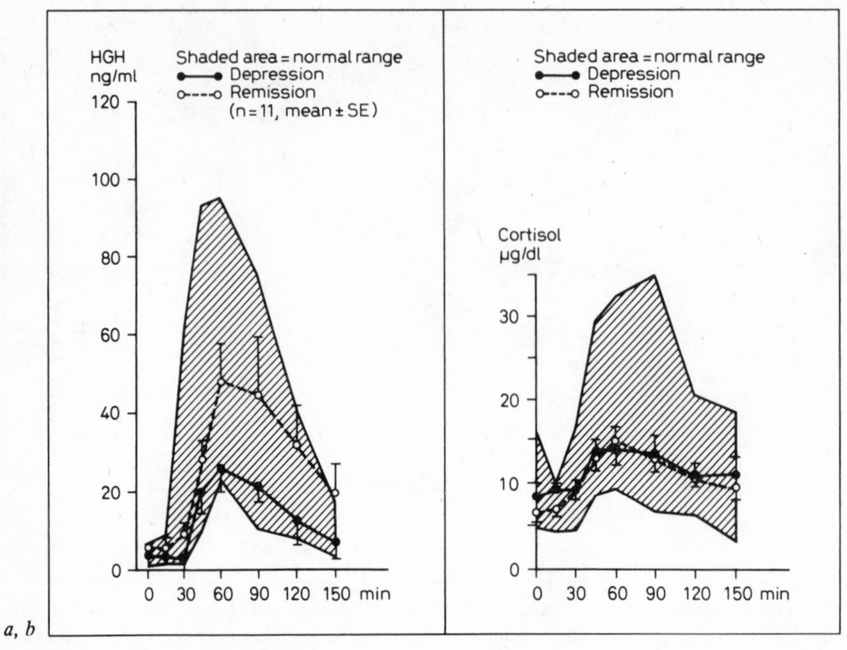

Fig. 2. Plasma GH (HGH) (a) and cortisol (b) responses to ITT of 11 depressive patients in terms of mean ± SEM. Solid circles and solid lines indicate levels during depression. Open circles and broken lines indicate levels after recovery. Shaded areas indicate the normal ranges.

was 91.2 ± 5.2 mg/dl during depression, and was 89.9 ± 1.9 mg/dl after recovery. Mean ± SEM of minimal blood sugar levels during ITT was 46.0 ± 5.9 mg/dl during depression, and was 41.5 ± 2.0 mg/dl after recovery.

Plasma Growth Hormone

Plasma GH responses to both stimuli, i.e. ITT and rapid ACTH test, tended to be attenuated during depression as compared with those after recovery (fig. 2a, 3a), although statistical significances were not proven because of wide scattering ($0.05 < p < 0.1$ at peak levels in ITT) and each mean fell within the normal range.

Plasma Cortisol

Mean ± SEM of morning plasma cortisol levels were within normal range at both clinical stages. The levels during depression tended to be slightly higher (8.2 ± 0.8 μg/dl) than after recovery (6.3 ± 1.0 μg/dl), though the difference was not statistically significant. Diurnal variation seemed to be maintained in most subjects (evening levels 3.8 ± 1.2 μg/dl during depression and 3.3 ± 0.4 μg/dl

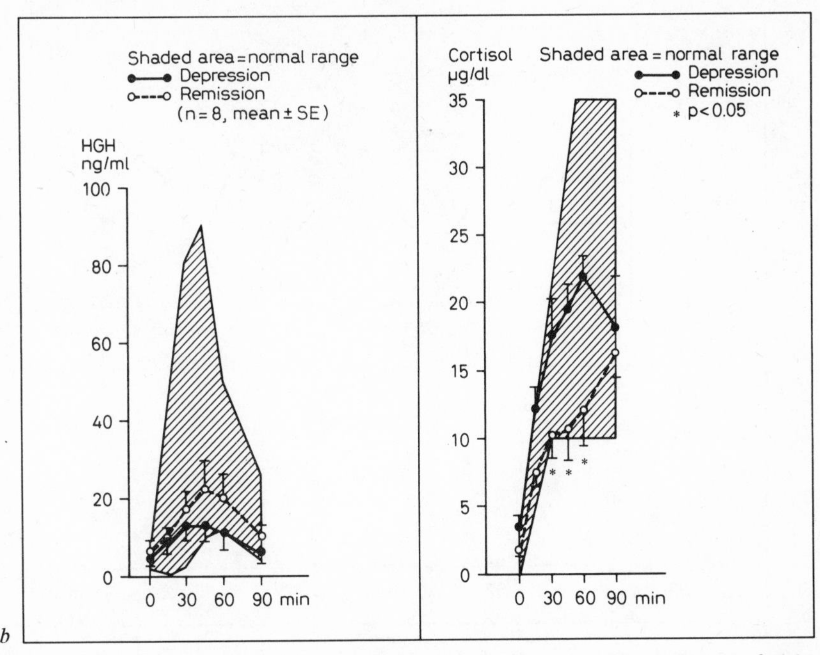

a, b

Fig. 3. Plasma GH (HGH) (a) and cortisol (b) responses to rapid ACTH test of eight depressive patients. Three patients did not receive rapid ACTH test. Symbols are the same as in figure 2.

after recovery), although high evening levels (> 5 μg/dl) were found in three patients during depression. Dexamethasone failed to suppress cortisol levels to the degree expected in three patients during depression. Mean ± SEM of morning levels after dexamethasone was 3.4 ± 0.8 μg/dl during depression and 1.4 ± 0.1 μg/dl after recovery ($0.05 < p < 0.1$). Levels during ITT were almost identical at both clinical stages (fig. 2b). On the other hand, the response to rapid ACTH test in eight patients was significantly higher during depression than after recovery (fig. 3b).

Plasma ACTH

Plasma ACTH levels at 8 a.m. before and after metyrapone were determined for seven patients who received rapid ACTH test as well. Basal levels were almost identical and were within normal range (23) at both clinical stages (73 ± 2 pg/ml during depression and 83 ± 11 pg/ml after recovery). The levels after metyrapone were also identical at both stages (228 ± 52 pg/ml during depression and 239 ± 41 pg/ml after recovery), and were within normal range (23).

Figure 4 illustrates a typical case.

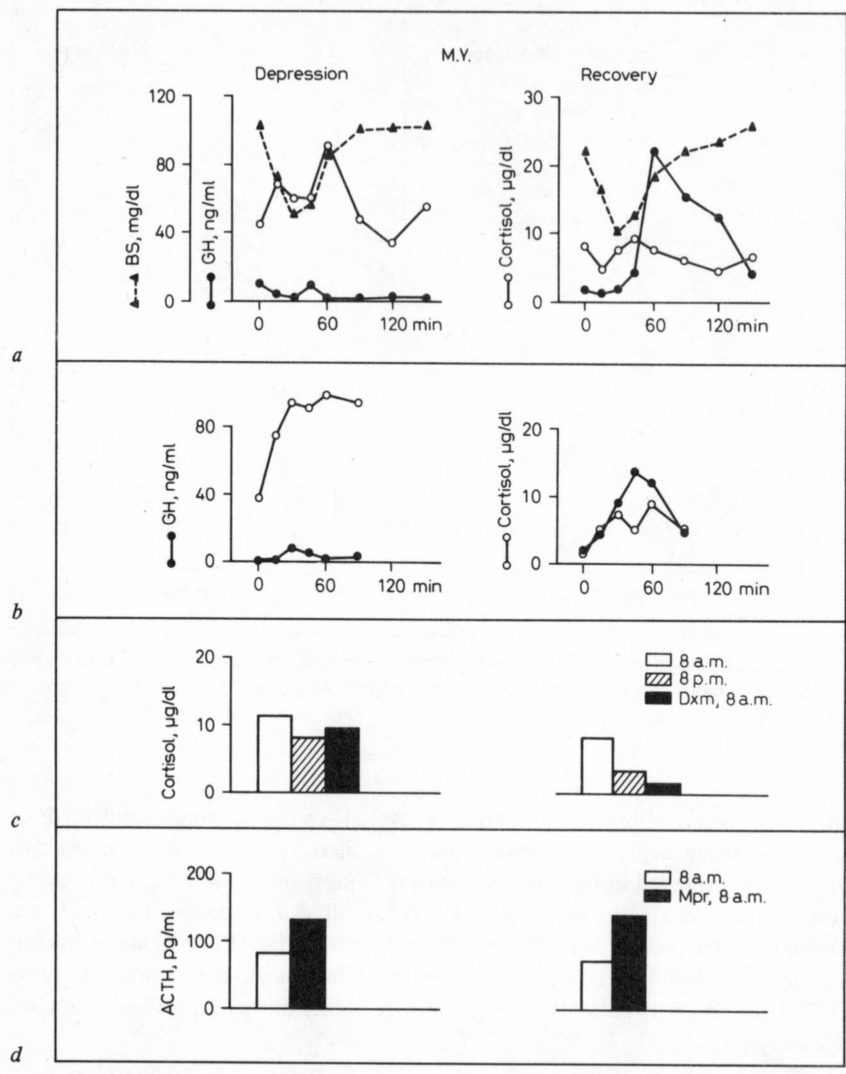

Fig. 4. ITT (a), rapid ACTH test (b), diurnal variation and dexamethasone test (Dxm) (c), and metyrapone test (Mpr) (d) of a 50-year-old male patient. Results on blood sugar (BS), GH, cortisol and ACTH obtained during depression (left) and after recovery (right). Six months before admission to hospital this patient had moved from a distant city to his present residence, and also obtained a more responsible position than before and got himself involved in over-work. Five months later he began to complain of insomnia and felt that he had lost all of his ability. He became agitated and was afraid that he might be ruined. When admitted, he was depressed and had psychomotor retardation, insomnia, self-reproach and suicidal ideas. He recovered after 2 months' treatment including antidepressant drug medication and electroconvulsive therapy.

Discussion

These results seem to indicate that GH response to some kinds of stimuli is slightly impaired during depression. *Sachar et al.* (30) and *Takahashi et al.* (34) also reported attenuated GH response in their depressive patients to L-dopa and L-5HTP, respectively. As a whole, these data suggest that mechanisms for secreting GH in response to various secretory stimuli are generally impaired during depression.

Results on plasma cortisol and ACTH levels seem to indicate some distortion in regulation of the hypothalamo-pituitary-adrenal system of depressive patients.

In our patients, pituitary ACTH secretion seems to be almost identical at both clinical stages, at least not so much heightened during depression, as judged by morning levels and their response to metyrapone. On the other hand, adrenocortical sensitivity to ACTH seems to be higher during depression than after recovery, as judged by rapid ACTH test. Results on dexamethasone also seem to indicate higher set-points in the hypothalamo-pituitary system of some depressive patients. Similar results on dexamethasone were reported by *Gibbons and Fahy* (19) and *Carrol et al.* (8).

Plasma cortisol levels during ITT are supposed to represent the summation of pituitary and adrenal responses to insulin hypoglycemia. Discrepancy between the plasma cortisol responses to ITT and rapid ACTH test could be explained by this reason. (There is a possibility that ACTH response to ITT is somewhat different from that to metyrapone, since the latter is thought to be the most potent stimulus for ACTH secretion.) This might also explain the difference between the other two reports (9, 29) on the cortisol response to ITT in depressive patients.

These factors might have to be kept in mind in interpreting the higher adrenocortical hormone levels or production rates (at least those higher than after recovery) during depression as reported by many other investigators (4, 7, 18, 22, 28, 31), unless all of our samples for ACTH measurement during depression were drawn at secretory intervals.

Similar results were also reported in an animal experiment by *Kawakami et al.* (20). They immobilized rabbits and measured corticosteroid production in terms of ^{14}C-incorporation and plasma levels, as well as the effect of ACTH on it. Six hours immobilization gave an increase of corticosteroid production. But after this stress was repeated for 7 days, corticosteroid production, as well as its response to immobilization, was slightly lowered. On the other hand, its response to administered ACTH was markedly increased. They concluded that sensitivity of the adrenal cortex to produce steroids in response to ACTH was very much heightened after prolonged stress in the presence of the possible depletion of pituitary ACTH secretion under that situation.

External stress is known to induce secretion of GH (5) and ACTH (1). In view of *Kawakami*'s data and ours, pituitary ACTH secretion might not be increased when the stress is sustained for a long time, but adrenocortical sensitivity might be increased. According to the variable set-point theory of *Yates et al.* (35), the bodily requirement for adrenocortical hormone is increased under stress, and the gap between this requirement and the actual plasma steroid levels perceived by the hypothalamic comparator element triggers more ACTH secre-

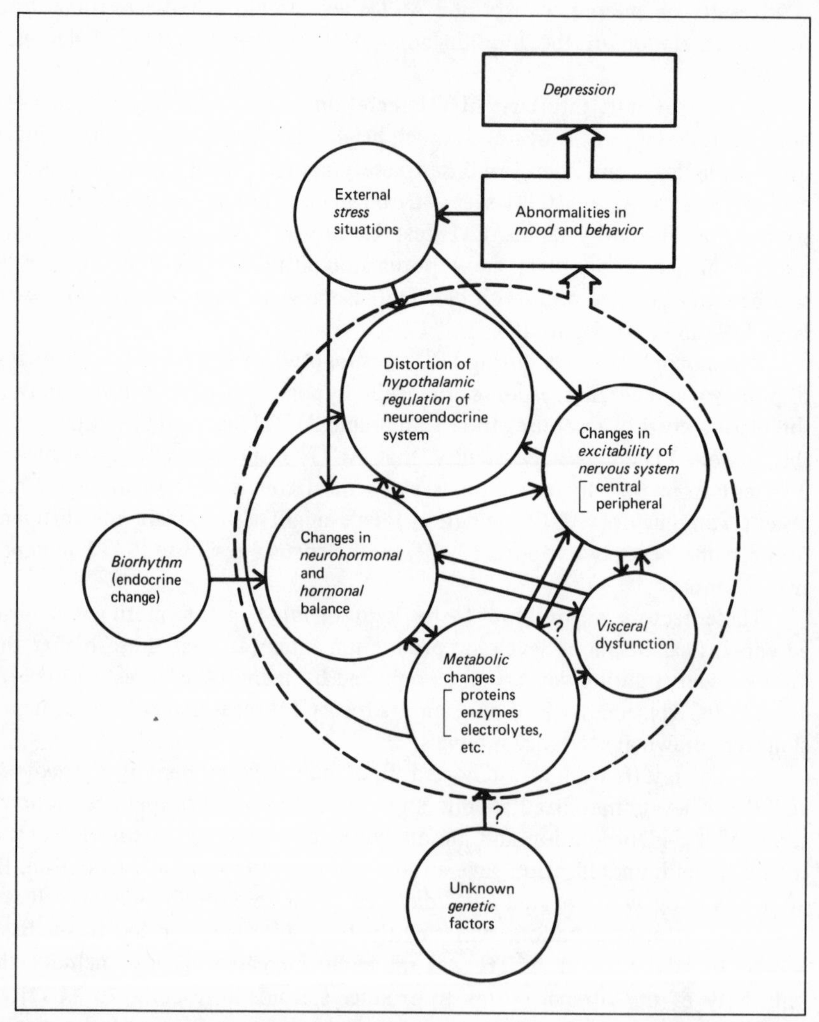

Fig. 5. A hypothetical model of the mechanisms which would finally precipitate the depressive state.

tion. But ample secretion of adrenocortical hormone as the result of increased adrenocortical sensitivity would partly compensate for this gap so that increased ACTH secretion would not further be required under sustained stress.

But in depressed patients, our data and others also suggest depleted GH secretion at the same time. If the same kind of mechanism is regulating GH secretion also, compensation for this hormone would not be very easy, since it does not have any particular target organ. In this case, an increased requirement for GH would not be sufficiently met. Thus, the gap between the degrees of fulfilment of requirements for GH and cortisol, in line with changes in the levels of other metabolic hormones (10, 14, 21), might cause subtle metabolic changes.

These metabolic changes would either, by themselves, or by affecting the excitability of the central and peripheral nervous systems or visceral function, serve to distort hypothalamic regulation and neurohormonal or hormonal balance further.

Figure 5 illustrates a hypothetical model of mechanisms which would finally precipitate the depressive state.

Many previously reported data by various investigators suggest changes in the neuroendocrine or endocrine factors in depressive patients including biogenic amines (3, 6, 12, 13, 33) or pituitary and peripheral hormones (11, 14, 22, 24, 30, 31), enzyme or co-enzymes (2, 25, 27), electrolyte balance (10) and the excitability of the nervous system (32), even though not infrequently diverse results are found.

Complicated vicious circles, as illustrated in figure 5, usually could be resolved when triggering elements subside, and their total effects are lowered to a certain level. But if they continue acting on the body for a long time, these circles would circulate so many times that they could not be stopped easily. It would cause inadequate response to external stimuli, which would be noticed as abnormalities in mood and behavior, in turn increasing stress situations, and finally precipitating a depressive state.

Most investigations of depressive illness thus far have likely dealt with only one or few aspects of these vicious circles. Many diverse results on any one aspect might not be so strange if one regards depression as a state in which these vicious circles are heightened to a degree that cannot be easily resolved.

Summary

Blood sugar, plasma GH, cortisol and ACTH levels were determined in 11 depressive patients during illness and after recovery. ITT, rapid ACTH test, dexamethasone test and metyrapone test were carried out at both clinical stages of each patient. Responses of blood sugar and cortisol to ITT and the response of ACTH to metyrapone were identical at both stages. During depression, the response of GH to ITT and rapid ACTH test tended to be attenuated, dexamethasone failed in some patients to suppress cortisol levels to the degree expected, and the cortisol response to rapid ACTH test was higher than that after recovery. The possible role of distortion in the neuroendocrine regulation in relation to other factors in precipitating depression was discussed.

Acknowledgements

We are indebted to Dr. *Teruya Yoshimi* and Dr. *Yoshikatsu Nakai*, Second Division, Department of Internal Medicine, Kyoto University for radioimmunoassay of cortisol and ACTH, respectively. We are also grateful to the staff of the Department of Psychiatry, Mie University for their cooperation and clinical assistance.

References

1 *Allen, J.P.; Allen, C.F.; Greer, M.A., and Jacobs, J.J.:* Stress-induced secretion of ACTH; in Brain-pituitary-adrenal interrelationships, pp. 99–127 (Karger, Basel 1973).
2 *Abdulla, Y.H. and Hamadah, K.:* 3′,5′-cyclic adenosine monophosphate in depression and mania. Lancet *i:* 378–381 (1970).
3 *Birkmayer, W. und Linauer, W.:* Störung des Tyrosin- und Tryptophanmetabolismus bei Depression. Arch. Psychiat. Nervenkr. *213:* 377–387 (1970).
4 *Brooksbank, B.W.L. and Coppen, A.:* Plasma 11-hydroxycorticosteroids in affective disorders. Brit. J. Psychiat. *113:* 395–404 (1967).
5 *Brown, G.M. and Reichlin, S.:* Psychological and neural regulation of growth hormone secretion. Psychosom. Med. *34:* 45–61 (1972).
6 *Bunney, W.E.; Davis, J.M.; Weil-Malherbe, H., and Smith, E.R.B.:* Biochemical changes in psychotic depression. Arch. gen. Psychiat. *16:* 448–460 (1967).
7 *Carpenter, W.T. and Bunney, W.E.:* Adrenal cortical activity in depressive illness. Amer. J. Psychiat. *128:* 31–40 (1971).
8 *Carrol, B.J.; Martin, F.I.R., and Davies, B.:* Resistance to suppression by dexamethasone of plasma 11-OHCS levels in severe depressive illness. Brit. med. J. *iii:* 285–287 (1968).
9 *Carrol, B.J.:* Hypothalamic-pituitary function in depressive illness: insensitivity to hypoglycemia. Brit. med. J. *iii:* 27–28 (1969).
10 *Coppen, A.:* The biochemistry of affective disorders. Brit. J. Psychiat. *113:* 1237–1264 (1967).
11 *Coppen, A.:* Pituitary-adrenal activity during psychosis and depression. Progr. Brain Res. *32:* 336–342 (1970).
12 *Coppen, A.:* Indoleamines and affective disorders. J. psychiat. Res. *9:* 163–171 (1972).
13 *Curzon, G. and Bridges, P.K.:* Tryptophan metabolism in depression. J. Neurol. Neurosurg. Psychiat. *33:* 698–704 (1970).
14 *Dewhurst, K.E.:* Blood-levels of thyrotrophic hormone, protein-bound iodine, and cortisol in schizophrenia and affective states. Lancet *ii:* 1160–1162 (1968).
15 *Endo, M.:* Plasma growth hormone measurement by talc precipitation method. 16th Gen. Meet. Jap. Endocrin. Soc., Western Division, Kumamoto 1968. Folia endocrin. jap. *44:* 1337–1338 (1969).
16 *Endo, M.:* Plasma growth hormone levels during insulin hypoglycemia in atypical psychosis. 43rd Annu. Meet. Jap. Endocrin. Soc., Osaka 1970. Folia endocrin. jap. *45:* 1295–1296 (1970).

17 *Endo, M.:* Radioimmunoassay of human growth hormone by adsorption to talc. Folia endocrin. jap. *49:* 1001–1010 (1973).
18 *Gibbons, J.L. and McHugh, P.R.:* Plasma cortisol in depressive illness. J. psychiat. Res. *1:* 162–171 (1962).
19 *Gibbons, J.L. and Fahy, T.J.:* Effect of dexamethasone on plasma corticosteroids in depressive illness. Neuroendocrinology *1:* 358–363 (1966).
20 *Kawakami, M.; Seto, K.; Negoro, H.; Yoshida, K.; Yanase, M., and Môri, M.:* Central regulation of pituitary-adrenal system. Jap. J. clin. Med. *27:* 1348–1363 (1969).
21 *Kreuz, L.E.; Rose, R.M., and Jennings, J.R.:* Suppression of plasma testosterone levels and psychological stress. Arch. gen. Psychiat. *26:* 479–482 (1972).
22 *Lingjærde, P.S.:* Plasma hydrocortisone in mental diseases. Brit. J. Psychiat. *110:* 423–432 (1964).
23 *Matsukura, S.; Matsuyama, H.; Nakai, Y., and Imura, H.:* ACTH; in Radioimmunoassay. Recent Progr. Prosp., Saishin-Igaku *26:* 1085–1090 (1971).
24 *Mueller, P.S.; Heninger, G.R., and McDonald, R.K.:* Insulin tolerance test in depression. Arch. gen. Psychiat. *21:* 587–594 (1969).
25 *Murphy, D.L. and Weiss, R.:* Reduced monoamine oxydase activity in blood platelets from bipolar depressed patients. Amer. J. Psychiat. *128:* 1351–1357 (1972).
26 *Nakai, Y.; Imura, H., and Matsukura, S.:* Paradoxical binding phenomenon in radioimmunoassay of adrenocorticotropic hormone. Folia endocrin. jap. *49:* 894–900 (1973).
27 *Paul, M.I.; Cramer, H., and Goodwin, F.K.:* Urinary cyclic AMP excretion in depression and mania. Arch. gen. Psychiat. *24:* 327–333 (1971).
28 *Sachar, E.J.; Hellman, L.; Fukushima, D.K., and Gallagher, T.F.:* Cortisol production in depressive illness. Arch. gen. Psychiat. *23:* 289–298 (1970).
29 *Sachar, E.J.; Finkelstein, J., and Hellman, L.:* Growth hormone responses in depressive illness. I. Response to insulin tolerance test. Arch. gen. Psychiat. *25:* 263–269 (1971).
30 *Sachar, E.J.; Mushrush, G.; Perlow, M.; Weitzman, E.D., and Sassin, J.:* Growth hormone responses to L-dopa in depressed patients. Science *178:* 1304–1305 (1972).
31 *Sachar, E.J.; Hellman, L.; Roffwarg, H.P.; Halspern, F.S.; Fukushima, D.K., and Gallagher, T.F.:* Disrupted 24-hour patterns of cortisol secretion in psychotic depression. Arch. gen. Psychiat. *28:* 19–24 (1973).
32 *Shagass, C.:* Evoked brain potentials in psychiatry (Plenum Press, New York 1972).
33 *Schildkraut, J.J.:* The catecholamine hypothesis of affective disorders: a review of supporting evidence. Amer. J. Psychiat. *122:* 509–522 (1965).
34 *Takahashi, S.; Kondo, H.; Yoshimura, M., and Ochi, Y.:* Growth hormone responses to administration of L-5-hydroxytryptophan in manic-depressive psychoses. Psychoneuroendocrinology, pp. 32–38 (Karger, Basel 1974).
35 *Yates, F.E. and Urquhart, J.:* Control of plasma concentrations of adrenocortical hormones. Physiol. Rev. *42:* 359–443 (1962).
36 *Yoshimi, T.; Endo, J., and Tachibana, S.:* Radioimmunoassay of plasma cortisol. 46th Annu. Meet. Jap. Endocrin. Soc., Kyoto 1973. Folia endocrin. jap. *49:* 325 (1973).

Authors' addresses: *Midori Endo,* MD and *Jiro Endo,* MD, Department of Neuropsychiatry and Central Clinical Laboratory, Faculty of Medicine, Kyoto University, *Kyoto 606; Mitsuhiro Nishikubo,* MD, *Takahisa Yamaguchi,* MD and *Noboru Hatotani,* MD, Department of Psychiatry, Mie University School of Medicine, *Tsu 514* (Japan)

Psychoneuroendocrinology. Workshop Conf. Int. Soc. Psychoneuroendocrinology, Mieken 1973, pp. 32–38 (Karger, Basel 1974)

Growth Hormone Responses to Administration of *L*-5-Hydroxytryptophan (L-5-HTP) in Manic-Depressive Psychoses

S. Takahashi, H. Kondo, M. Yoshimura and Y. Ochi

Department of Psychiatry and Neurology and Second Department of Internal Medicine, Kyoto Prefectural University of Medicine, Kyoto

There is increasing evidence of an abnormality of indoleamine metabolism in the affective disorders (4). Decreases in serotonin and 5-hydroxyindoleacetic acid (5-HIAA) concentrations in the brains of depressive suicides (17), as well as in the cerebrospinal fluid concentration of 5-HIAA in depressive patients are reported (1, 5, 12, 18). *L*-tryptophan, a precursor of serotonin, is reported to have an antidepressant action similar to imipramine with or without MAOI administration (3).

Recently, clinical trials with *L*-5-hydroxytryptophan (L-5-HTP), which is believed to be converted directly to serotonin, were made by *Sano* (15), who found the agent to have a potent antidepressant effect.

The hypothalamic regulation of human growth hormone (HGH) release from the anterior pituitary appears to be closely related to the brain biogenic amines. Previous studies in animals (10, 11) revealed that both adrenergic and dopaminergic mechanisms in the brain seem to be involved in regulating pituitary HGH release. Oral administration of L-dopa stimulates HGH release both in patients with Parkinson's diesease and in normal young adults (2).

The authors of this paper have reported that the oral administration of L-5-HTP also causes a rise in plasma HGH in normal adults (21). Although serotoninergic structure is known to be highly distributed in the hypothalamus, the role of serotonin in regulating human pituitary function remains unsolved.

Certain depressed patients failed to release HGH adequately both in response to insulin-induced hypoglycemia (13) and to oral administration of L-dopa (14), and some of them had persistent insensitivity of pituitary function after their recovery.

Thus, it may be hypothesized that depressed patients manifest an impairment of HGH release when L-5-HTP is administered, and that antidepressive efficacy of L-5-HTP could be evaluated by the endocrine response, which might be involved in defective indoleamine metabolism.

Materials and Methods

Fourteen patients with manic-depressive psychoses, eight males and six females, aged 17–65, who were admitted to the psychiatric ward, Kyoto Prefectural University Hospital, were studied. Unequivocal diagnosis of the patients was established by the agreement of psychiatrists at the Department of Psychiatry and Neurology. They were classified into three groups by their psychic symptoms and their previous histories of manic and depressive episodes; i.e. five cases were classified as manic-depressive (or bipolar), five as endogenous-depressive (or unipolar) and four as neurotic-depressive. The protracted depressive group consisted of two cases of endogenous-depressives and four neurotic-depressives, who had not responded to any antidepressants and electroconvulsive therapy in 2 or more years. Global judgement on the severity of the illness was evaluated, and assessment of depressive symptoms using *Hamilton*'s Rating Scale (7) was carried out on the day of an endocrine examination.

Summary of psychiatric data is given in table I.

None of the patients were obese or cachectic. No particular findings in endocrine and metabolic functions were found during hospitalization.

Twenty-four endocrine examinations were carried out, and the majority of the patients were tested again for replication in accord with the change of psychic state. All medications were withdrawn for at least 7 days prior to the experimental day.

After overnight fasting and at resting state, 200 mg of L-5-HTP was orally administered at 8.30 a.m. Blood was taken at 30-min intervals for 3 h before and after the administration of L-5-HTP.

Table I. Psychiatric data for 14 manic depressive patients

Patient No.	Age, years	Sex	Diagnosis[1]	Previous history	Global severity[2]	*Hamilton*'s depression score
1	24	f	MD, manic[3]	2	2	–
2	31	f	MD, manic	1	2	–
3	60	f	MD, manic and depressive	4	1	18, 5
4	17	f	MD, manic and depressive	2	3	25 >, 25 >, 21 >
5	58	m	MD, depressive	3	2	29
6	51	f	ED	3	1	22, 11
7	30	m	ED	no	2	–
8	35	m	ED	no	3	24
9	50	m	ED, protracted for 3 years[4]	no	3	29
10	65	m	ED, protracted for 2 years	2	2	23, 18
11	45	m	ND, protracted for 6 years	5	2	13, 23
12	45	f	ND, protracted for 3 years	no	1	10, 13
13	55	m	ND, protracted for 2 years	no	2	19
14	47	m	ND, protracted for 2 years	no	1	11

1 MD = manic depressive; ED = endogenous depression; ND = neurotic depression.
2 Global severity: 0 = not ill; 1 = mild; 2 = moderate; 3 = marked.
3 Psychic state where HGH measurements were done.
4 Resistant depressive symptoms protracted over 2 years or more.

An aliquot of blood was used for the assay of blood glucose by means of Auto Analyzer according to *Hoffman*'s method (8). The plasma of remaining blood was separated by centrifugation, samples were frozen and subsequently analyzed in duplicate for HGH by double antibody radioimmunoassay of *Schalch and Parker* (16). This method gives an accuracy of the measurements of less than 0.5 ng/ml level. Plasma cortisol levels just prior to the L-5-HTP administration were done by radioimmunoassay as described by *Yoshimi et al.* (20).

Results

Pituitary HGH response to L-5-HTP in four patients in the manic state (patients No. 1, 2, 3 and 4) was sufficiently positive, showing a peak value of 8.2–38.8 ng/ml. The minimum plasma HGH response with oral administration of L-5-HTP, considered as adequate, was above 5 ng/ml, a figure which was regarded acceptable in our previous report on normal subjects and patients with various endocrine disorders (21). Mean value of maximum HGH levels in these patients was 17.7 ± 7.1 ng/ml (mean ± SEM), that was not significantly different from the value of seven controls – 10.5 ± 1.6 ng/ml, ranging from 6.4–18.8 ng/ml, referred from the authors' previous report (21) – though it seemed to be much higher than the latter (see table II).

Table II. Summary of the results 1. Base line and maximum levels of plasma HGH and glucose

Diagnostic group	No. of patients	Age, years	HGH, ng/ml		Glucose, mg/dl	
			base line	maximum	base line	maximum
Control[1]	7 (7)[2]	33 ± 2	2.3 ± 0.8	10.5 ± 1.6	85 ± 2	97 ± 3
Manic (MD)	4 (4)	33 ± 9	11.5 ± 8.4	17.7 ± 7.1	77 ± 4	86 ± 2
Depressive (MD and ED)	6 (10)	42 ± 7	2.4 ± 0.3	4.7 ± 1.4[3,5]	82 ± 3	97 ± 3[5]
Protracted depressive (ED and ND)	6 (10)	51 ± 3	1.5 ± 0.3	2.1 ± 0.5[4]	91 ± 4	102 ± 5

1 Referred from the authors' previous report (24).
2 Number of determinations.
3 $p < 0.05$.
4 $p < 0.01$: significantly different from the value of control group.
5 $p < 0.05$: significantly different from the value of manic group.
Figures are shown by mean ± SEM.

Base-line levels of plasma HGH just prior to L-5-HTP administration, exceeded 5 ng/ml level in two of these manic patients. This premature HGH release was presumably due to manic overactivity or lack of depressive stress which may inhibit the hypothalamic function. However, these two obviously responded to the stimulation of L-5-HTP showing a peak of HGH concentration, which occurred 60–90 min after the ingestion of the drug.

Blood glucose rose slightly from 77 ± 4 mg/dl (mean ± SEM) to 86 ± 2 mg/dl after administration of L-5-HTP. These values appeared to be lower than the control values, but were statistically insignificant.

In contrast, manic-depressive bipolar patients in depressive state failed to secrete HGH adequately. When studied in both manic and depressive states, two of these patients (patients No. 3 and 4) showed a striking contrast.

Of five examinations on three endogenous depressive patients (patients No. 6, 7 and 8), three achieved an adequate response. Including these, statistics of ten determinations from six depressives are given in table II. Both maximum levels of HGH and of glucose were significantly different from the values of manic group ($p < 0.05$ and $p < 0.05$, respectively). Peaks of plasma HGH concentrations, occurring 90–150 min after L-5-HTP ingestion, tended to be delayed as compared with the reaction of manic patients.

Ten examinations, except for one from six patients of protracted depressive group aged 45–65, resulted in a deficient response of HGH to L-5-HTP. Base-line levels of HGH were 1.5 ± 0.3 ng/ml (mean ± SEM), reaching the maximum levels of only 2.1 ± 0.5 ng/ml, which were significantly lower than control values ($p < 0.01$) (table II).

Blood glucose levels were higher in the non-protracted depressives than in the manics, and still higher in the protracted depressive group. Plasma cortisol levels were observed as rather high in each of the three groups – manics: 17.3 ± 1.8 μg/100 ml; non-protracted depressives: 16.3 ± 4.9 μg/100 ml; protracted depressives: 15.1 ± 5.2 μg/100 ml, mean ± SEM; normal range 5–12 μg/100 ml (20). Increments in blood glucose levels following L-5-HTP administration in the protracted depressives were less marked than those in the controls, so that HGH insensitivity was presumably irrelevant to blood glucose changes.

Discussion

Manic-depressive psychotics raise unique problems as subjects for controlled endocrine studies. Table III shows response of pituitary HGH to L-5-HTP administration achieved by the psychic states and the three diagnostic categories. Open circles show adequate HGH responses with a magnitude of more than 5 ng/ml. This table demonstrates that oral administration of L-5-HTP raises plasma HGH levels in the manic state. Depressed patients are likely to have no such response.

Table III. Summary of the results 2. Responses of HGH secretion to L-5-HTP administration

Manic state	Diagnostic group	Depressive state
OOOO	MD	XXXXX ($p < 0.05$)[1]
	ED	OXXOO
	Protracted	XXXOXXXXXX

O = Adequate HGH response (magnitude more than 5 ng/ml).
X = Deficient HGH response.

1 Significance level $p < 0.05$, manic versus depressive in MD group.

Our results, however, were not obtained frequently enough to contribute to the current concept of biological differences between unipolar and bipolar depressives. Of our patients, each of three bipolar depressives failed to respond to L-5-HTP, whereas two of three unipolars were found to be responders. This observation indicates further investigation.

Protracted depressives responded to L-5-HTP quite differently. Each of them failed to show an adequate HGH response. The mode of nonresponders to L-5-HTP was replicated in three of four patients retested. Although we have no age-matched controls, difference from the normal values was significant ($p < 0.01$). When the data were analyzed in terms of percentage of deficient responses, the difference was also obviously significant ($p < 0.01$, $\chi^2 = 13.4$).

Age may be a factor in the response of HGH to L-5-HTP administration. It was reported that among the 14 older normals, aged 48–68, only five failed to have an adequate response of HGH to L-dopa administration, although difference from the younger normals, aged 20–32, was significant, suggesting that the pituitary HGH response to an amine precursor diminishes with age (14).

Several other factors may account for the diminished HGH response in depressed patients. For example, ingested L-5-HTP may have been poorly absorbed because of the delayed passage in the gastrointestinal tract, or may have failed to reach the brain in adequate amounts because of retarded metabolic dysfunction. The action of L-5-HTP in releasing HGH from the anterior pituitary also may be non-specific, like an amino acid effect by arginine.

Of course, in depressions, interference of high cortisol levels which antagonize the HGH secretion is important. Several depressed patients who failed to release HGH in response to the insulin tolerance test have been reported, although the insulin-induced hypoglycemia was low enough to stimulate the anterior pituitary hormones in normal subjects (9, 13). The high cortisol levels were interesting as being due to depressive stress rather than the central nervous system mechanism involved.

The most intriguing possibility is that the failure to release HGH after L-5-HTP administration is due to a central neurochemical disturbance in indoleamine metabolism, perhaps a failure to convert L-5-HTP to serotonin. Protracted depressed patients in the present study had been quite resistant to antidepressive agents and electroconvulsive therapy for years, and failed repeatedly to have an endocrine response to L-5-HTP as well, suggesting an enduring factor that is possibly concerned with their vulnerability to prolonged depressive symptoms. Four of these resistant depressives had no favorable clinical reactions to treatment with L-5-HTP, 300 mg daily for 2 weeks, which was recently prescribed as a new magical drug for depressions. Our patients in protracted depression may not fit the recommended indication for L-5-HTP, as they were nonresponders of HGH to L-5-HTP, suggesting certain disorders in brain serotonin biosynthesis, which inhibit the utilization of exogenous L-5-HTP.

Summary

Fourteen hospitalized patients with manic-depressive psychoses received *L*-5-hydroxytryptophan (L-5-HTP), a serotonin precursor, which has been postulated as a potent antidepressive agent. Plasma human growth hormone (HGH) and glucose levels, were measured at 30-min intervals after oral administration of 200 mg of L-5-HTP. Plasma cortisol levels prior to L-5-HTP administration were also measured.

Five manic-depressive (bipolar) patients studied in the manic state showed an adequate HGH response of more than 5 ng/ml. While those in the depressed state failed to secrete HGH adequately ($p < 0.05$), nine of ten tests from six patients with protracted depression for 2–6 years, aged 45–65, had deficient responses of HGH ($p < 0.01$).

Deficiency of pituitary HGH secretion appeared to correlate neither with the score of *Hamilton*'s Depression Scale nor with the global judgement on the severity of the illness. Increments in the blood glucose levels after L-5-HTP administration were mild and the HGH insensitivity may be irrelevant.

Because there is evidence suggesting pituitary hormone release related to brain biogenic amines, the deficient HGH responses to L-5-HTP in depressed patients may be due to a neurochemical defect that has been hypothesized in the manic-depressive psychoses. The HGH responses were most distinctly diminished in patients with protracted depression, suggesting an endocrine hypofunction which may cause the fixation of the depressive symptoms.

References

1 *Ashcroft, G.W.; Crawford, T.B.B.; Eccleston, D.; Sharman, D.F.; MacDougall, E.J.; Stauton, J.B., and Binns, J.K.:* 5-Hydroxyindole compounds in the cerebrospinal fluid of patients with psychiatric or neurological disease. Lancet *ii:* 1049–1052 (1966).

2 *Boyd, A.E. III; Lebovitz, H.E., and Pfeiffer, J.B.:* Stimulation of human-growth-hormone secretion by L-dopa. New Engl. J. Med. *283:* 1425–1429 (1970).

3 *Coppen, A.; Shaw, D.M., and Farrell, J.P.:* Potentiation of the antidepressant effect of a monoamine oxidase inhibitor by tryptophan. Lancet *i:* 79–80 (1963).

4 *Coppen, A.:* The biochemistry of affective disorders. Brit. J. Psychiat. *113:* 1237–1264 (1967).
5 *Coppen, A.; Prange, A.J., jr.; Whybrow, P.C., and Noguera, R.:* Abnormalities of indoleamines in affective disorders. Arch. gen. Psychiat. *26:* 474–478 (1972).
6 *Dunner, D.L.; Goodwin, F.K.; Gershon, E.S.; Murphy, D.L., and Bunney, W.E., jr.:* Excretion of 17-OHCS in unipolar and bipolar depressed patients. Arch. gen. Psychiat. *26:* 360–363 (1972).
7 *Hamilton, M.:* A rating scale for depression. J. Neurol. Neurosurg. Psychiat. *23:* 56–62 (1960).
8 *Hoffman, W.S.:* A rapid photoelectric method for the determination of glucose in blood and urine. J. biol. Chem. *120:* 51–55 (1937).
9 *Mueller, P.S.; Henninger, G.R., and McDonald, R.K.:* Insulin tolerance test in depression. Arch. gen. Psychiat. *21:* 587–594 (1969).
10 *Müller, E.E.; Sawano, S.; Arimura, A., and Schally, A.V.:* Blockade of release of growth hormone by brain norepinephrine depletors. Endocrinology, Springfield *80:* 471–476 (1967).
11 *Müller, E.E.; Pra, P.D., and Pecile, A.:* Influence of brain neurohumors injected into the lateral ventricle of the rat on growth hormone release. Endocrinology, Springfield *83:* 893–896 (1968).
12 *Van Praag, H.H. and Korf, J.:* Endogenous depressions with and without disturbances in the 5-hydroxytryptamine metabolism. A biochemical classification? Psychopharmacologia, Berl. *19:* 148–152 (1971).
13 *Sachar, E.J.; Finkelstein, J., and Hellman, L.:* Growth hormone responses in depressive illness. I. Response to insulin tolerance test. Arch. gen. Psychiat. *25:* 263–269 (1971).
14 *Sachar, E.J.; Mushrush, G.; Perlow, M.; Weitzman, E.D., and Sassin, J.:* Growth hormone responses to L-dopa in depressed patients. Science *178:* 1304–1305 (1972).
15 *Sano, I.:* 'Precursor-therapy' with active amines. Part I. Treatment of depression by L-5-HTP (L-5-hydroxytryptophan). Psychiat. Neurol. Jap. *73:* 809–815 (1971).
16 *Schalch, D.S. and Parker, M.L.:* A sensitive double antibody immunoassay for human growth hormone in plasma. Nature, Lond. *203:* 1141–1142 (1964).
17 *Shaw, D.M.; Camps, F.E., and Eccleston, E.G.:* 5-Hydroxytryptamine in the hindbrain of depressive suicides. Brit. J. Psychiat. *113:* 1407–1411 (1967).
18 *Sjöström, R. and Roos, B.E.:* 5-Hydroxyindoleacetic acid and homovanillic acid in cerebrospinal fluid in manic-depressive psychosis. Europ. J. clin. Pharmacol. *4:* 170–176 (1972).
19 *Takahashi, S.; Kondo, H., and Kato, N.:* Effect of *L*-5-hydroxytryptophan on brain monoamine metabolism and evaluation of its clinical effect in depressed patients. Presented at the Fourth Congress of ISPNE, September 1973, Berkeley, USA.
20 *Yoshimi, T.; Endo, J., and Tachibana, K.:* A simple technique of radioimmunoassay for plasma cortisol. Presented at the 46th Annu. Meet. Jap. Endocrinol. Soc., April 1973, Kyoto, Japan.
21 *Yoshimura, M.; Ochi, Y.; Miyazaki, T.; Shiomi, K., and Hachiya, T.:* Effect of L-5-HTP on the release of growth hormone, TSH and insulin. Endocrinol. Japon. *20:* 135–141 (1973).

Authors' addresses: Dr. *S. Takahashi* and Dr. *H. Kondo,* Department of Psychiatry and Neurology, Kyoto Prefectural University of Medicine, *Kyoto;* Dr. *M. Yoshimura* and Dr. *Y. Ochi,* 2nd Department of Internal Medicine, Kyoto Prefectural University of Medicine, *Kyoto* (Japan)

Psychoneuroendocrinology. Workshop Conf. Int. Soc. Psychoneuroendocrinology, Mieken 1973, pp. 39–47 (Karger, Basel 1974)

Humoral Regulation of Gonadotropin Secretion in the Female

O. Tanizawa, T. Aono, J. Minagawa, A. Miyake, K. Kawamura, N. Fukada, H. Ichii, K. Yamaji and K. Kurachi

Department of Obstetrics and Gynecology, Osaka University Medical School, Osaka

Introduction

Since radioimmunoassay has been introduced for measurement of serum gonadotropins, it has become possible to make accurate measurement of serum LH and FSH levels in various conditions in women (5). The serum LH and FSH levels gave us much fundamental knowledge which was related to regulate the hypothalamo-pituitary-ovarian axis.

Moreover, *Schally et al.* (6) succeeded in isolation and synthesis of LH-releasing hormone (LH-RH) in 1971, and now the synthetic LH-RH is clinically available for the pituitary reserve function test (11). On the other hand, it is a well-known fact that estrogen has negative and positive feedback effects on the hypothalamo-pituitary axis.

The present study summarized the serum LH and FSH levels of Japanese women with normal menstrual cycle and examined the dynamic effects of LH-RH and conjugated estrogen (Premarin) on the serum LH and FSH levels of women at the different phases of the menstrual cycle. In order to further study the feedback mechanism of the sex steroid hormones, the serum LH levels of adult female rabbits were measured by radioimmunoassay following injection of LH-RH and cupric sulfate with or without pretreatment of estrogen and progesterone.

Variation of Serum LH and FSH Levels in Normal Women by Age

The serum LH and FSH levels were measured by the radioimmunoassay-kit of human LH and FSH from the Carbiochem (2). The purified LH and FSH were labelled with ^{125}I and the double antibody technic was used for separation of bound and free. The second IRP-HMG was used as standards for LH and FSH.

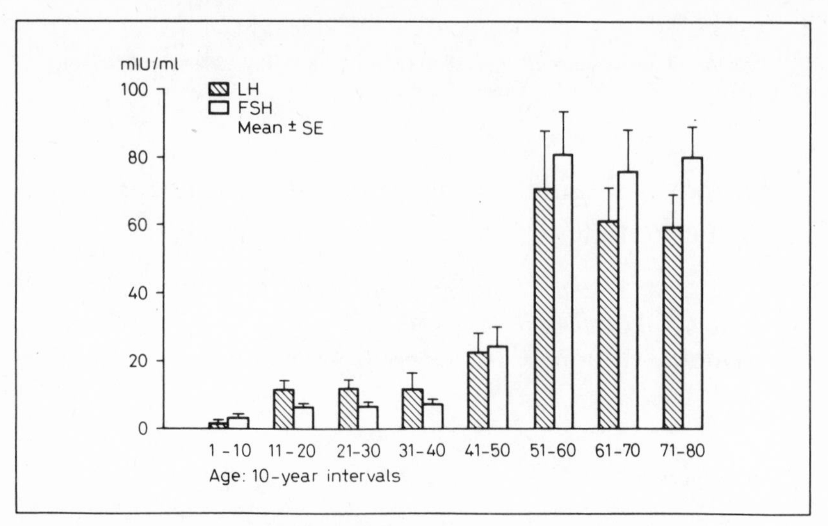

Fig. 1. Serum LH and FSH levels in normal females in relation to age.

The range of measurement was 0.5–250 mIU/ml for LH and 1–500 mIU/ml for FSH. The coefficient of variations was satisfactory for both LH and FSH being 14.3 and 11.6 %, respectively.

Measurement was made on 104 Japanese women aged from 1 to 80 years by dividing them by 10 years (fig. 1). The FSH levels increased gradually from ages 10 to 20 years and for those in their 40s it increased to 3.4 times the FSH levels of those in their 30s. Around the fiftieth year, when menopause occurs, it increased to over 10 times. The variations of LH levels were roughly the same with FSH. The LH levels of those in their 40s was 1.9 times of those in their 30s. For the 50–80-year-olds, it increased 5.2–6.8 times that of those aged between 20–30.

Variation of Serum LH and FSH Levels during the Menstrual Cycle

The daily serum levels of LH and FSH were determined in seven women with normal menstrual cycle (7). The data of LH and FSH levels during menstrual cycles were arranged in relation to the day of LH surge and averaged (fig. 2).

The mean level of LH in the first half of follicular phase (10.2 ± 5.6 mIU/ml) increased towards the second half (20 ± 5.6 mIU/ml). At midcycle, the abrupt surge of LH level (113.1 ± 20 mIU/ml) was demonstrated. The value then decreased to a lower level (11.8 ± 1.5 mIU/ml) throughout the luteal phase. The serum FSH level stayed high (5.9 ± 0.3 mIU/ml) during the first two-thirds of the follicular phase, followed by the transitory drop (4.9 ± 0.4 mIU/ml) in the last one-third of the follicular phase. The pattern of FSH showed the peak (14.1 ± 1 mIU/ml) which coincided with LH surge, then decreased to the lower level (3.7 ± 0.4 mIU/ml) in the early luteal phase, followed by the gradual increase towards the end of this phase.

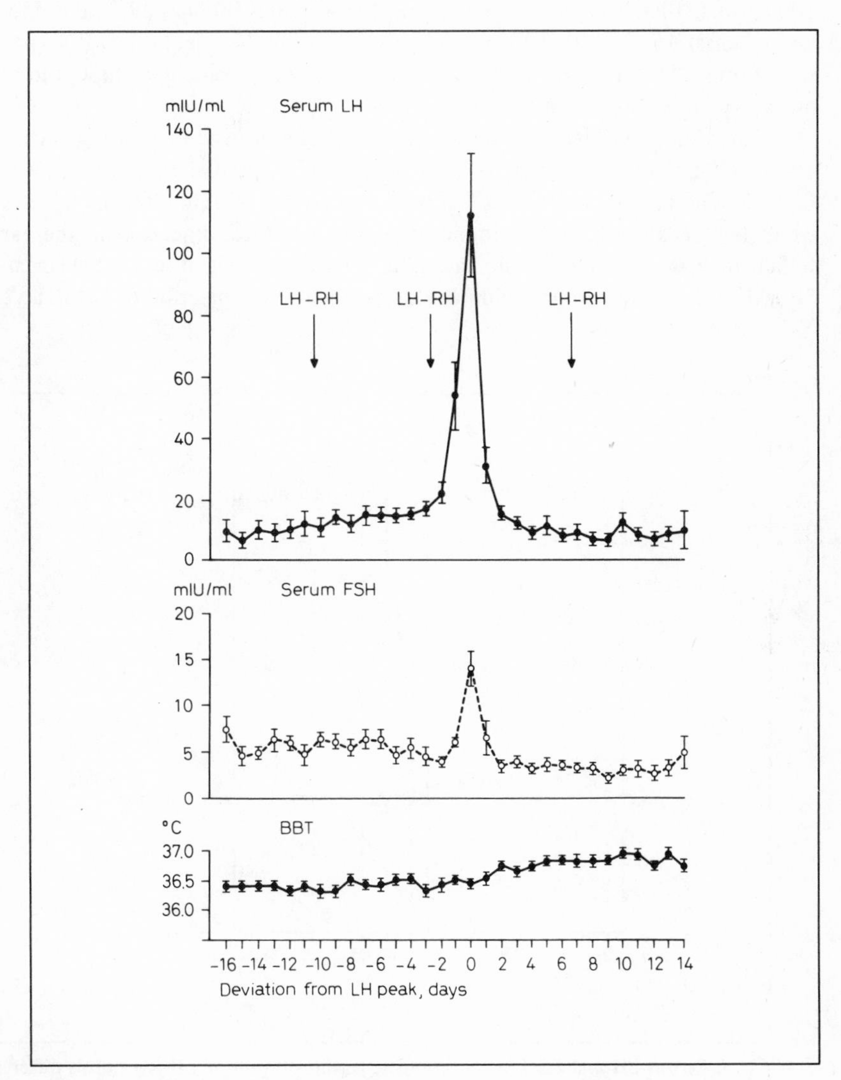

Fig. 2. Serum LH and FSH levels during seven normal menstrual cycles.

Effect of LH-RH and Estrogen on the Serum LH and FSH Levels of Women

Responses of the Serum LH and FSH Levels to LH-RH

The LH-RH preparation used in this study was synthetized by Dr. *Sakakibara* and generously supplied by the Daiichi Pharmaceutical Co. (Tokyo, Japan) and the peptide content was approximately 85 % by weight. Subjects were kept fasting overnight and 5 ml of blood was drawn at around 9 a.m. 100 μg of LH-RH was intravenously injected in about 30 sec and 5 ml of blood was taken at 15, 30, 60, 120 and 180 min following the injection. Eight women with normal menstrual cycle aged 21–30 at the early follicular phase, the preovulatory phase and mid-luteal phase volunteered for the study.

The mean (± SE) levels of serum LH and FSH before and following injection of LH-RH in normal women at three phases of menstrual cycle are shown in figure 3. The maximum levels of LH were observed at 30 min after the injection, while the peak of FSH was found at 60 min after the injection in the early follicular phase and mid-luteal phase. The maximum levels of serum LH (mean ± SE mIU/ml) and maximum fold-increase following the injection of LH-RH were

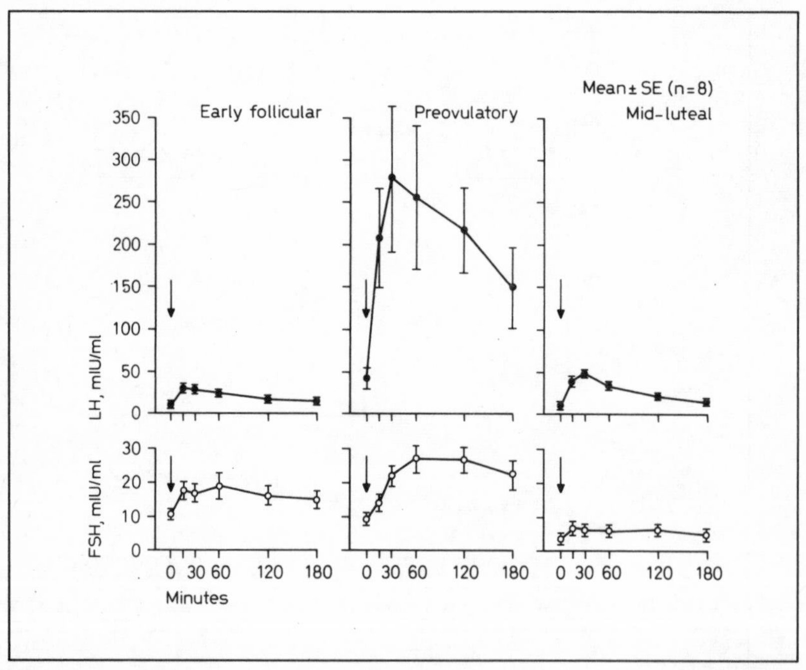

Fig. 3. Serum LH and FSH levels following administration of LH-RH during different phases of menstrual cycles.

in the order of 279.4 ± 87.6 mIU/ml (6.7-fold) at the preovulatory phase, 48.2 ± 3.5 mIU/ml (5.6-fold) at the mid-luteal phase and 29.9 ± 4.9 mIU/ml (2.8-fold) at the early follicular phase. The peak value of serum FSH (mean ± SE mIU/ml) with maximum fold-increase was in the order of 26.7 ± 3.8 mIU/ml (3-fold) in the preovulatory phase, 18.5 ± 3.6 mIU/ml (1.7-fold) in the early follicular phase and 7 ± 1.7 mIU/ml (1.8-fold) in the mid-luteal phase. *Yen et al.* (11) also recently reported the same variation pattern of pituitary responsiveness to LH-RH during different phases of cycle.

Responses of the Serum LH and FSH Levels following Injection of LH-RH in Menopausal Women

Fifteen menopausal women were divided into three groups according to the condition of menstrual irregularity. That is: (1) pre-menopausal: patients who complained of functional bleeding during recent months; (2) peri-menopausal: women who complained of amenorrhea since more than 2 years, and (3) post-menopausal: women who have been amenorrhea since more than 2 years. The responses of serum LH and FSH to LH-RH were summarized in figure 4.

The pre- and post-administration levels of LH and FSH were the lowest in pre-menopausal women among three groups. The resting levels of LH and FSH in peri-menopausal women were already high and the most remarkable increases of LH and FSH were observed following LH-RH injection. This high responsiveness of LH and FSH was somewhat reduced in post-menopausal women. These data,

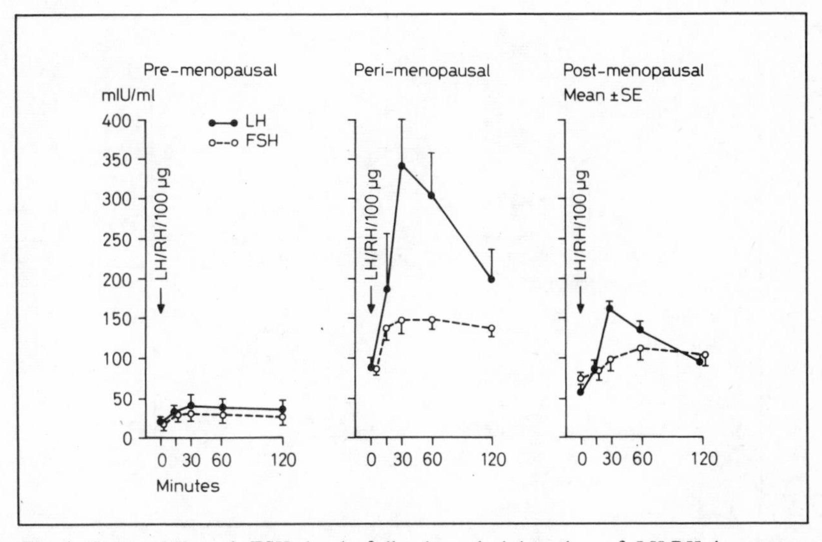

Fig. 4. Serum LH and FSH levels following administration of LH-RH in women approaching the menopause and postmenopausal women.

interpreted as the pituitary reserve function of gonadotropin secretion, are closely related to the ovarian function during the menopause.

Responses of the Serum LH and FSH Levels following Injection of Conjugated Estrogen

In order to clarify the effect of rapid increase of serum estrogen on the LH release, 20 mg of conjugated estrogen (Premarin) was intravenously injected in the women in various conditions. The serum LH levels were determined by radioimmunoassay before and following the injection of Premarin up to 120 h. For the purpose of using this schedule as the hypothalamic function test, the pituitary reserve of gonadotropin, previously checked by injection of synthetic LH-RH in a dose of 100 μg (fig. 5), served during the normal menstrual cycle. The maximum latent time (24 h) and the highest LH level (4.6-fold) were observed at the preovulatory phase. The sensitivity of the hypothalamo-pituitary complex to the feedback action of estrogen varied during the menstrual cycle.

Effect of LH-RH and Cupric Sulfate on the Serum LH Levels of Rabbits

Radioimmunoassay for Rabbit LH

The serum LH levels of adult female rabbit were measured following injection of LH-RH and cupric sulfate with or without pretreatment of estrogen and

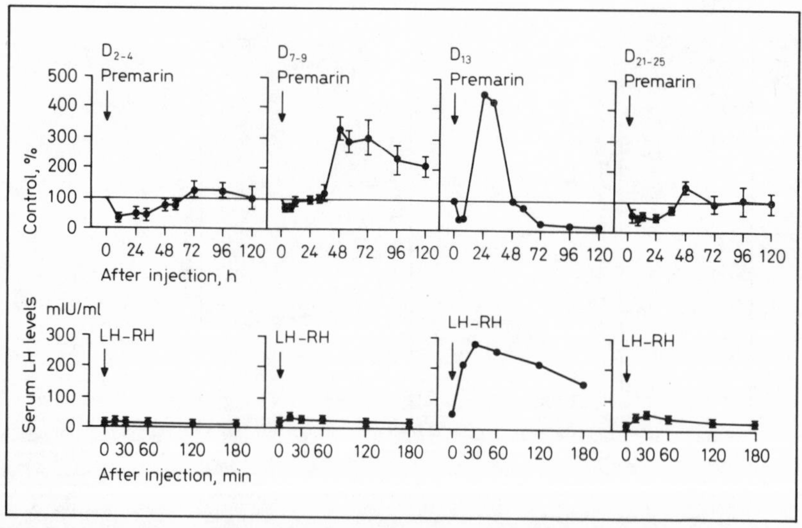

Fig. 5. Changes in serum LH levels following injection of Premarin or LH-RH during different phases of menstrual cycles.

progesterone. The serum LH levels of rabbit were measured by radioimmunoassay (10) as a modified method described by *Scaramuzzi et al.* (9). An ovine LH (NIH-LH-S8) with complete Freund's adjuvant was used for immunization of guinea pigs. The NIAMD-rat-LH was iodinated with ^{125}I using the method of *Greenwood et al.* (3). In the ovine-rat LH radioimmunoassay system (10), binding of ^{125}I-LH to antibody was reduced by addition of increasing doses of crude LH extract of rabbit, NIAMD-rat-LH-RP-1 and rabbit serum. These dose-response curves were parallel between themselves, in the range of 20–80 % bound (0.1–10 μg/ml NIAMD-rat-LH-RP-1). The adult female rabbits (2.5–3 kg) were used for experiments. All the animals were surgically fitted with polyethylene catheters at the femoral vein.

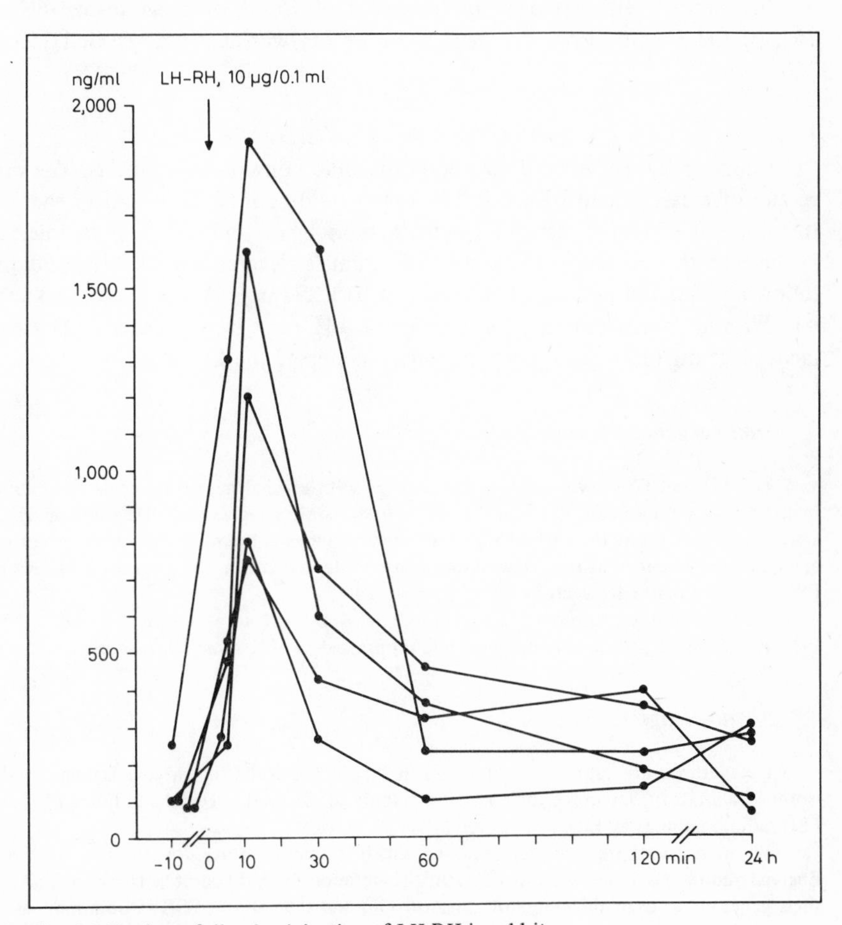

Fig. 6. LH release following injection of LH-RH in rabbits.

Effect of LH-RH on the Serum LH Levels of Rabbits

Five rabbits were injected 10 μg of LH-RH in 0.1 ml of saline from ear vein. 3-ml blood samples were collected via catheter prior to the injection and 5, 10, 30, 60, 120 min, and 24 h following injection.

The injection of 10 μg of LH-RH always resulted in a significant rising of serum LH levels. The serum LH levels peaked at 10 min, following injection (1,250 ± 203 ng/ml) and then sharply decreased to basal levels during 60–120 min (fig. 6).

The time-course pattern of LH levels was quite similar to that of human. All rabbits ovulated at 24 h following injection of 10 μg of LH-RH injection.

The LH surge was clearly enhanced by pretreatment of 200 μg of estradiol benzoate. The peak value of LH showed 4,000 to 2,300 ng/ml.

In contrast with estrogen pretreatment of 25 mg of progesterone before 24 h of LH-RH injection, the peak value of LH was decreased from 1,150 to 260 ng/ml.

Effect of Cupric Sulfate on the Serum LH Levels of Rabbits

Cupric sulfate was used as a hypothalamic stimulant on rabbits. The procedure of measurement of serum LH levels was described above. After the pretreatment of 200 μg of estradiol benzoate, 6 mg/kg of cupric sulfate was injected in the rabbits. The serum LH levels of all animals elevated remarkably at 30 min following injection and high LH levels (1,850 ± 294 ng/ml) were continued until 4 h. Without pretreatment of estrogen, the LH surge was decreased. Pretreatment of 25 mg of progesterone completely supressed the LH surge.

Summary and Comment

The LH and FSH levels of Japanese women with normal menstrual cycle were determined by radioimmunoassay. From the results of dynamic effects of LH-RH and conjugated estrogen (Premarin) at the various phases of menstrual cycle, the maximum secretion of LH and FSH was obtained at the preovulatory phase. After ovulation, the magnitude of LH and FSH surge was markedly decreased.

In the rabbit experiments, estrogen enhanced the LH surge which was induced by LH-RH and cupric sulfate, and progesterone suppressed the LH surge remarkably.

Acknowledgements

The authors are grateful to the Daiichi Pharmaceutical Co. and the Daiichi Radioisotope Laboratory, Japan for their generous supply of synthetic LH-RH and human LH and FSH radioimmunoassay kits, respectively.

We wish to express appreciation to the National Institute for Medical Research, England and the National Institute of Arthritis, Metabolism, and Digestive Diseases, USA for their generous gifts of the standard gonadotropin and rat LH and FSH radioimmunoassay kits, respectively.

References

1 *Arimura, A. and Schally, A.V.:* Physiological and clinical studies with natural and synthetic luteinizing hormone-releasing hormone. Med. J. Osaka Univ. *23:* 77–100 (1972).

2 *Aono, T.; Minagawa, J.; Kawamura, J.; Tanizawa, O., and Kurachi, K.:* Radioimmunoassay of human pituitary FSH and LH. J. jap. obstet. gynaec. Soc. *24:* 821–829 (1972).

3 *Greenwood, F.C.; Hunter, W.M., and Glover, J.S.:* The preparation of ^{131}I-labelled human growth hormone of high specific radioactivity. Biochem. J. *89:* 114–123 (1963).

4 *Kupperman, H.S.; Epstein, J.A.; Blatt, M.H.G., and Stone, A.:* Induction of ovulation in the human; therapeutic and diagnostic importance. Amer. J. Obstet. Gynec. *75:* 301–309 (1958).

5 *Midgley, A.R., jr. and Jaffe, R.B.:* Human luteinizing hormone in serum during the menstrual cycle: determination by radioimmunoassay. J. clin. Endocrin. *26:* 1375–1381 (1966).

6 *Matsuo, H.; Baba, Y.; Nair, R.M.G.; Arimura, A., and Schally, A.V.:* Structure of the porcine LH- and FSH-releasing hormone. Biochem. biophys. Res. Commun. *43:* 1334–1339 (1971).

7 *Minagawa, J.; Aono, T.; Kawamura, K.; Tanizawa, O., and Kurachi, K.:* Radioimmunoassay of serum FSH and LH during normal menstrual cycle and anovulatory conditions. Folia endocrin. Jap. *48:* 396–407 (1972).

8 *Miyake, A.; Aono, T.; Kinugasa, T.; Minagawa, J.; Kawamura, K.; Miyazaki, M.; Tanizawa, O., and Kurachi, K.:* Changes in serum LH levels following intravenous injection of Premarin. J. jap. obstet. gynaec. Soc. (in press, 1973).

9 *Scaramuzzi, R.J.; Blake, C.A.; Papkoff, H.; Hilliard, J., and Sawyer, C.H.:* Radioimmunoassay of rabbit LH. Serum levels during various reproductive states. Endocrinology, Springfield *90:* 1285–1291 (1972).

10 *Tanizawa, O.; Fukada, N.; Kobayashi, Y.; Yamaji, K.; Ichii, H.; Aono, T., and Kurachi, K.:* Bioassay of LH-RH in rats and rabbits. Luteinizing hormone-releasing hormone (in press, 1973).

11 *Yen, S.S.C.; Van den Berg, G.; Rebar, R., and Ehara, Y.:* Variation of pituitary responsiveness to synthetic LRF during different phases of the menstrual cycle. J. clin. Endocrin. *35:* 931–934 (1972).

Author's address: Dr. *O. Tanizawa,* Department of Obstetrics and Gynecology, Osaka University Medical School, *Osaka 553* (Japan)

Psychoneuroendocrinology. Workshop Conf. Int. Soc. Psychoneuroendocrinology, Mieken 1973, pp. 48–55 (Karger, Basel 1974)

Psychoneuroendocrine Aspects in Gynecology and Obstetrics

Y. Okamura, M. Kitazima, K. Arakawa, H. Tateyama, M. Nagakawa, T. Goto, A. Kurano, M. Nakamura and Y. Maruki[1]

Department of Obstetrics and Gynecology (Director: Prof. *I. Taki*), Kyushu University Medical School, Fukuoka

The limbic system and the hypothalamus appear to mediate the somatic manifestations of emotion by acting as a bridge between the neocortex, the autonomic nervous system and the endocrine glands. In this manner, emotions may cause a multitude of endocrine and autonomic reactions, such as increased or decreased secretion of hormones, rise in blood pressure, increased smooth muscle activity of uterus, and other organic or functional changes.

The purpose of this presentation was to investigate the psychoneuroendocrine aspects in gynecology and obstetrics, namely: endocrine diseases, vegetative disturbance and childbirth.

Psychoneuroendocrine Aspects in Endocrine Diseases

The patients of endocrine diseases, who had anovulatory cycles, primary or secondary amenorrhea, were studied. In an attempt to diagnose in which organs the lesions existed (the diencephalonhypophysis, ovaries, or adrenal glands), we made various endocrine stimulating tests in a total of 46 patients, comprising six patients with anovulatory cycle, and 40 patients of amenorrhea. We analyzed the results as is shown in this presentation, and further examined the involvement of the psychogenic factor in these endocrine functions.

1 The authors would like to express hearty thanks to Prof. *I. Taki* and Prof. *Y. Ikemi* for their support and encouragement.

New Concept of Endocrine Disorders

Systematic considerations have been made as to the mechanism of development of endocrine disorders from three points of view, viz., psychology, endocrinology, and neurology, in an attempt to get hold of endocrine disorders as a whole from a new concept as is shown in table I.

The f_1 stands for psychogenic fitness, f_2 for somatic fitness, x for genetic factor, y for situational factor, t for time factor, and z for endocrine disorder.

The endocrine disorder that develops from the x under f_1 is of congenital nature, and anorexia nervosa belongs to this group. Mental conflict at home and anxiety at work may be cited as the y under f_2. The endocrine disorder by the x under f_2 is the case of chromosomal disorder, such as Turner's syndrome and testicular feminization syndrome. Inadequate use of hormones, dietary restrictions (for weight reduction), and massive postpartum hemorrhage may be cited as the y under f_2. The t under f_1 means the period after a psychological load (mental stress) has been applied, and the t under f_2 indicates the period after a somatic situational load has been applied.

Diagnosis of Endocrine Diseases

Electroencephalogram (EEG) was made to examine cerebral cortex-diencephalon, LH-RH test, metopyrone test, and vasopressin test, to examine the diencephalon-pituitary function (2, 4), the gonadotropin-dexamethasone test (G-D test) to examine the ovarian function (3), and the rapid ACTH test to examine the adrenocortical function (6), as shown in table II. Abnormalities of EEG were classified as slowing, slow burst, and 6 and 14 c.p.s. positive spike. The analysis of psychogenic factor was made by the interview, and psychological tests (namely: Cornell medical index; manifest anxiety scale; Yatabe-Guildford's test; Minnesota multiphasic personality inventory; Rosenzweig's picture frustration study, and Rorschach test) based on a new concept of f_1 and f_2 classification.

Table I. New concept of endocrine disorders

$$Z = \left[\begin{array}{l} f_1 (x \cdot yt) \\ f_2 (x \cdot yt) \end{array} \right.$$

Z	=	endocrine disorder.
f_1	=	psychogenic fitness.
f_2	=	somatic fitness.
x	=	genetic factor.
y	=	situational factor.
t	=	time factor.

When we classified the 46 cases into the group f_1 and group f_2, 19 patients belonged to the group f_1 in which the involvement of psychogenic factor was strongly implied, and 22 patients to the group f_2 in which there was no involvement of psychogenic factor, and five patients to neither of the groups. Abnormal functional tests of the diencephalonhypophysis were found in three of the 19 cases, which is 15.9 %; in the group f_1, the result of the ovarian function test found abnormality in nine cases, which is 47.7 %; the result of the adrenocortical function test found abnormality in one case, which is 5.3 %, and no abnormality was found in 7 of the 19 cases, which is 47.7 %, as is shown in table III.

On the other hand, the results of the functional tests of the diencephalon-hypophysis found abnormality in 11 of 22 cases, which is 50 % in the group f_2; abnormal ovarian function was found in 14 cases, which is 63.6 %; abnormal adrenocortical function was found in seven cases, which is 31.8 %, and in only one case from 22 was no abnormality found, which is 4.5 %.

When we examined these findings by classification into the group f_1 and the group f_2, the primary lesions could not be estimated in 18 of the 19 patients, which is 94.7 % in the group f_1, except for one case in the primary lesion which

Table II. Function test for endocrine diseases

Cerebral cortex – diencephalon	EEG
Diencephalon-pituitary	LH-RH test Metopyrone test Vasopressin test
Ovary	Gonadotropin-dexamethasone test
Adrenal cortex	Rapid ACTH test

Table III. Abnormal reaction

	f_1	f_2
	No. of cases, %	No. of cases, %
Diencephalon-pituitary	3/19 (15.9)	11/22 (50.0)
Ovary	9/19 (47.7)	14/22 (63.6)
Adrenal cortex	1/19 (5.3)	7/22 (31.8)
Unknown	9/19 (47.7)	1/22 (4.5)

f_1 : n = 19; f_2 : n = 22; unclassified: n = 5.

Table IV. Primary lesion

	f_1		f_2	
	No. of cases	%	No. of cases	%
Diencephalon-pituitary	0	0	5	22.7
Ovary	1	5.3	4	18.2
Adrenal cortex	0	0	3	13.6
Unknown	18	94.7	10	45.5
Total	19	100.0	22	100.0

was found to exist in the ovaries, the primary lesion could be estimated to exist in the diencephalonhypophysis in five of 22 cases, which is 22.7 % in the group f_2; in the ovaries in four cases which is 18.2 %, and in the adrenal cortex in three cases, which is 13.6 %, but the primary lesions could not be estimated in the rest, viz., ten cases, which is 45.5 % as is shown in table IV.

The primary lesions could not be estimated in 18 cases in the group f_1, except for one patient, but could be estimated in about half of the patients in the group f_2. The results of the above-mentioned tests suggest that mild disharmony of overall endocrine situation is significantly related to psychogenic factor.

Psychoneuroendocrine Aspect of Vegetative Disturbance

Most methods of examinations of autonomic nervous system (ANS) and reading of the results were complicated, and patients were in pain during the examinations. Therefore, we tried to develop a simple method which was easily accepted by patients.

The patients of psychosomatic disorder in the fields of gynecology and obstetrics were checked with 40 questions concerning the ANS.

Plethysmogram (PTG) was taken for 5–10 min during rest, then a tolerance test with noradrenaline was performed. Since October 1971 we preferred intravenous injection of the drug 1 μg/kg in 20 ml of 5-percent glucose solution for 1 min. The pronounced change of PTG was chiefly alteration of amplitude.

Base line was almost horizontal and no variation was found in healthy women.

With ANS score and PTG, in the fields of gynecology and obstetrics, 61 cases of psychosomatic patients were examined and a few interesting findings are described below in two cases of patients with an ANS score above 15.

Table V. Influence of AT on plasma level of free 11-OHCS, length of labor and asphyxia of the newborn

No.	Name	Age	*-para*	Presentation	Body weight of infant, g	AT	Apgar	11-OHCS, μg/dl	Length of labor	
									stage I, h	stage II, min
Means of control			prim.	cephalic → OA	3,105.5	no	–	95.16	20.12	49.7
			mult.	cephalic → OA	3,203.3	no	–	73.24	9.00	26.2
1	R.M.	19	prim.	cephalic → OA	3,190.0	no	10	59.70	10.30	10.0
2	S.H.	27	prim.	cephalic → OA	3,320.0	no	10	63.70	5.50	25.0
3	H.H.	29	mult.	cephalic → OA	3,625.0	epidural	7	87.50	11.20	29.0
4	M.O.	25	mult.	cephalic → OA	3,460.0	epidural	10	83.70	19.30	53.0
5	K.Y.	32	prim.	cephalic → OA	2,820.0	pudendal	10	91.80	5.30	41.0
6	H.H.	25	prim.	cephalic → OA	3,230.0	AT 1 ×	10	67.40	6.00	25.0
7	M.K.	27	prim.	cephalic → OA	3,300.0	AT 1 ×	10	70.50	15.80	20.0
8	A.K.	26	prim.	cephalic → OA	3,260.0	AT 1 ×	10	93.50	18.30	18.0
9	S.K.	23	prim.	cephalic → OA	2,520.0	AT 3 ×	10	109.30	13.00	29.0
10	H.U.	28	mult.	cephalic → OA	3,430.0	AT 4 ×	10	85.40	3.80	22.0
11	M.T.	29	mult.	cephalic → OA	3,120.0	AT 2 ×	9	53.60	8.00	10.0
12	I.Y.	30	mult.	cephalic → OA	3,360.0	AT 4 ×	9	62.40	2.00	25.0
13	N.H.	22	prim.	cephalic → OA	3,010.0	AT 4 ×	10	93.80	3.00	55.0

Case 1: N.K., aged 40 years, married.
ANS score was 19. Marked degree of imbalance was shown in the PTG during rest, and after tolerance test with noradrenaline, base line became more irregular and amplitude became smaller. Her blood pressure rose from 124/78 to 142/80 mm Hg.

Case 2: Y.I., aged 35 years, married.
ANS score was 27. The PTG during rest showed also a high degree of imbalance, and after tolerance test with noradrenaline, some pronounced responses of very small amplitude and irregular base line were found.

Psychoneuroendocrine Aspect in Childbirth

Autogenic training for childbirth requires considerable time and effort on the part of both the patient and the doctor, but observations of women in labor and at the time of delivery have given the impression that such training has had a most beneficial effect.

Three 'levels' in which relaxation may operate are distinguished as muscular, central and interpersonal. A woman may benefit from relaxation on one or more of these levels, according to the susceptibility of her fundamental personality.

Therefore, the main part of autogenic training of pregnant women is education prior to labor. Through this education, pregnant women come to understand the physiological aspects of pregnancy, labor and its pains, so that fear and anxiety can be removed (1, 7, 8). At the same time, pregnant women receive instruction in antenatal bodily exercises, that is, supplemental gymnastics. The fluorimetric method for the estimation of free 11-hydroxycorticoids in human plasma is given in reference 5.

Consideration should be directed to what influence autogenic training has on the plasma concentration of 11-OHCS immediately after delivery. Table V shows them in detail.

Cases 1 and 2 received neither autogenic training nor any kind of obstetric anesthesia and yet they never complained of pain, had no fear and anxiety and looked very comfortable throughout labor and delivery. In these cases the length of labor was of short duration and the levels of plasma 11-OHCS immediately after delivery were low when compared with the control women. These two cases seem to suggest an ideal form of autogenic training and form the basis of such training.

While cases 3 and 4 received epidural anesthesia and suffered no pain, they expressed fear and anxiety, complained of pains and the level of plasma 11-OHCS was higher than the average level of the control women. These cases also seem to suggest the effectiveness of autogenic training from an opposite position. At this point, from the observation of the above four cases, there is an impression that the plasma 11-OHCS levels immediately after delivery are correlated with fear and anxiety.

The same impression was also given by observation of case 5 who received a pudendal block at the moment of full dilatation of the cervix. Cases 6–8 received autogenic training only once and cases 9–13 received training two to four times.

Cases 6 and 7 were able to obtain considerable relaxation and were comfortable throughout labor and delivery, expressing no fear and anxiety. In these cases, support during labor was necessary because they had received autogenic training only once, but plasma levels of 11-OHCS were lower than corresponding levels of the control women and the length of labor was shorter.

Case 8 expressed fear and anxiety in spite of medical and psychological support during labor and the level of plasma 11-OHCS was high even though the length of labor was shortened.

Cases 9 and 10 also expressed fear and anxiety and especially case 9 who could hardly exercise any supplemental movements in spite of support during labor. In both cases the levels of plasma 11-OHCS were high.

In cases 11 and 12, fear and anxiety were removed to some degree and the plasma levels of 11-OHCS were somewhat lower than the levels of the controls.

Case 13 did not complain of pains but expressed fear and anxiety during labor and delivery and the plasma level of 11-OHCS was high.

From the above observations it was concluded that for women who had attained adequate relaxation and had been released from fear and anxiety, the levels of plasma 11-OHCS remained low immediately after delivery, and the length of labor was shortened, indicating a beneficial influence from autogenic training on the mechanics of childbirth.

Conclusion

As the female reproductive tract is extremely susceptible to the physiological expressions of emotional conflicts, the specialties of obstetrics and gynecology lend themselves to the psychosomatic approach more than any other specialty.

Since a successful therapy must be directed to the emotional factors, one must possess a good knowledge of the patient's physical and mental processes as they interact to produce symptoms.

References

1 *Buxton, C.L.:* A study of psychophysical methods for relief of childbirth pain (Saunders, Philadelphia 1962).

2 *Gwinup, G.:* Test for pituitary function using vasopressin. Lancet *18:* 572–573 (1965).
3 *Laitinen, O. and Pesonen, S.:* Some aspects of the function of ovaries in secondary amenorrhea. Acta endocrin., Kbh. *50:* 254–260 (1965).
4 *Liddle, G.W.; Island, D.; Lance, E.M., and Harris, A.P.:* Alterations of adrenal steroid patterns in man resulting from treatment with a chemical inhibitor of 11β-hydroxylation. J. clin. Endocrin. *18:* 906–911 (1958).
5 *Mattingly, D.:* A simple fluorimetric method for the estimation of free 11-hydroxycorticoids in human plasma. J. clin. Path. *15:* 374–379 (1962).
6 *Moncloa, F.; Velezco, I., and Gutierrez, L.A.:* One-hour intravenous ACTH test. J. clin. Endocrin. *26:* 482–483 (1966).
7 *Prill, H.J.:* Methoden psychischer Geburtsschmerzerleichterung. Z. Geburtsh. Gynäk. *146:* 211–229 (1956).
8 *Roth, F.:* Schmerzlose Geburt durch Psychoprophylaxe (Thieme, Stuttgart 1959).

Author's address: Dr. *Y. Okamura,* Department of Obstetrics and Gynecology, Faculty of Medicine, Kyushu University, *Fukuoka-shi* (Japan)

Stress and Biological Rhythm

Psychoneuroendocrinology. Workshop Conf. Int. Soc. Psychoneuroendocrinology, Mieken 1973, pp. 58–66 (Karger, Basel 1974)

Emotional Stress and Biochemical Responses of Manic-Depressive Patients

R. Takahashi, T. Nakahara and Y. Sakurai

Department of Neuropsychiatry, Nagasaki University School of Medicine, Nagasaki

There is fairly good evidence which is consistent with abnormal amine metabolism in manic-depressive disease. However, many of the reported data concerning etiological roles of catecholamines (CA), 5-hydroxytryptamine and amine precursors are controversial. Recent important results of the studies on pathogenesis of manic-depressive disease are the biochemical findings which are common to manic and depressive phases of patients. These are the following: (a) the decreased level (2) of cerebrospinal fluid 5-hydroxyindoleacetic acid (5-HIAA); (b) the decreased accumulation (12) of this compound after probenecid; (c) the decreased level of CSF tryptophan (3), and (d) the decreased level of plasma tyrosine and tryptophan (13). It is worth pointing out that the lower level of 5-HIAA and the latter amino acids were observed also in patients during the complete recovery period (2, 13). This fact suggests there might be a relationship between the abnormal amine metabolism and the predisposition of manic-depressive disease. It is well known that in the recovered patients, mania or depression is easily induced by slight emotional stress inherent in everyday life. In this connection, it seems to be important in the study of pathogenesis of manic-depressive disease to clarify the biological mechanisms which are involved in induction of mania and depression. In such a study, it is necessary to obtain data concerning the biological characteristics of the recovered patients in stressful situations. For this purpose, neuroendocrine responses of the recovered patients during unpleasant emotional state were studied.

Materials and Methods

Eleven patients with an endogenous depressive illness were tested during their complete recovery period. Three of them had a history of manic illness. They were not receiving any medication for at least 2 weeks prior to being used in this study. The patient population consisted of seven females and four males, ranging in age from 21 to 61 years. Eleven controls were normal volunteers who had no known history of psychiatric disorder nor were they receiving any medication. The controls were selected so that their age range and sex distribution were similar to those of the patients.

CPRG rating scale of depression for patient's use (7) was recorded by the controls and the patients before the test. All subjects were explicitly told to refrain from smoking, drinking alcoholic and caffeine-containing beverages, eating bananas, vanilla ice creams, etc., from the previous morning. They participated in a 6-hour experimental procedure – similar to *Levi*'s method (8) – which involved three consecutive 2-hour periods, the second one being the stress period. The test began at 10.00 a.m., and the subjects took no lunch. The stressor included two parts: (a) a 15-min Kraepelin's continuous adding test, and (b) viewing a 30-min film, eliciting feelings of fear and fright. This film showed not only the bloody scenes of traffic accidents, and the resulting tragic and unhappy families of the drivers, but also scenes of brain surgery. This film was produced by the Traffic Safety Association of Japan. During the control periods, preceding and following the stress period, the subjects were allowed to relax, and read a weekly magazine in a room in which soft music was played. During each period the subjects ingested 300 ml of cold tap water. Urine samples were collected at the end of each 2-hour period. Each subject was given the *Taylor*'s manifest anxiety scale (MAS) test (14) and their blood pressures and heart rates were also measured at the end of each period. Volumes and specific gravity of the urine samples were measured and a 20-ml aliquot of each sample was taken for the determination of 3-methoxy-4-hydroxyphenylethylene glycol (MHPG). Added was 10 mg of sodium metabisulfite as preservative and the samples were stored at –25 °C. The remainder of each sample was acidified by adding 1 ml of 6 *N* HCl per 100 ml of urine and kept at –25 °C until the catecholamines were determined. Norepinephrine (NE) and epinephrine (EP) were measured by the method of *Crout* (4) whereas MHPG was measured by the method of *Dekirmenjian* (5) using gas-liquid chromatography with electron capture detector. Urinary levels of catecholamines and MHPG were expressed as μg and ng per mg creatinine.

Results

There was no significant difference in the mean age of the controls and patients, in their sex, nor in the mean of their scores of the depression rating scale (see table I). After the test ended, the subjects reported experiencing feelings of tension and fear during the stress period.

Table I. Scores of depression rating scale in depressive patients during the complete recovery period and in controls

	No.	Age, years		Sex		Score of depression rating scale	
		mean	SE	male	female	mean	SE
Controls	11	36.5	3.5	2	9	127.1	12.9
Recovered depressive patients	11	44.8	4.1	4	7	113.6	10.9

MAS

As expected, a rise occurred in the score of the MAS, both in the controls and in the patients. Although there was no significant difference between the means of MAS scores during the stress period of the controls and the patients, the means of MAS scores during the second control period for the patients did not fall as sharply as those in the control subjects (see fig. 1). When the profile of each subject in the two groups was examined, only three of eleven patients showed lower MAS scores in the second control period than in the first, while nine of the control subjects showed lower values. This difference was statistically significant at the level of $p < 0.025$. These findings suggest the mental stress reaction continued longer in the recovered patients than in the control subjects.

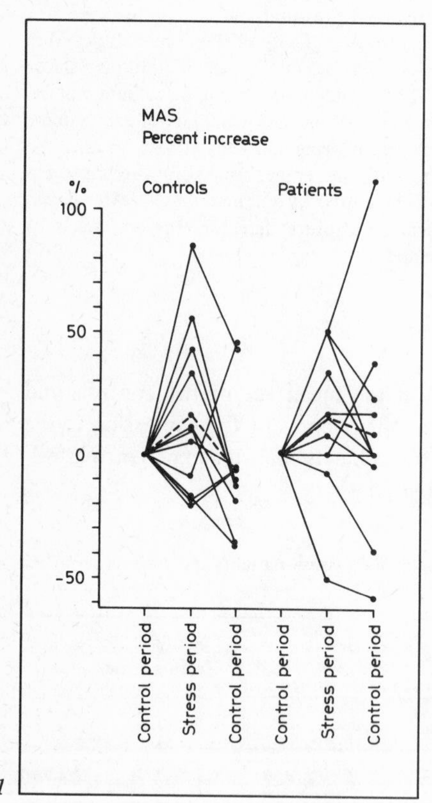

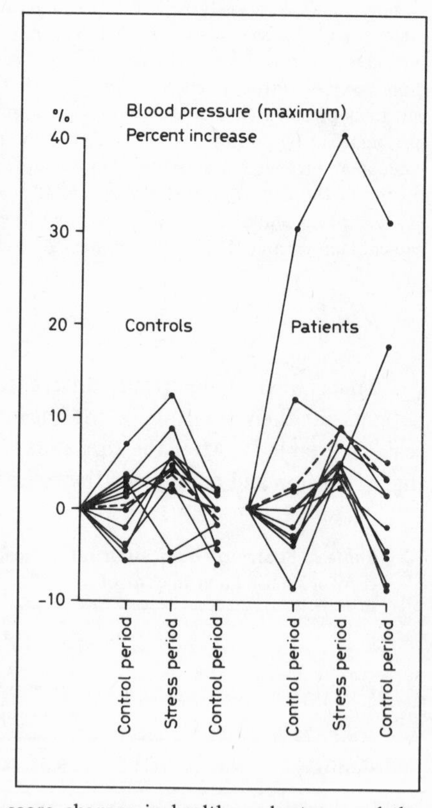

Fig. 1. Taylor's Manifest Anxiety Scale score changes in healthy volunteers and the recovered patients before, during and after emotional stress. Dotted line expresses means.

Fig. 2. Systolic blood pressure changes during the conditions indicated in figure 1.

Table II. Correlation between systolic blood pressure and MAS

		Control period	Stress period	Control period
Controls	r	0.682 $p < 0.025$	0.631 $p < 0.05$	0.595 $p < 0.1$
Patients	r	0.216 NS	0.328 NS	0.558 NS

Blood Pressure

The stress-induced change in systolic blood pressure occurred both in the control and patient groups (fig. 2). The mean of change in the patient group was not significantly different from that in the control group during the stress and the control periods.

There was no significant difference in the change of diastolic blood pressure and heart rate between the controls and the patients during all three periods. Although the emotional responses were different between the two groups (as mentioned above), these results indicate that the patients react to the emotional stress with the physical activity of the same degree as the control subjects. There was significant correlation between systolic blood pressure and MAS score during the stress period in the controls, but no significant correlation was found in the patients (table II). Thus, the dissociation between blood pressure response and emotional response in the patients was confirmed again. Furthermore, the correlation between blood pressure and MAS score in the patients was not significant in the first control period. This finding might indicate either that the control period used in this study was already stressful to the patients or that the patients have a dissociation between physical activity and emotional state in the usual situation, like everyday life.

Catecholamine Excretion

Contrary to our expectation, the mean value of urinary NE did not increase during the stress period in the control group, although it showed a slight but not significant increase in the second control period (fig. 3). However, the mean value of urinary NE in the patients during the stress period was significantly lower than that during the first control period. Furthermore, when the profile of each subject was examined, five of ten individuals in the control showed a rise in NE excretion during the stress period as compared with the first control period.

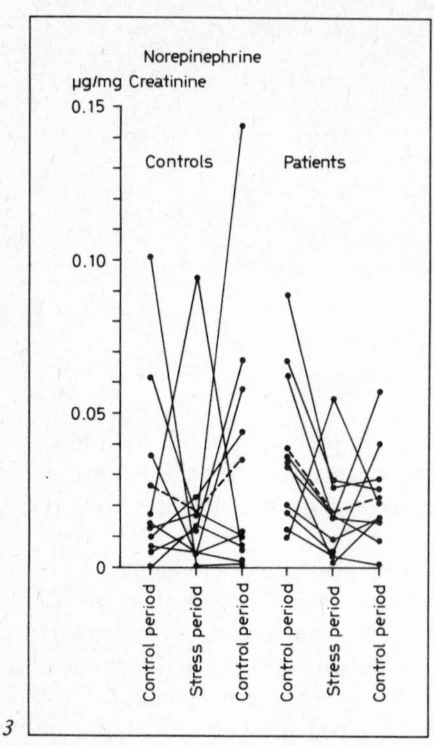

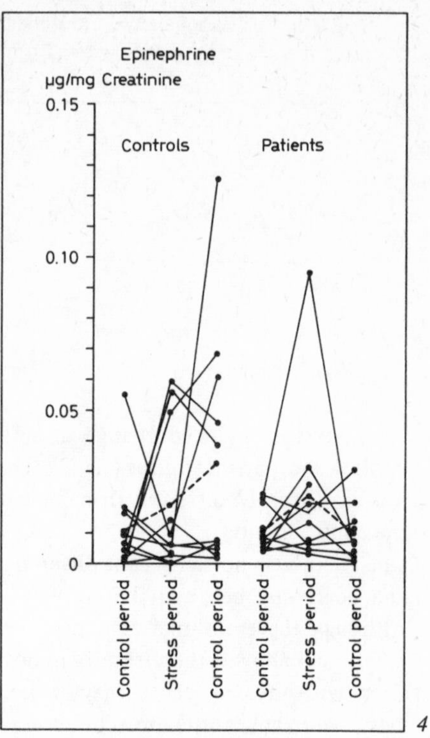

Fig. 3. Urinary excretion of norepinephrine (NE) during the conditions indicated in figure 1.

Fig. 4. Urinary excretion of epinephrine (EP) during the conditions indicated in figure 1.

In the patients, only one subject showed a rise in NE excretion during the stress period. This difference was statistically significant at the level of $p < 0.03$.

There was no significant correlation between the urinary NE level and MAS score in the controls or in the patients during the three periods.

The data in figure 4 show that EP levels in urine from controls increased during the stress period and during the following control period. However, this rise was not statistically significant, although the value of the second control period was close to significance ($t = 1.94$). In the patient group, there was no significant change in the mean value of EP level in urine between the stress and the control periods. However, only four patients excreted more EP in the second than in the first control period, while in the controls, eight subjects excreted more EP in the second than in the first control period. This difference was statistically significant at the level of $p < 0.03$. This finding indicates that the adrenomedullary response to an emotional stress in the patients was different

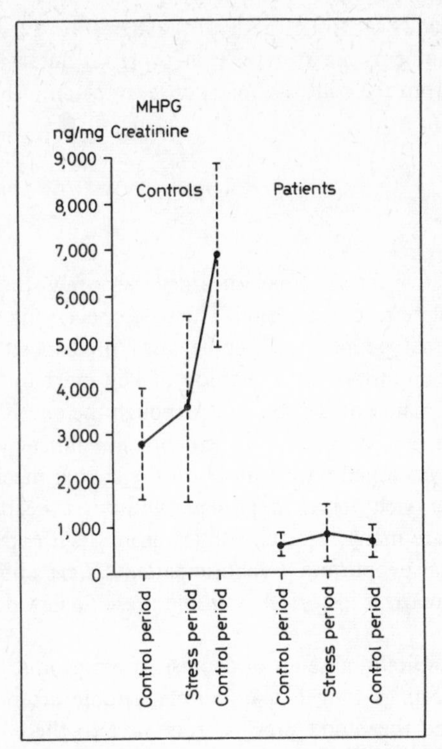

Fig. 5. Urinary excretion of 3-methoxy-4-hydroxyphenylethylene glycol (MHPG) during the conditions indicated in figure 1. Solid line expresses means. Dotted line expresses standard error of mean.

from that in the controls. Correlation between the urinary EP level and MAS score was not significant during three periods either in the controls or in the patients.

MHPG Excretion

There was a significant difference in levels of MHPG in urine between the controls and the patients during the control periods. The patients excreted much less MHPG than the controls as shown in figure 5. The stress-induced change of MHPG excretion was not found in the patient group, while there was a tendency of increase of MHPG level following the stress in the control subjects. However, the stress-induced increase of MHPG excretion in the controls was not statistically significant because of two reasons: the large variation of each value and the small number of determinations made (control 6, patient 8). Therefore, further studies on a large population of subjects are necessary in order to determine whether the patients are different from the controls in response to the emotional stress (in terms of MHPG excretion).

Discussion

As far as biochemistry of affective disorders is concerned, very few studies deal with the patients during complete recovery period. The data concerning the biological difference between normal people and people with predisposition toward manic-depressive illness are important for the study of key mechanisms which are involved in precipitating mania or depression. Although the excellent paper (1) on a biochemical switch mechanism of manic-depressive illness was recently reported, the latter was concerned with studies on the switch mechanism from depression to mania or vice versa, in patients without a healthy period. On the other hand, there are many research studies dealing with stress and biochemical responses in normal people and psychotic patients. *Levi*'s book (9) is a good reference since it summarizes historical facts and presents new data in this field.

In our study, we found that in spite of absence of depressive symptoms, the recovered patients have anxiety feelings longer than normal people after responding to unpleasant stress. Since the blood pressure responses to the emotional stress were similar in the patients and the controls, it is presumed that the recovered patients have dissociation between physical activity and emotional state under the unpleasant situations. The blood pressure response to infused NE has been reported to be greater after recovery from depression than during depression (11). This finding does not seem incompatible with the present result.

The stress-induced response of urinary NE and EP was different between the patients and the controls. Since nearly all of the urinary NE comes from sympathetic nerves and EP from adrenal medulla, lower responses of NE and EP indicate that the patients have impaired capacity in sympathico-adrenomedullary response to emotional stress. On the basis of good evidence (10), urinary MHPG level is now thought to reflect some of the adrenergic activity in the brain. Therefore, the determination of urinary MHPG excretion was undertaken to study the central adrenergic activity of the recovered patients. Because of a small number of determinations of urinary MHPG, it was not possible to conclude that the patients had reduced response of central adrenergic activity. However, it should be pointed out that the recovered patients excreted much less MHPG than the control subjects during the control period. Only one paper (6) described the longitudinal study of urinary MHPG level in depressed patients and reported that decreased urinary MHPG returned to normal level upon remission from depression. Although our results seem to be contrary to that report, a comparison cannot be meaningful, since the patients in this study were not receiving any medication during complete recovery and all subjects were placed on a 6-hour experimental procedure which was in many ways different from usual life. Collecting the data of urinary MHPG in large populations of recovered patients are needed.

Summary

Eleven subjects with an endogenous depressive illness during their complete recovery period are compared with 11 normal volunteers. The groups were matched with respect to age and sex. Measurement of anxiety scales, blood pressure and urinary catecholamine responses to emotional stress were made. The results indicate that the recovered patients have dissociation between physical activity and emotional state and impaired capacity of sympatho- adrenomedullary response to emotional stress.

References

1 *Bunney, W.E., jr.; Goodwin, F.K., and Murphy, D.L.:* The switch process in manic depressive illness. II. Relation to catecholamines, REM sleep and drugs. III. Theoretical implication. Arch. gen. Psychiat. *27:* 304–317 (1972).

2 *Coppen, A.; Prange, A.J., jr.; Whybrow, D.C., and Noguera, R.:* Abnormalities of indoleamines in affective disorders. Arch. gen. Psychiat. *26:* 474–478 (1972).

3 *Coppen, A.; Brooksbank, B.W.L., and Peets, M.:* Tryptophan concentration in the cerebrospinal fluid of depressive patients. Lancet, June 24: 1393 (1972).

4 *Crout, J.R.:* Catecholamines in urine; in *Seligson* Standard methods of clinical chemistry, vol. 3, pp. 62–80 (Academic Press, New York 1961).

5 *Dekirmenjian, H. and Maas, J.W.:* An improved procedure of 3-methoxy-4-hydroxyphenylethylene glycol determination by gas-liquid chromatography. Analyt. Biochem. *35:* 113–122 (1970).

6 *Greenspan, K.; Schildkraut, J.J.; Gordon, E.K.; Baer, L.; Aronoff, M.S., and Durell, J.:* Catecholamine metabolism in affective disorders. III. MHPG and other catecholamine metabolites in patients treated with lithium carbonate. J. Psychiat. Res. *7:* 171–183 (1970).

7 *Kurihara, M.; Itoh, H.; Kawakita, Y.; Kudo, Y.; Satoh, Y.; Takahashi, R., and Tanimukai, H.:* A double-blind controlled study on the clinical efficacy of four anti-depressant drugs – clomipramine, dimethacrine, nortriptyline and protriptyline. Clin. Evaluat. *1:* 27–51 (1972).

8 *Levi, L.:* Neuro-endocrinology of anxiety. Brit. J. Psychiat. Special publ. *3:* 40–52 (1968).

9 *Levi, L.:* Society, stress and disease, vol. 1. The psychosocial environment and psycho-somatic diseases. (Oxford Univ. Press, New York 1971).

10 *Maas, J.W. and Landis, D.H.: In vivo* studies of metabolism of norepinephrine in central nervous system. J. Pharmacol. exp. Ther. *163:* 147–162 (1968).

11 *Prange, A.J., jr.; McCurdy, R.L., and Cochrane, C.M.:* The systolic blood pressure response of depressed patients to infused norepinephrine. J. psychiat. Res. *5:* 1–13 (1967).
12 *Sjöström, R. and Roos, B.E.:* 5-Hydroxyindoleacetic acid and homovanillic acid in cerebrospinal fluid in manic-depressive psychosis. Europ. J. clin. Pharmacol. *4:* 170–176 (1972).
13 *Takahashi, R.:* Discussion at symposium 'biological mechanisms of schizophrenia and schizophrenia-like psychoses', Kyoto Conference on Clinico-Biological Psychiatry, Kyoto, Japan, 1973.
14 *Taylor, J.A.:* A personality scale of manifest anxiety. J. abnorm. soc. Psychol. *48:* 285–290 (1953).

Authors' address: Prof. *R. Takahashi,* Dr. *T. Nakahara* and Dr. *Y. Sakurai,* Department of Neuropsychiatry, Nagasaki University School of Medicine, *Nagasaki* (Japan)

Psychoneuroendocrinology. Workshop Conf. Int. Soc. Psychoneuroendocrinology, Mieken 1973, pp. 67–76 (Karger, Basel 1974)

Growth Hormone and Cortisol Secretion during Nocturnal Sleep in Narcoleptics and in Dogs

K. Takahashi, Y. Takahashi, S. Takahashi and Y. Honda
Departments of Neurochemistry and Psychology,
Tokyo Metropolitan Institute for Neurosciences, Tokyo;
Department of Psychiatry and Neurology,
Kyoto Prefectural University of Medicine, Kyoto, and
Department of Neuropsychiatry, University of Tokyo, Tokyo

It has been well established that human growth hormone (HGH) release occurs immediately after sleep onset. This time-relationship was clearly demonstrated by the fact that when sleep onset was shifted, HGH secretion followed the shift (2, 6, 11, 15, 17). Cortisol secretion was suggested also to be related to sleep-wakefulness rhythm. Reversal of sleep time and daily activity for 1–2 weeks was followed by a phase reversal of the circadian pattern of plasma cortisol level (4, 7). *Weitzman et al.* (21) demonstrated the main secretory phase of cortisol during the sixth, seventh and eighth h of sleep. Prolactin (12), testosterone (1) and luteinizing hormone (5) were reported also to be released during nocturnal sleep.

We were encouraged by these findings to study the secretory patterns of hormones in subjects with a disturbance of the sleep-wakefulness rhythm. One of the best known diseases with a disruption of normal sleep-wakefulness rhythm is narcolepsy. It is characterized by frequent attacks of sleep in the daytime and frequent awaking during nocturnal sleep. Direct entry to REM sleep from wakefulness is another characteristic feature in narcolepsy (8, 13). It was of interest, therefore, to see how pituitary hormones are released during nocturnal sleep in this pathological condition.

Dogs were used as the experimental model for the spontaneous disruption of the sleep-wakefulness rhythm in humans, as the dog shows polyphasic sleep pattern over a period of 24 h (18). In addition, dogs are convenient for the purpose of taking serial blood samples and because of their similarity to humans in terms of their secretory patterns of GH and cortisol. Not only are the diurnal patterns of cortisol comparable in dogs and humans (10) but also dogs respond to a number of external stimuli such as insulin hypoglycemia, L-dopa and cyclic

AMP with a marked increase in levels of plasma canine GH (16, 20). These stimuli are well known to produce HGH release in humans, but not in rats and cats (3, 14).

Materials and Methods

Studies in Humans

Ten typical male narcoleptics aged 24–43 years were used and matched 12 normal adults selected as controls. Normal sleep-wakefulness rhythm was maintained in narcoleptic patients by medication which was withdrawn 3 days before the experiment. On the day of the actual experiment subjects were not restricted in terms of normal activities, food and water intake; after 6 p.m. of the same day the subjects were only allowed to take water. They were instructed to go to sleep at their customary time in a soundproof, dark and electrically shielded room.

Sleep was monitored on a polygraph and classified by means of the standardized scoring criteria (9). Blood was sampled every 20–30 min by means of an indwelling catheter placed in the antecubital vein. HGH was determined in plasma by radioimmunoassay. Plasma cortisol was measured by competitive protein binding method. A secretory phase was defined as the period between two sampling points during which plasma level of hormones rose more than 3 ng/ml in HGH or 3 μg/100 ml in cortisol and the concentration at the sampling point subsequent to the two points was also higher than the initial one.

Studies in Dogs

Ten male mongrel dogs of adult age weighing 8–10 kg were used. EEG electrodes were implanted stereotaxically into the motor and visual cortex and dorsal hippocampus. Screw electrodes for electrooculogram (EOG) were implanted near the lateral orbital ridges. Electrodes for electromyogram (EMG) were implanted also in the neck muscle. All electrodes were connected to a female plug which was fixed firmly on the skull with dental cement. A silicone catheter was inserted into the right atrium through the jugular vein and connected to a plastic two-way connector on the skull embedded in the dental cement, so that connecting tubing was adaptable for blood sampling during the experiment.

Sleep stages were classified into the following four stages of W (wakefulness); L-sleep (light sleep); D-sleep (deep sleep), and REM sleep. L-sleep is characterized by humps, sleep spindles and K-complexes. D-sleep is defined by high voltage, δ-waves covering more than 20 % of the record. REM-sleep is characterized by low voltage fast neocortical and rhythmic 4–6 c/sec hippocampal waves, accompanied by rapid eye movements and abolishment of EMG in the neck muscle.

Canine GH was determined by the radioimmunoassay developed by *Tsushima et al.* (20). Plasma cortisol was measured by competitive protein binding method. The hormone secretory phase was determined in the same manner as described in the study in humans.

In the studies with dogs two different designs were used as follows:

(1) Dogs under uncontrolled conditions. Dogs were housed in a room where no strict light control was made. Dogs were fed once in the morning and permitted to sleep as desired. A few hours before the experiment, each dog was transferred to a soundproof, electrically shielded box and the female plug on its skull was connected to a polygraph recorder. Experiments were conducted for a period of 4 h in the early evening.

(2) Dogs under controlled conditions. Dogs were housed in a light-controlled room (light on 9 a.m. to 5 p.m.) and fed once in the morning. One week before the experiment, each dog was transferred into a dog cage placed in an electrically shielded, light-controlled

and soundproof room. The dog was kept awake by artificial stimuli between 1 p.m. and 5 p.m. every day for a week until the day of the experiment. Polygraphic recording and blood sampling were performed during the period of forced wakefulness and the subsequent 4 h in the darkness.

Results

Studies in Humans

In all of the 12 normal subjects, a marked HGH secretion occurred within 30 min after the onset of sleep. One to five secretory phases were observed during a night sleep (fig. 1). As shown in figure 1, the initial peak of plasma HGH level was the largest in all cases except one, and most of the secretory phases appeared in the first half of a night's sleep.

One to five secretory phases of cortisol were observed during a night's sleep in normal subjects as shown in figure 2. Low level was maintained in the first half of sleep and most of cortisol secretion occurred in the latter half. The highest level of plasma cortisol was reached towards the end of sleep (fig. 2).

No HGH secretion was observed in one case of postencephalic narcolepsy. In the other nine cases, one to five secretory phases appeared during a night's sleep, as shown in figure 3. Distribution of secretory phases was very irregular and the largest HGH secretion, occurring within 2 h after sleep onset, was observed in only three cases. On the other hand, the largest peak appeared in the latter half of a night's sleep in five cases (fig. 3).

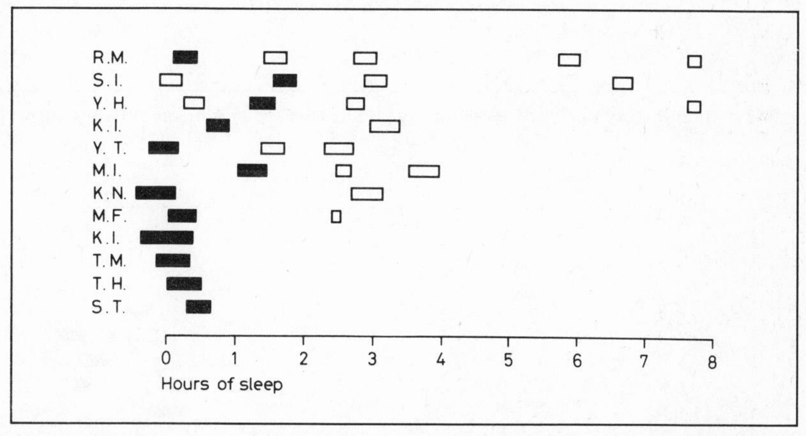

Fig. 1. Secretory phases of HGH during nocturnal sleep in normal subjects. For this and subsequent figures (fig. 2–4), bars represent duration and timing of individual secretory phases. A solid bar shows the secretory phase during which plasma hormone level reached the highest value in each individual. The abscissa shows time in hours after the onset of sleep.

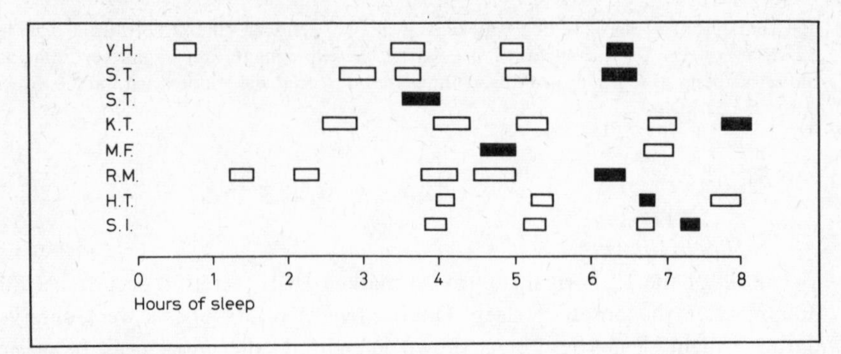

Fig. 2. Secretory phases of cortisol during nocturnal sleep in normal subjects.

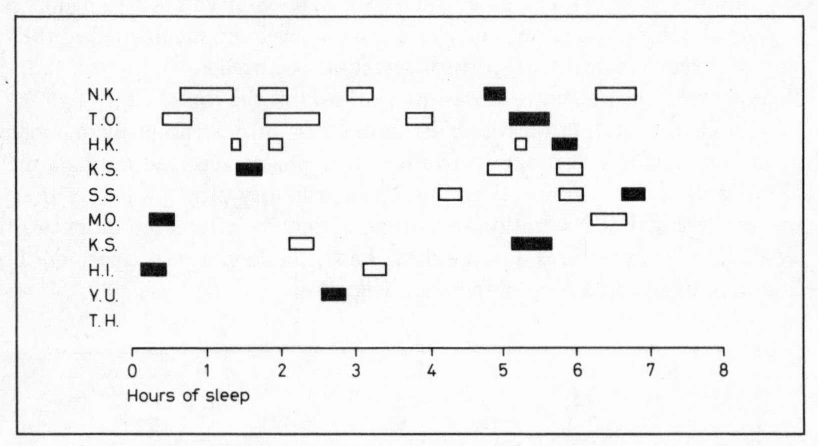

Fig. 3. Secretory phases of HGH during nocturnal sleep in subjects with narcolepsy.

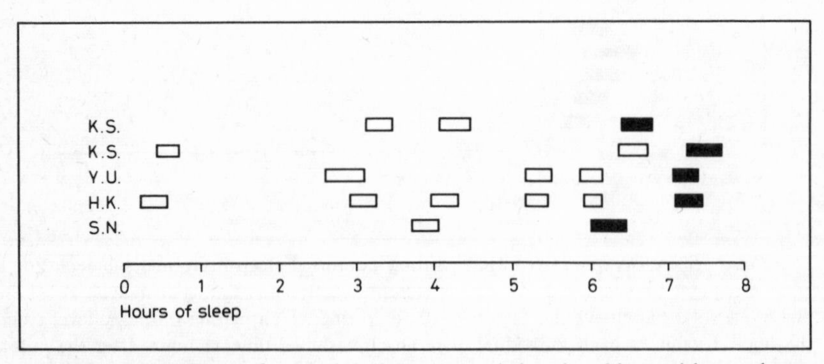

Fig. 4. Secretory phases of cortisol during nocturnal sleep in subjects with narcolepsy.

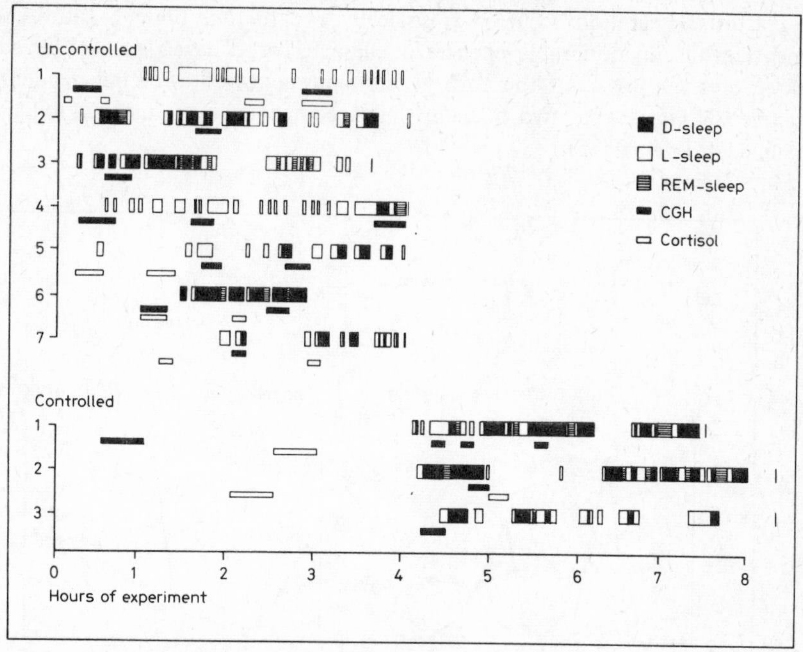

Fig. 5. Patterns of sleep and hormone secretion in dogs under uncontrolled conditions (top) and in dogs under controlled conditions (bottom). Thick solid bars, thick open bars and thick striped bars represent D-sleep, L-sleep and REM-sleep, respectively. Thin solid bars and thin open bars represent the secretory phases of canine GH and cortisol, respectively. The abscissa shows time in hours after the start of experiment.

Contrary to the abnormal secretory patterns observed in HGH, cortisol was released in the manner similar to normals. As shown in figure 4, secretory patterns of cortisol in the narcoleptics resembled that in normals. Most of cortisol secretion occurred in the latter half of a night's sleep and the last peak was the largest in all subjects tested.

Studies in Dogs

(1) Dogs under uncontrolled conditions. Sleep pattern of dogs in the present study were very irregular and varied from dog to dog (fig. 5). Canine GH level increased episodically, forming one to three peaks during the experimental hours. Canine GH peaks observed in the whole series of experiments were 12. Six of them appeared during sleep, while three of them appeared during wakefulness. Three other peaks were difficult to correlate to any sleep stage because of frequent changes in stages (fig. 5). Some examples of canine GH and cortisol secretion are shown in figure 6.

Cortisol secretion occurred episodically two to four times in four experiments, while no noticeable changes in plasma cortisol levels were observed in three of seven dogs. All the cortisol secretions appeared to be independent of canine GH release, but two of them were accompanied by an increase in plasma canine GH level (fig. 5).

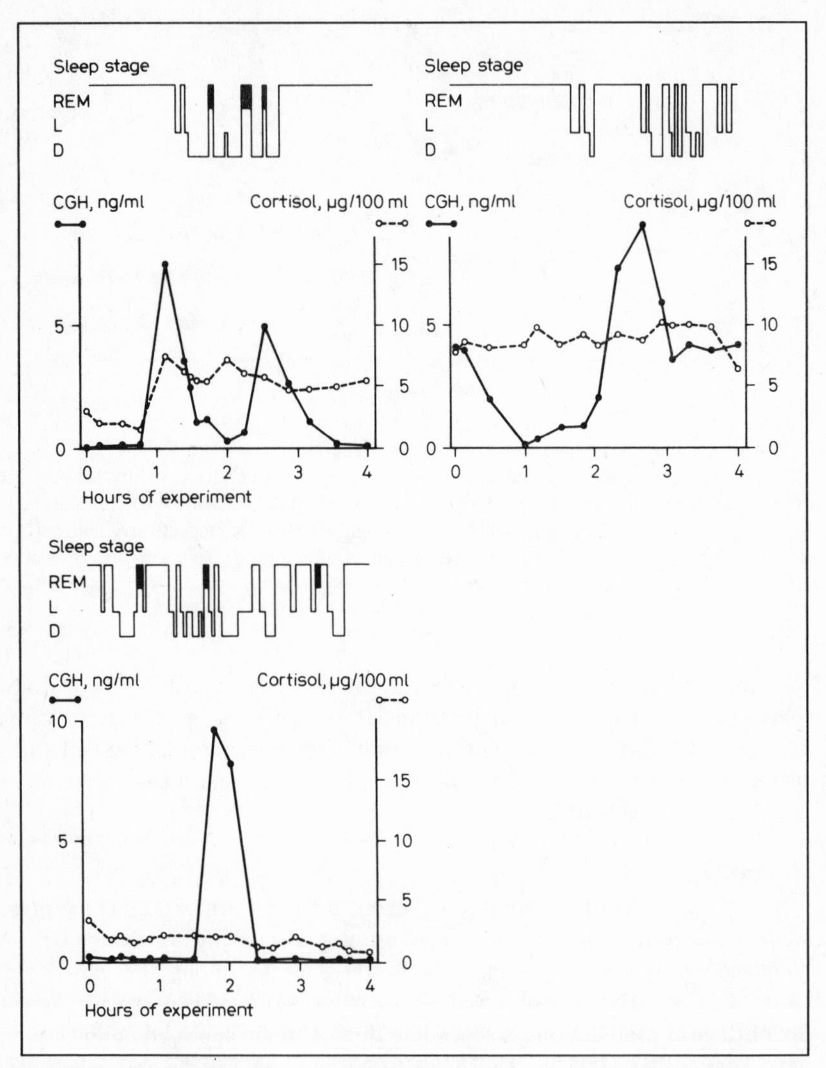

Fig. 6. Canine GH and cortisol secretion in dogs under uncontrolled conditions. For this and figure 7, solid circles represent plasma level of canine GH and open circles plasma level of cortisol. The abscissa shows time in hours after the start of experiment.

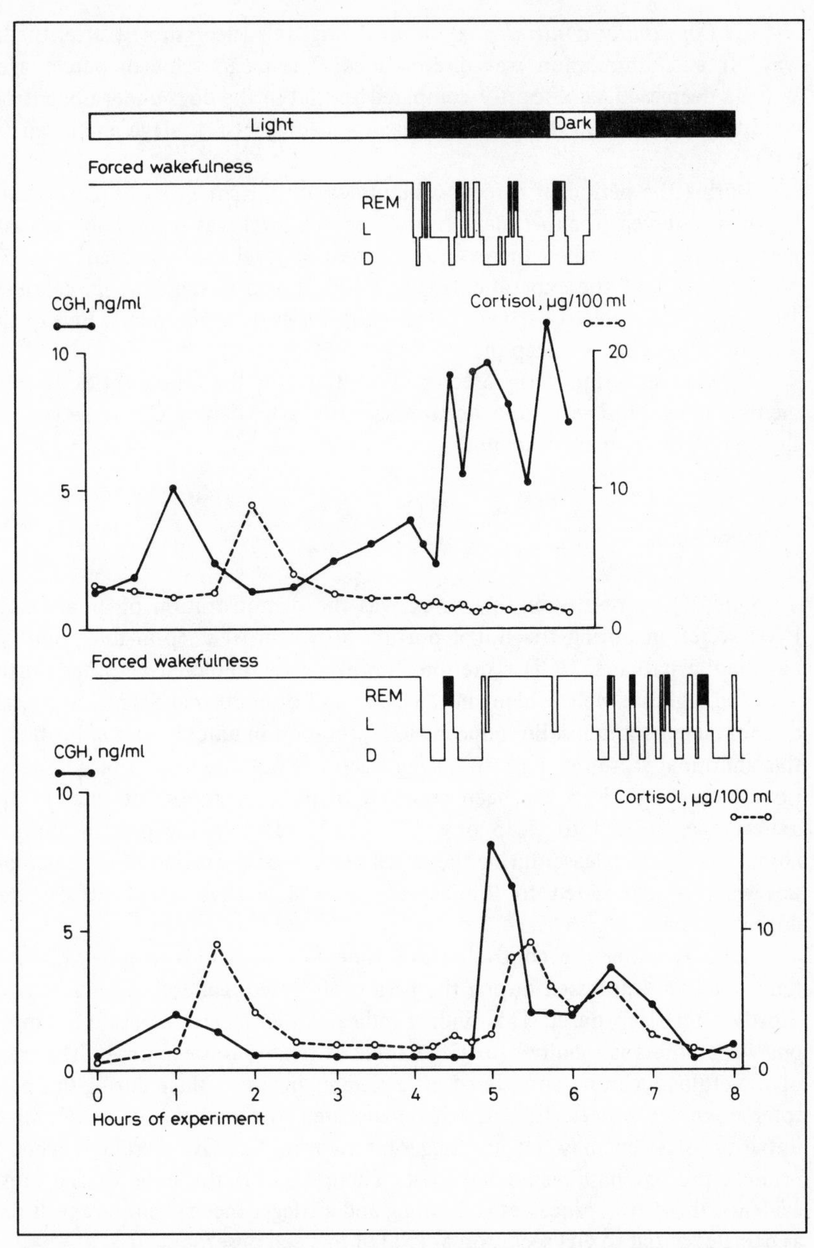

Fig. 7. Canine GH and cortisol secretion in dogs under the controlled condition. Dogs were kept awake by external stimuli for the first 4 h of the experiment. The external stimulation was discontinued and light was turned off 4 h after the start of experiment. The abscissa shows time in hours after the start of experiment.

(2) Dogs under controlled conditions. Dogs fell asleep shortly after the light was off and stimulation was discontinued. The mean value of percentage of W stage decreased significantly compared to that in the dogs under uncontrolled conditions ($p < 0.005$), and the mean percentage of D stage increased ($p < 0.005$).

During the period of forced wakefulness, only a small rise of plasma canine GH was observed in experiment 1 and base-line level was maintained in experiment 2, while a marked increase in canine GH level was observed after sleep onset in both of the experiments (fig. 7). As shown in figure 5, six canine GH secretory phases were observed in the whole series of experiments. Five of them were observed close to sleep onset.

Cortisol secretion was observed three times in the whole series of experiments. One of the secretions occurred shortly after canine GH secretion when the dog woke up in experiment 2.

Discussion

One of the results of this study was the demonstration of an absence of HGH secretion during the initial portion of nocturnal sleep in the majority of narcoleptic patients. HGH secretion, however, was detected at other portions with the highest levels evident in the latter half of nocturnal sleep. The presence of spontaneous and insulin-induced HGH secretion in narcoleptics indicated that this abnormal secretory pattern during sleep was not due to pituitary hypofunction. Slow-wave sleep has been reported to play an important role for HGH secretion as related to sleep onset (2, 11, 17, 19). In the present study the absence of HGH release during the initial portion of nocturnal sleep in narcoleptics was not correlated to this aspect, since slow-wave sleep was detectable during this period.

In dogs under the controlled experimental conditions, canine GH release tended to be suppressed during the period of forced wakefulness and occurred shortly after sleep onset. This finding indicates that a certain period of continuous wakefulness is required for GH release as related to sleep onset. This period of wakefulness can be considered as 'a priming process', since during this period some unknown process for GH release appeared to be in progress. Alternatively, initiation of sleep may act as 'trigger-mechanism' for GH secretion, when the priming process has been completed. Therefore, on the basis of the present evidence these two processes (a priming and a trigger-mechanism) are postulated as being essential to GH secretion as related to sleep onset.

It is tempting to speculate that in narcolepsy there is a disturbance of the priming and/or trigger-mechanism leading to a defect in HGH secretion as related to sleep onset. One of the most characteristic symptoms of narcolepsy is frequent attacks of sleep during the daytime. A disruption of normal sleep-wakefulness cycle caused by attacks of sleep might result in a disturbance of the priming process. On the other hand, the pathological phenomena at sleep onset, such as direct entry to REM sleep, hypnagogic hallucinations, sleep paralysis, cataplexy and so on are characteristic of narcolepsy. Such abnormalities in the process of sleep onset might affect the trigger-mechanism for HGH secretion during the initial portion of nocturnal sleep. It still remains to be clarified which process, the priming or triggering process, is more responsible for the lack of HGH secretion as related to sleep onset.

In normal subjects HGH secretion occurred scarcely in the latter half of a night's sleep. Marked secretion of HGH in the latter half of nocturnal sleep in narcoleptics may reflect a loss of the relationship between slow-wave sleep and HGH secretion, and this could account for the subsequent random HGH release.

Contrary to the abnormality of HGH secretion, the pattern of cortisol secretion in narcolepsy is similar to that in normals, indicating that ACTH regulating system is intact and independent of the dysfunction of HGH regulation. Furthermore, a disruption of normal sleep-wakefulness cycle and the pathological phenomena at sleep onset did not appear to influence the process of ACTH secretion like that of HGH secretion.

Summary

A marked secretion of HGH as related to sleep onset, which was consistent in normal subjects, was not observed in the majority of narcoleptics. Contrary to the abnormality of secretory pattern of HGH, normal secretory pattern of cortisol was observed in all narcoleptics tested.

In dogs under uncontrolled conditions, secretion of canine GH was difficult to correlate to any stages of sleep, because of frequent changes of sleep and wakefulness. In dogs under controlled conditions, in which the animals were kept awake by external stimuli for 4 h in the afternoon every day for a week before the experiment, canine GH release tended to be suppressed during the period of forced wakefulness and occurred shortly after sleep onset.

References

1 *Evans, J.I.; MacLean, A.W.; Ismail, A.A.A., and Love, D.:* Concentrations of plasma testosterone in normal men during sleep. Nature, Lond. *229:* 261–262 (1971).

2 *Honda, Y.; Takahashi, K.; Takahashi, S.; Azumi, K.; Irie, M.; Sakuma, M.; Tsushima, T., and Shizume, K.:* Growth hormone secretion during nocturnal sleep in normal subjects. J. clin. Endocrin. *29:* 20–29 (1969).

3 *Kokka, N.; Garcia, J.F.; Morgan, M., and George, R.:* Immunoassay of plasma growth hormone in cats following fasting and administration of insulin, arginine, 2-deoxyglucose and hypothalamic extract. Endocrinology, Springfield *88:* 359–366 (1971).

4 *Krieger, D.T.; Kreuzer, J., and Rizzo, F.A.:* Constant light: effect on circadian pattern and phase reversal of steroid and electrolyte levels in man. J. clin. Endocrin. *30:* 1634–1638 (1969).
5 *Nankin, H.R. and Troen, P.:* Overnight pattern of serum luteinizing hormone in normal men. J. clin. Endocrin. *35:* 705–710 (1972).
6 *Parker, D.C.; Sassin, J.E.; Mace, J.W.; Gorlin, R.W., and Rossman, L.G.:* Human growth hormone release during sleep: Electroencephalographic correlation. J. clin. Endocrin. *29:* 871–873 (1969).
7 *Perkoff, G.T.; Eik-Nes, K.; Nugent, C.A.; Fred, H.L.; Nimer, R.A.; Rush, L.; Samuels, L.T., and Tyler, F.H.:* Studies of the diurnal variation of plasma 17-hydroxycorticosteroids in man. J. clin. Endocrin. *19:* 432–438 (1959).
8 *Rechtschaffen, A.; Wolpert, E.; Dement, W.; Mitchell, S., and Fischer, C.:* Nocturnal sleep of narcoleptics. Electroenceph. clin. Neurophysiol. *15:* 559–569 (1963).
9 *Rechtschaffen, A. and Kales, A.:* A manual of standardized techniques and scoring system for sleep stages of human subjects (Public Health Service, US Govt. Printing Office, Washington, D.C. 1968).
10 *Rijnberk, A.; Kindern, P.J., and Thijssen, J.H.H.:* Investigations on the adrenal function of normal dogs. J. Endocrin. *41:* 387–395 (1968).
11 *Sassin, J.F.; Parker, D.C.; Mace, J.W.; Goltin, R.W.; Johnson, L.C., and Rossman, L.G.:* Human growth hormone release: Relation to slow-wave sleep and sleep-waking cycles. Science *165:* 513–515 (1969).
12 *Sassin, J.F.; Frantz, E.D.; Weitzman, E.D., and Kapen, S.:* Human prolactin: 24-hour pattern with increased release during sleep. Science *176:* 177–178 (1972).
13 *Suzuki, J.:* Narcoleptic syndrome and paradoxical sleep. Folia psychiat. neurol. jap. *20:* 123–134 (1966).
14 *Takahashi, K.; Daughaday, W.H., and Kipnis, D.M.:* Regulation of immunoreactive growth hormone secretion in male rats. Endocrinology, Springfield *88:* 909–917 (1971).
15 *Takahashi, K.; Takahashi, S.; Azumi, K.; Honda, Y., and Utena, H.:* Changes of plasma growth hormone during nocturnal sleep in normal and in hypersomnic patients. Adv. neurolog. Sci. (in Japanese) *14:* 743–754 (1971).
16 *Takahashi, K.; Tsushima, T., and Irie, M.:* Effect of catecholamines on plasma growth hormone in dogs. Endocrin. jap. *20:* 323–330 (1973).
17 *Takahashi, Y.; Kipnis, D.M., and Daughaday, W.H.:* Growth hormone secretion during sleep. J. clin. Invest. *40:* 2079–2090 (1968).
18 *Takahashi, Y.; Honda, Y.; Takahashi, K.; Tsushima, T., and Irie, M.:* Growth hormone secretion during sleep in dogs. Folia endocrin. jap. (in Japanese) *49:* 502 (1973) (abst.).
19 *Takahashi, Y.:* Growth hormone secretion during sleep. A review; in *Kawakami* Biological rhythms in neuroendocrine activity (Igakushoin 1974).
20 *Tsushima, T.; Irie, M., and Sakuma, M.:* Radioimmunoassay for canine growth hormone. Endocrinology, Springfield *89:* 685-693 (1971).
21 *Weitzman, E.D.; Fukushima, D.; Nogeire, C.; Roffwarg, H.; Gallagher, T.F., and Hellman, L.:* Twenty-four-hour pattern of the episodic secretion of cortisol in normal subjects. J. clin. Endocrin. *33:* 14–22 (1971).

Author's address: Dr. *K. Takahashi,* Department of Neurochemistry, Tokyo Metropolitan Institute for Neurosciences, 2–6 Musashidai, *Fuchu-city, Tokyo* (Japan)

Psychoneuroendocrinology. Workshop Conf. Int. Soc. Psychoneuroendocrinology, Mieken 1973, pp. 77–82 (Karger, Basel 1974)

Stress and Cerebro-Hepatic Axis with Special Reference to Steroids Metabolism

J. Nomura, K. Hisamatsu, M. Maeda, S. Kamiya, H. Hattori and N. Hatotani
Department of Psychiatry, Mie University School of Medicine, Tsu

Introduction

We have studied many cases of acute or atypical psychoses from the endocrinological point of view (3). During the psychotic period, metabolic disturbances of androgen and estrogen were found frequently. These disturbances may be regarded as a result of hepatic dysfunctions. However, such abnormalities were also found in the psychotic state induced by *d*-lysergic acid diethylamide (LSD) and in the diencephalo-pituitary disorders, and were normalized upon remission without any protective treatment of the liver. Consequently, such hepatic dysfunctions during the psychotic period are not primary dysfunctions, but seem to indicate a breakdown of the cerebro-hepatic homeostasis. Although such a homeostatic breakdown is a non-specific bodily response, it may be an important factor in causing acute mental disorders. In order to further examine experimentally the cerebro-hepatic relationship, we have studied the effects of various stress conditions on the activities of several enzymes in rat liver (6). In this research, we examined the effects of electric stimulation of hypothalamus, administration of LSD, and forced running and immobilization, on the metabolism of androgen and estrogen in the liver of rabbits and rats.

Materials and Methods

Male and female Wistar rats and white male rabbits were used in all experiments. The ability of estrogen inactivation by the liver was measured by *Jailer*'s method (5). A 1 mg of estradiol pellet was transplanted into the spleen of the ovariectomized female rats. When the amount of estradiol in the spleen gradually decreased to the extent that the liver could completely inactivate the hormone, the cornified cells in the vaginal smear would disappear. The reappearance of cornified cells in the vaginal smear was regarded as a decrease in the ability of the liver to inactivate estrogen.

The metabolism of androgen was determined as follows. The liver of a male rat or a male rabbit was homogenized with a glass homogenizer in 5 volumes of cold 0.25 *M* sucrose. The homogenate was centrifuged at 1,000 *g* for 10 min. The resulting supernatant equivalent to 5 g of liver was incubated with Δ^4-androstane-3,17-dione 0.7 μM, reduced triphosphopyridine nucleotide (NADPH) 1.2 μM, $MgSO_4$ 100 μM and Tris-HCl buffer (pH 7.4) 200 μM for 90 min at 37 °C. The mixture was extracted by 150 ml of ethyl acetate. The ethyl acetate layer was washed with 15 ml of 10 % NaOH and water successively, and evaporated to dryness. The residue was dissolved by 15 ml of methanol and left overnight at –20 °C to remove excess lipids. The precipitate was filtered and the filtrate was evaporated. The residue was dissolved by 2 ml of benzene and loaded on an aluminum column (10 × 0.7 cm). The column was washed by 4 ml of benzene and eluted stepwise by the mixture of benzene 97:ethanol 3, benzene 84:ethanol 16 and benzene 65:ethanol 35. The fractions were combined and evaporated. Δ^4-androstane-3,17-dione, androsterone and etiocholanolone were analyzed quantitatively by gas chromatography (column XE 20, 210 °C, carrier gas N_2 1.2 kg, H_2 0.7 kg, air 1 kg).

In the experiment of the electric stimulation of hypothalamus, bipolar stainless electrodes were stereotaxically implanted unilaterally in the ventromedial nucleus of the hypothalamus (VMH) of male adult rabbits. The stimulating procedures were started 2 weeks after the electrode implantation. The animals had complete freedom of movement in their observation boxes and were stimulated for 20 sec, with 2-min rest intervals, for durations totaling 2 h. Monophasic square-wave pulses of 1 msec duration, 100 cycle/sec, with an amplitude of 3–5 V were delivered from an electronic stimulator. At the conclusion of the experiment, the disposition of the electrode tips was confirmed histologically using cresyl violet stains.

In the experiment of forced running, the rats were placed in a running drum 30 cm in diameter, revolving at a speed of 3 times per min. In the experiment of immobilization, the rats were placed face down upon a board with their legs fastened. Carbon tetrachloride and LSD were administered intramuscularly.

Results

A. Estrogen Metabolism

The results are summarized in table I. When the liver was injured by the injection of 0.2 ml of carbon tetrachloride, the cornified cells reappeared in 48 h. Forced running for 96 h and immobilization for 72 h also resulted in the reappearance of cornified cells. The result showed that stress conditions, such as forced running and immobilization, as well as the liver injury by carbon tetrachloride, decreased the ability of the liver to inactivate estrogen.

B. Androgen Metabolism

1. Electric Stimulation of VMH in Rabbits

The results are summarized in table II. In the control rabbits, 40 % of Δ^4-androstane-3,17-dione was converted to androsterone and etiocholanolone,

Table I. Effect of carbon tetrachloride, stress and LSD on estrogen metabolism in rat liver

	Rat No.	Appearance of cornified cells in vaginal smear				
		24 h	48 h	72 h	96 h	120 h
Carbon tetrachloride, 0.2 ml	1	–	+	+	+	
	2	–	+	+	+	
Forced running	1	–	–	–	+	
	2	–	–	–	±	+
Immobilization	1	–	–	+	+	
	2	–	±	+	+	
LSD, 25 μg	1	–	–	–	–	
	2	–	–	–	–	

Table II. Effect of electric stimulation of VMH on androgen metabolism in rabbit liver

	No. of rabbit	A/E	$(A + E/\Delta^4) \times 100$
Control	5	0.51 (± 0.13)	40 (± 7)
VMH stimulation, 2 h	11	0.07* (± 0.02)	27 (± 8)

A = androsterone; E = etiocholanolone; Δ^4 = Δ^4-androstane-3,17-dione; mean (± SE); * = $p < 0.01$.

and the ratio of androsterone and etiocholanolone was 0.51. When VMH was stimulated, the rabbits stopped moving, squatted and then began a running or circling movement. The motor reactions were stimulus-bound and it appeared that they were expressions of fear or rage. Following 2 h sessions of stimulation of VMH, the conversion ratio of Δ^4-androstane-3,17-dione to androsterone and etiocholanolone was 27 % and the ratio of androsterone and etiocholanolone was 0.07. The decrease of conversion ratio was not statistically significant, but the decrease of the ratio of androsterone and etiocholanolone was highly significant. The result indicates that the 5α-hydrogenation of Δ^4-androstane-3,17-dione to androsterone in rabbit liver is strongly inhibited by the stimulation of VMH.

Table III. Effect of LSD and forced running on androgen metabolism in rat liver

	No. of rat	A/E
Control	4	1.61 (± 0.30)
LSD administration, 25 μg, 6 h	4	0.75* (± 0.10)
Forced running, 6 h	3	0.79* (± 0.08)

A = androsterone; E = etiocholanolone; mean (± SE); * = $p < 0.05$.

2. Administration of LSD and Forced Running in Rats

The results are summarized in table III. In the rat, the activity of Δ^4-3-ketosteroid hydrogenase was much higher than that in the rabbits and only the ratio of androsterone and etiocholanolone was determined. In the control rats, the ratio of androsterone and etiocholanolone was 1.61. When the rats were killed 6 h after the administration of LSD (25 μg), the ratio decreased to 0.75. The addition of LSD *in vitro* (6×10^{-6} *M*) had no effect on the 5α-hydrogenation. When the rats were forcedly run for 6 h, the ratio decreased to 0.79. The results indicate that LSD and forced running inhibit the 5α-hydrogenation of Δ^4-androstane-3,17-dione to androsterone in rat liver probably through the cerebro-hepatic axis.

Discussion

We studied the fraction pattern of urinary estrogen and 17-ketosteroids in the cases of acute psychoses (3). As to the estrogen fraction pattern, the ratio between estradiol, estrone and estriol was approximately 1:2:3, respectively. During the psychotic phase, estrone was much higher than estriol. The result suggested that the inactivation of estrogen in the liver was disturbed in the psychotic period. We showed experimentally that the ability of the liver to inactivate estrogen was decreased by stress, such as forced running and immobilization.

Concerning the fraction pattern of 17-ketosteroids, fraction 4 contains the products of 5α-hydrogenation including androsterone and fraction 5 contains the products of 5β-hydrogenation including etiocholanolone. The ratio of fraction 4 and fraction 5 was over 1.5 in normal individuals or in the lucid phase of psychotic patients. During the psychotic episodes, the ratio was below 1.5 or even below 1. The decrease of this ratio may reflect the decrease of 5α-hydrogenation of testosterone or Δ^4-androstane-3,17-dione in the liver. If the total activity of Δ^4-3-ketosteroid hydrogenase was determined by *Tomkins*' method (9) or by our method, neither stress nor LSD effected the enzyme activity. In this research, we determined 5α- and 5β-hydrogenation of Δ^4-3-ketosteroid separately, and demonstrated that electric stimulation of VMH, administration of LSD and forced running inhibited the 5α-hydrogenation, and decreased the ratio of androsterone and etiocholanolone. We did not obtain a definite result, when the other areas of brain, such as the lateral hypothalamus, the centre median, the hippocampus and the amygdala, were stimulated. However, the ratio of androsterone and etiocholanolone tended to decrease with the stimulation of these areas and the inhibition of 5α-hydrogenation may not be specific for VMH.

These results indicate that various stress conditions, which may affect the brain and hypothalamus, cause the metabolic disturbances of sex steroids in the liver probably through the cerebro (hypothalamo)-hepatic axis. Such a cerebro-hepatic relationship is supported by some other reports. *Higashimura et al.* (4) reported that the electric stimulation of VMH decreased the activities of arginase and arginine synthetase in the rabbit liver, and increased the level of blood ammonia. The hyperammonemia which is often found in the psychotic patients may be explained by the same mechanism (1). *Shimazu* (8) reported that the electric stimulation of hypothalamus increased tryptophan pyrrolase activity in the normal and the adrenalectomized rabbit liver and postulated some neural factors which control enzyme activity in the liver. *Nomura* (7) reported that the activity of tryptophan pyrrolase in rat liver was significantly increased by various stress conditions. Adrenalectomy did not prevent the increase in enzyme activity, but hypophysectomy eliminated the enzyme induction by stress. The result suggested that some unknown hypophyseal factor was involved in the regulation of rat liver tryptophan pyrrolase activity. *Colby* (2) recently reported that the effects of the gonadal hormones on hepatic Δ^4-steroid hydrogenase activity were manifested only in the presence of the pituitary gland. It is obvious that some endocrine as well as neural factors are involved in the regulation of some enzymatic activities in the liver. However, the exact nature of the cerebro-hepatic axis remains to be elucidated.

Summary

The ability to inactivate estrogen in rat liver was decreased by forced running and immobilization. The 5α-hydrogenation of Δ^4-androstane-3,17-dione to androsterone in rabbit and rat liver was inhibited by electric stimulation of the ventromedial nucleus of the hypothalamus, by administration of LSD and by forced running. These results indicate that the metabolism of sex steroids in the liver was influenced by the various stress conditions probably through the cerebro-hepatic axis. This cerebro-hepatic axis may play an important role in the pathogenesis of acute psychoses.

References

1 *Chatagnon, C. et Chatagnon, P.-A.:* Au sujet de l'ammoniémie au cours d'états psychotiques et névrotiques. Ann. méd.-psychol. *121:* 468–473 (1963).

2 *Colby, H.D.; Gaskin, J.E., and Kitay, J.I.:* Requirement of the pituitary gland for gonadal hormone effects on hepatic corticosteroid metabolism in rats and hamsters. Endocrinology, Springfield *92:* 769–774 (1973).

3 *Hatotani, N.; Ishida, C.; Yura, R.; Maeda, M.; Kato, Y.; Nomura, J.; Wakao, T.; Takekoshi, A.; Yoshimoto, S.; Yoshimoto, K., and Hiramoto, K.:* Psychophysiological studies of atypical psychoses. Endocrinological aspect of periodic psychoses. Folia psychiat. neurol. jap. *16:* 248–292 (1962).

4 *Higashimura, T.; Masuda, Y., and Kamiya, S.:* Changes in blood ammonia levels induced by electro-stimulation of the hypothalamus in rabbits. Folia psychiat. neurol. jap. *24:* 71–74 (1970).

5 *Jailer, J.W.:* The effect of inanition on the inactivation of estrogen by the liver. Endocrinology, Springfield *43:* 78–82 (1948).

6 *Nomura, J.; Maeda, M.; Nakazawa, K., and Hatotani, N.:* Experimental studies on cerebro-hepatic relationship. Effects of several conditions on enzyme activities in rat liver and brain, and on estrogen inactivation in rat liver. Folia psychiat. neurol. jap. *19:* 156–166 (1965).

7 *Nomura, J.:* Effects of stress and psychotropic drugs on rat liver tryptophan pyrrolase. Endocrinology, Springfield *76:* 1190–1194 (1965).

8 *Shimazu, T.:* The effect of electric stimulation of hypothalamus on rabbit liver tryptophan pyrrolase. Biochim. biophys. Acta *65:* 373–375 (1962).

9 *Tomkins, G.M.:* The enzymatic reduction of Δ^4-3-ketosteroids. J. biol. Chem. *225:* 13–24 (1957).

Authors' addresses: Dr. *J. Nomura,* Dr. *S. Kamiya,* Dr. *H. Hattori* and Dr. *N. Hatotani,* Department of Psychiatry, Mie University School of Medicine, *Tsu, Mie 514;* Dr. *K. Hisamatsu,* Tsu National Hospital, *Hisai, Mie 514-12;* Dr. *M. Maeda,* Otsu Red Cross Hospital, *Otsu, Shiga 520* (Japan)

Psychoneuroendocrinology. Workshop Conf. Int. Soc. Psychoneuroendocrinology, Mieken 1973, pp. 83–101 (Karger, Basel 1974)

Stress and Biological Rhythm: Their Impact on Pathophysiologic Research Strategy into Schizophrenic Illnesses

T. Fukuda[1], *Y. Hirota, T. Shimizu, T. Ohiwa and N. Iida*
Department of Neuropsychiatry, Osaka Medical College, Osaka-Takatsuki, and Neuropsychiatric Policlinic and Division of Mental Hygiene, Doshisha University, Kyoto

The cerebro-hypothalamo-pituitary-gonadal system, generally known as the system regulating the homeostatic equilibrium (*Bernard,* 1859; *Cannon,* 1939), is one of the research targets that seem to have provided us with manifold productive information, especially since the advent of the concept 'Stress' and 'General Adaptation Syndrome' (*Selye,* 1953). Instructive as they are, all the information presently available seems to render the impression that the sympathico-adrenal axis is the only hero in action. Uncertainty of such an impression may be questioned considering the works, for example of *Hess* (1949), *Gellhorn* (1957, 1967), which show integrated functionings of both 'ergotropic' (sympathetic) and 'trophotropic' (parasympathetic) system under various conditions. Information pertaining to the latter system, in relation to the subject matter to be dealt with here, are surprisingly scanty.

The present communication deals with synthetic presentation of a series of investigations, both basic and clinical, which were conducted primarily with the hope of obtaining some insights as to certain psychiatric problems. Our results, in addition to the data on cholinergic system, especially at correlative level, seem to urge a better physiological delineation of the concept 'stress'.

Animal Experiments

In table Ia, experimental design is shown, in which reproduction of slow EEG known to accompany electroconvulsive shock treatment (EST) on mental

1 (Honorary) Clinical Professor, Missouri Institute of Psychiatry, State University of Missouri-Columbia, School of Medicine, St. Louis Division, St. Louis, Mo.

Table I. Experimental design

a)	*Material and method*		
1.	Animal	Adult albino rats (200–300 g)	
2.	ES	55–58 ma for 0.2 sec, through electrodes applied to the ears bilaterally once every 1 or 2 days	
3.	EEG recording	Bipolar recording via permanently implanted surface electrodes	
b)	*Experimental conditions to produce slowing of EEG*	*c)*	*Means to block the slowings*
1.	Cerebral damage	1.	Injection of atropine
2.	Preloading stress, overloading stress (ES)	2.	Injection of acetylcholine
3.	Hypophysectomy, overloading stress (ES)	3.	Administration of hormones (?)

patients was of principal concern. Electroconvulsive shock (ECS or ES) did not produce EEG slowings.

Faced with such unexpected result, a serious challenge to find means to produce EEG slowings was made, which was in fact the impetus to utilize 'stressed' animals. The animals, preloaded with stress of one type or another, showed invariably EEG slowings.

The injection of acetylcholine and adrenaline given after preloading stress ('enforced activity' in a running drum) exerted more or less opposite expected effects in terms of desynchronization and synchronization of slow EEG.

The same was not seen, except in the brain-damaged rats, with atropine which showed a blocking effect against slowings of EEG in human material (*Ulett and Johnson,* 1957). These findings are summarized in table Ib and c.

Some of the more important inferences made from the foregoing experimental observations (for full account, see: *Fukuda et al.,* 1959; *Fukuda and Funasaka,* 1961), the following seems to warrant mention in relation to the context of this communication:

(1) Though species differences must be taken into account, EEG slowings as function of a series of EST may not develop in human material, possibly depending upon the conditions and/or types of the individuals; the slowings would then be anticipated in individuals who suffer, and/or preloaded with, stresses of some sort.

(2) Most, if not all, of 'stressful' conditions exert identical influences onto the regulatory system, and hence give rise to similar responses as measured in terms of EEG slowings.

The first-mentioned has nevertheless led us to re-examine the clinical materials with some positive findings, as will be dealt with later. The second issue, however, was subjected further to exploratory investigation in search of structures and/or systems within the brain, other than the regulatory system in view of the multiplicity of dynamic function of the brain, by employing two different methods or parameters, i.e. intra-arterial methylene blue vital stain (*Fukuda,* 1957, 1958, 1971) and the modified thiocholine method known to date as the only reliable histochemical technique for acetylcholinesterase (AChE), the cholinergic enzyme (*Fukuda and Koelle,* 1959; *Fukuda,* 1959, 1960, 1962, 1971; *Fukuda et al.*, 1965).

(For figures see pages 86–89)

Fig. 1a–d. Animals are adult albino rats weighing 120–300 g. Frontal sections were cut between 80–140 μ as serial as possible after the methylene blue vital staining had been completed. Unless indicated otherwise, the areas of these photographs are mainly of the reticular formation at the level of the lower brain stem. *a* Daily injection (i.p.) of methamphetamine 10 mg/kg for 10 days. Though some 'autistic' behaviors (see: *Utena,* 1974; *Machiyama et al.*, 1974) were noticeable, neuronal features display no recognizable peculiarity. Findings 60 days after the injection remained the same. (*Fukuda and Hirota,* 1971, unpublished.) × 120. *b* Three days after single injection (i.p.) of reserpine 10 mg/kg. Neural elements appear morphologically intact, although perinuclear pallor as well as some fainter staining of cytoplasmic and dendritic processes are present, hence an increased number of the methylene blue 'weak positive' (*Fukuda,* 1971) neurons. Such findings are considered tentatively due to the hypotensive effects of the drug. These features are observed from 3 h up to 2 weeks after the single injection. Animals remained 'immobile' and withdrawn throughout the observation period. × 180. *c* Vital stain as counterstained with the thiocholine method (*Fukuda,* 1971). Yellowish precipates are seen partly overlapped with the 'strong positive' neurons of *N. oculomotorius* in the thalamus. Cholinergic fibers are also stained, which stand out relatively clearly in contrast with the bluestained structures. × 40. *d* Nucleus rubber stained in the same manner as in figure 1c under higher magnification. Amidst 'strong-positive' and 'weak-positive' neurons with the vital stain, 'moderate', 'faint' as well as 'unstained' ones for AChE are recognizable. The thickness does not allow analysis of further detail, however. × 180.

Fig. 2. a 30 min following administration (i.p.) of 6-AN. Though morphologically 'intact', there is a pronounced increase of methylene blue 'strong positive' neuronal elements; both dendritic and axonal arborizations are clearly discernible, in addition to glial elements which usually remain 'weak-positive' and much less in number. These features may be called a sort of 'alarming reactions' of neuronal tissue against noxious agents. × 80. *b* 90 min after 6-AN. Cell bodies and processes are undergoing a variety of disconcerting changes; displacement of nucleus, perinuclear pallor as much as distortion of cellular contours. Dendritic elongations appear in the process of losing structural integrity in general. × 80. *c* Three days after 6-AN. Neurons are seen with various types of disconcerting features, not only perinuclear pallor and other disintegrating features, but also peculiar morphology, e.g. unusual knob-like formations noticeably with the smaller neurons in the middle (from *Fukuda,* 1971; through the courtesy of Folia psychiat. neurol. jap.). × 80. *d* 14 days after 6-AN. Most of the neural tissues are seen to be taken over by glial ones at this stage. Dendritic processes are fading away, though some neurons do remain, apparently unattacked, even 2 months later. × 80.

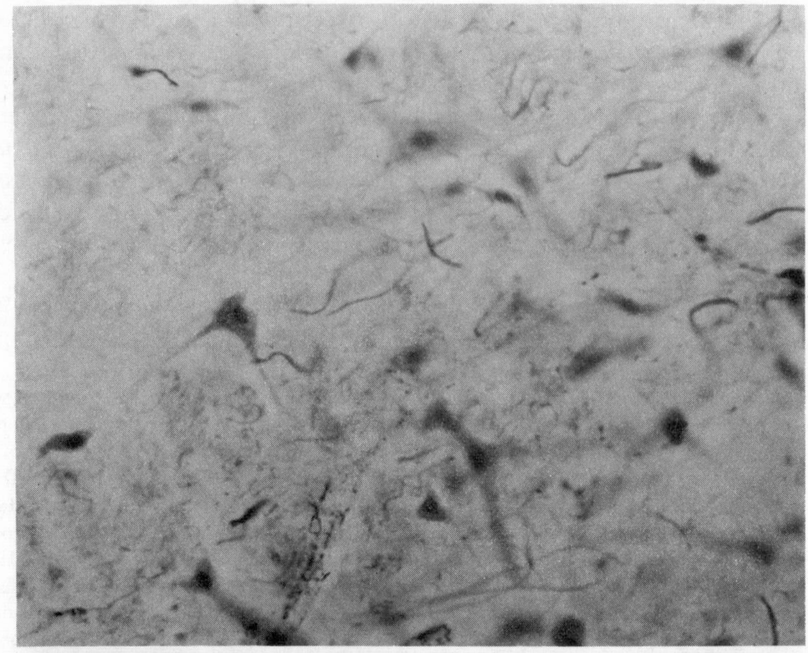

1a

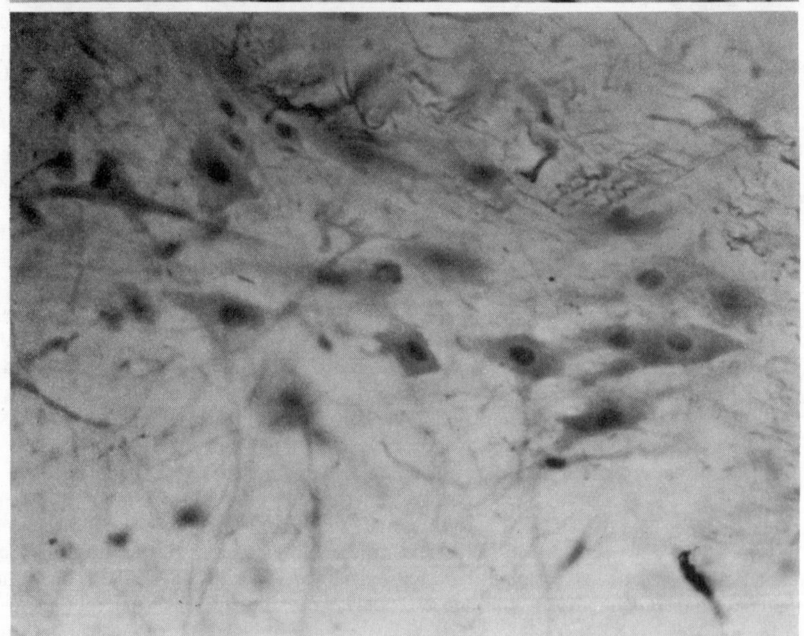

1b

(For legends see page 85)

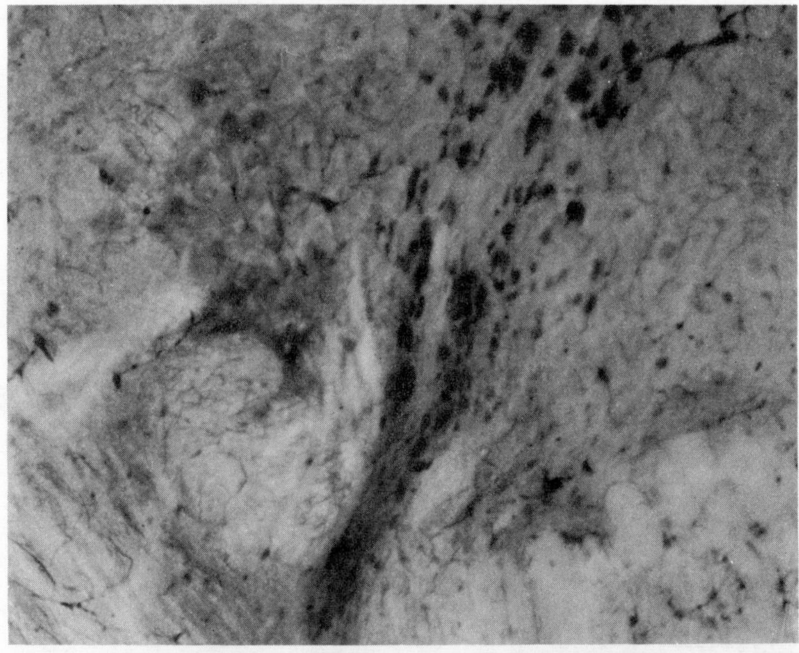

1c

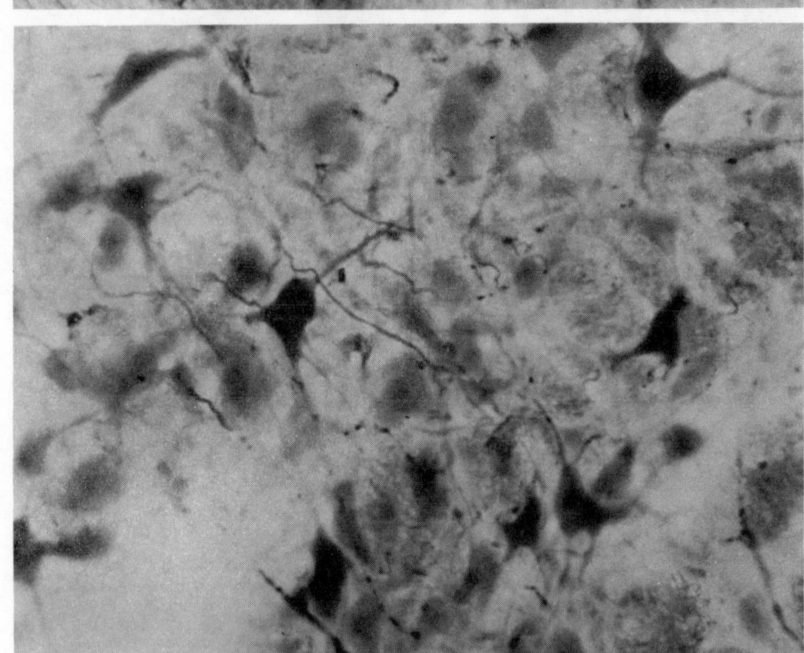

1d

(*For legends see page 85*)

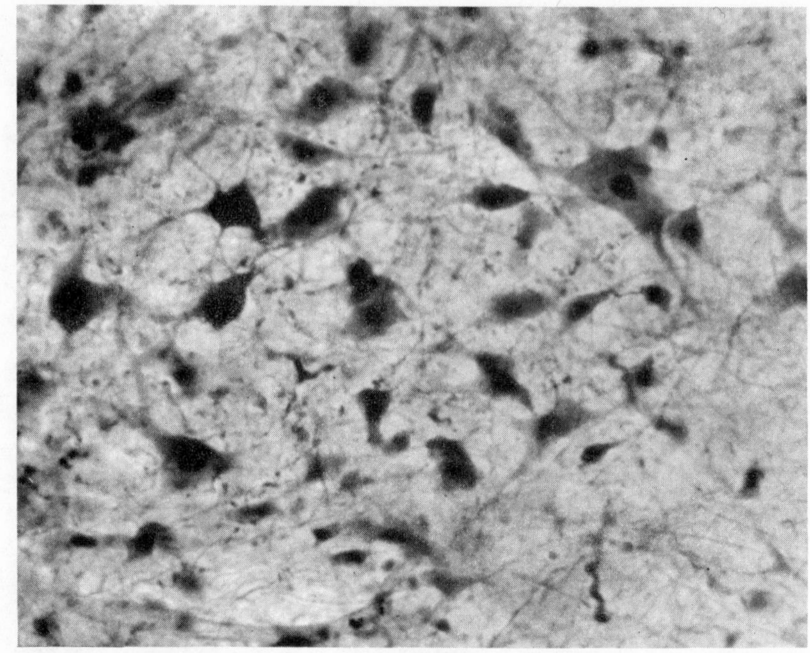

2a

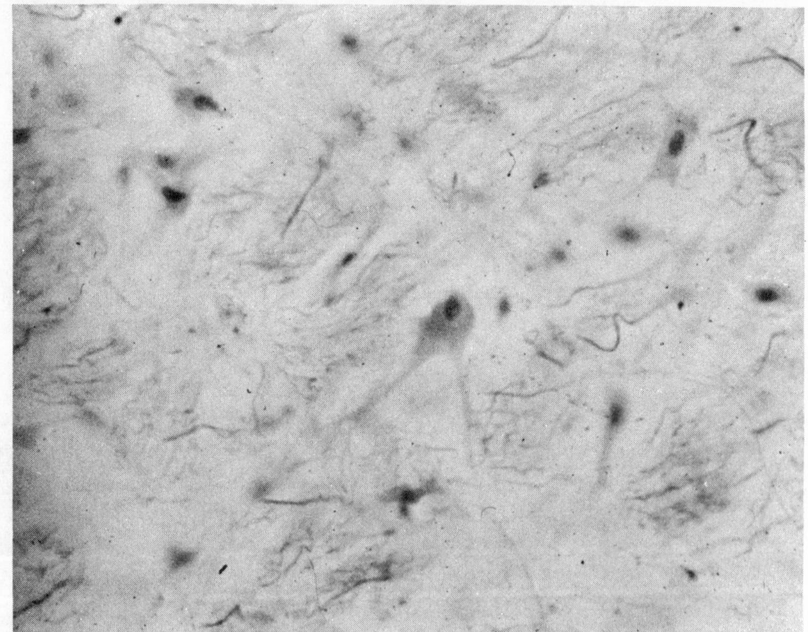

2b

(For legends see page 85)

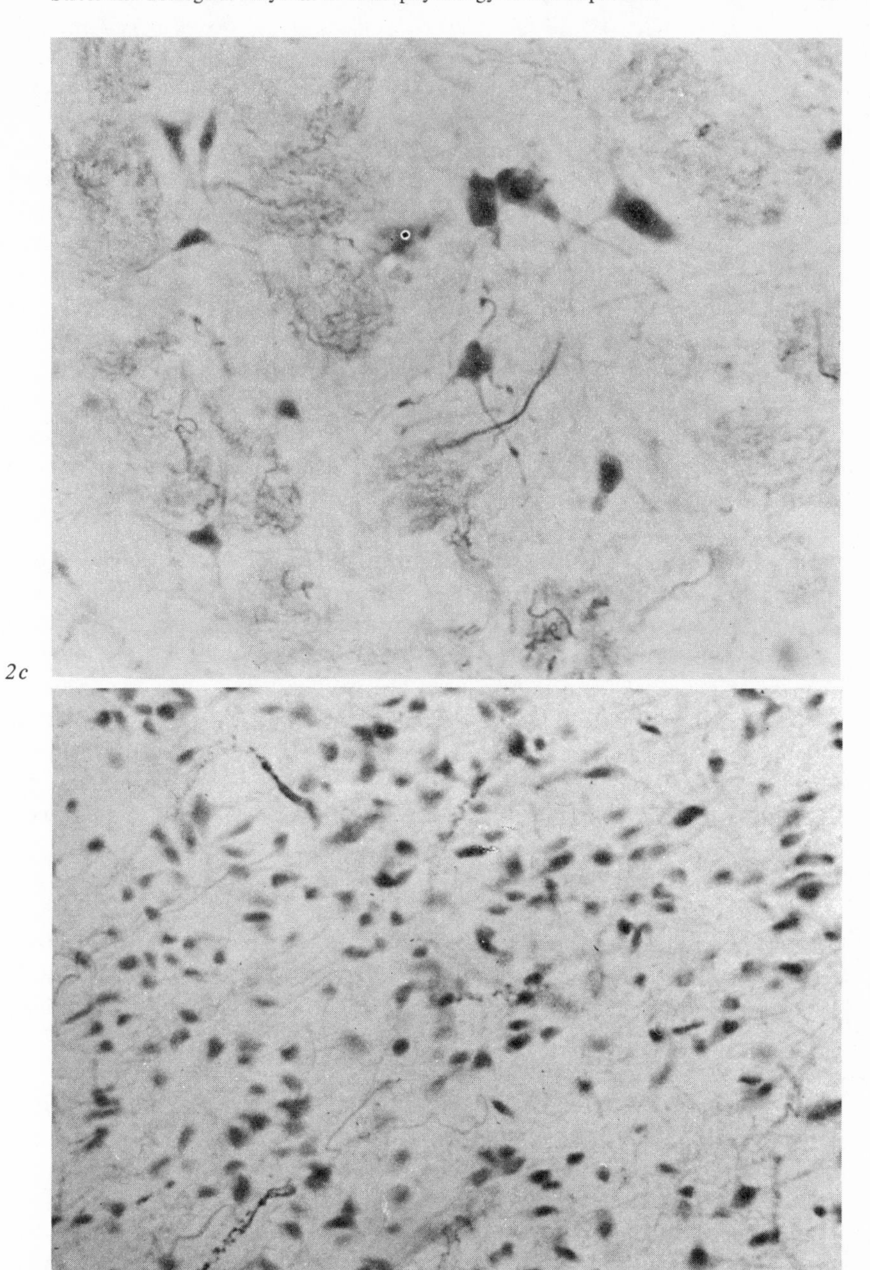

2c

2d

(For legends see page 85)

Negative and Trivial Morphologic Findings with the Vital Stain

ECS (or ES) alone did not reveal recognizable changes of the central neurons with the vital stain technique, which is nevertheless a finding endorsing the view generally held on this subject. Questionable changes of neuronal features, difficult to define according to the 'text-book' description of histopathology, involving mainly cell bodies and dendritic structures were observed in the brain of the rats treated with the preloading stress + ES that were examined post-mortem immediately following termination of the slow EEG experiment. Though such trivial changes were seen to be spread over almost the entire brain, the interpretation necessarily remained quite conservative (see: *Fukuda,* 1958). Such findings would imply that morphologic aspects demonstrable with this particular stain are of little or no significance and that more sensitive methods, or better, other parameters, are more appropriate in dealing with the matter here in focus. Nonetheless, some efforts have been made more recently in order to prove or disprove definitely the aforementioned indefinable features. A few drugs, such as acute and chronic administration of reserpine, methamphetamine have been tried so far with negative findings (fig. 1 a and b), even in the face of the known 'stressful' effects, i.e. depletion of amines and production of 'schizophrenic' animals, respectively, as reported by various investigators. Though a full account of the work along this line will be published shortly, a list of peculiar changes of neuronal features have been observed with the vital stain following intraperitoneal administration of 6-amino-nicotine amide (6-AN), an antimetabolite of nicotine amide supposed to produce the false co-enzyme NADP; the principal lesions are allocated and identified exclusively to the lower brain stem with this particular substance as the effect of intoxication-stress, leaving manifest neurologic deficits for several days or longer (fig. 2 a, b, c, d) (*Schneider and Coper,* 1968; see also: *Fukuda,* 1971).

Thus, it is only to be hoped that, by listing up all seizable changes of neuronal features in a spectrum starting from well-defined ones, the indefinable features become eventually 'definable' as well, so as to be of some value in understanding how the entire brain would operate under the 'stress' conditions.

Quantitative and Histochemical Findings (AChE)

The cholinergic neurons in the brain are mapped out by use of the modified thiocholine method. Criteria used to identify cholinergic, cholinoceptive and non-cholinergic neurons are already accounted for (*Fukuda,* 1960, 1962). To be brief, using the peripheral autonomic ganglia as the model, 10μ sections of the brain tissue were incubated, from 5–30 min, thus minimizing non-specific artefacts. With such refined histochemical method, four types of neurons are identified with reasonable constancy; heavily stained; moderately stained; faintly stained, and unstained. The heavily stained neurons are called 'cholinergic' whereas the faintly stained and unstained are identified 'cholinoceptive' and 'non-cholinergic', respectively. The moderately stained, on the other hand, are difficult to label, while some seem to be cholinergic and the remainder more cholinoceptive (table IIa and b). The findings were more or less confirmatory of the earlier work of *Koelle* (1954), except in that, unlike the peripheral autonomic ganglia such as ciliary ganglion, the cellular localizations of AChE are seen to be confined to the cytoplasm without the excess deposit in and along the cell membrane, the only exception which applies as well to the otherwise admirable works of the Cambridge group where the criteria of cholinergic neurons are taken from the peripheral ganglia without reservation (see: *Shute,* 1970). At any event, AChE was investigated in animals treated similarly with ES and with preloading stress + ES. Worthy of note are two findings: first, quantitative changes of the enzyme as measured photocolorimetrically are definitely significant between the cortex and brain-stem, while the subcortex comprising the hippocampus and hypothalamus displays the changes somewhere in between; second, cytologic localization of AChE became peculiarly confined to the nuclear portion, where the enzyme is normally not demonstrable, when the cholinergic neurons at the brain stem level seem to be run out of the cytoplasmic AChE (*Fukuda,* 1971; *Fukuda et al.,* 1965). A detailed mapping of cholinergic neurons as recognizable under such experimental conditions is presently under way (*Fukuda and Ukawa,* in preparation), which may elucidate the possible role of cholinergic system in such stressful conditions.

In this connection, it may be pointed out that the pituitary AChE cannot be demonstrated in the adenohypophysis, but only in the neurohypophysis, the strategic implication thereof at functional level remaining a matter of conjecture at present (*Koelle,* 1971; see also: *Harris,* 1955). Nevertheless, there is evidence pertinent to the important role of the cholinergic system, not necessarily limited to rather discrete areas of the brain (see: *Jasper,* 1969). Combined techniques of both vital stain and electronmicroscope (see: *Fukuda,* 1971) as well as both vital stain and the thiocholine method (fig. 1c and d) are being utilized currently to obtain further information along this line.

Clinical Investigations

Effect of EST on EEG of Typical and Atypical Schizophrenics

On the basis of the inferences already mentioned, a systematic investigation was carried out with clinical materials, by dividing schizophrenic patients into two groups, typical and atypical, according principally to the classification system of *Mitsuda* (1967). Resting EEGs were graded into six (I–VI) stages between the normal and seizure pattern; likewise, effect of hyperventilation (HV) was roughly identified with three types (A–C) (see: *Fukuda and Matsuda,* 1969).

Table IIa. Heavily stained regions (AChE) (*Fukuda*, 1962)

	Region	Stained structures	Classification[1]
1.	Anterior horn	Motoneurons and axons, scattered fibers	M-1
2.	Cr. XI	Most axons in transverse section	M-1
3.	Nucleus of Cr. XI	Small motoneurons	M-1
4.	Nucleus and root of Cr. XII	Larger cells	M-1
5.	Reticular formation (medulla)	Scattered neurons and fibers	Misc.
6.	Ventral nucleus of reticular formation	Neurons and fibers	C (?)
7.	Dorsal motor nucleus of Cr. X	Few large neurons and fibers	M-1
8.	Nucleus and motor root of Cr. VII	Large neurons and fibers	M-1
9.	Motor nucleus of Cr. V	Large neurons and fibers	M-1
10.	Nucleus of trapezoid body	Numerous scattered neurons and fibers	S-3 (?)
11.	Cerebellar cortex	Scattered cells in granular layer; few stellate cells in molecular layer	C
12.	Nucleus of Cr. VI	Few small neurons	M-1
13.	Pontine nucleus	Numerous scattered neurons and fibers	Misc.
14.	Edinger-Westphal nucleus	Few small, rounded neurons	M-1
15.	Nucleus of Cr. III	Spindle-shaped neurons	M-1
16.	Superior colliculus	Densely packed neurons of stratum griseum, scattered fibers of non-optic layer of Str. opt.	M-2, C
17.	Anterior nucleus of thalamus	Few scattered neurons and numerous fibers of ventrolateral portion	C
18.	Lateral nucleus of thalamus	Neurons and fibers interspersed with unstained tract	S-3
19.	Caudate nucleus	Closely-packed neurons and fibers	C
20.	Putamen	Densely-packed neurons and fibers	C
21.	Amygdala (central nucleus)	Scattered, fusiform neurons and fibers	C
22.	Amygdala (lateral nucleus)	Scattered, smaller globular neurons and fibers	C

1 M = motoneuron; S = sensory neuron; C = correlation center; Misc. = miscellaneous. (The figures 1, 2 and 3 stand for primary, secondary and tertiary, respectively.)

Table IIb. Moderately stained regions (*Fukuda*, 1962)

	Regions	Stained structures	Classification[1]
1.	Spinal cord; dorsal gray column	Occasional neurons and numerous fibers	S-2, Misc.
2.	Nucleus gracilis and cuneatus	Most neurons and fibers	S-2
3.	Olivary nucleus	Few neurons and fibers	C
4.	Nucleus of Deiters	Numerous large neurons and fibers	S-2, M-2
5.	Dentate nucleus dorsal and lateral	Diffuse staining of most neurons and fibers	C
6.	Tegmental nucleus	Numerous small, globular neurons	C
7.	Pontine nucleus	Occasional neurons, numerous fibers	Misc.
8.	Chief vestibular nucleus	Diffuse staining of most neurons and fibers	S-2
9.	Inferior colliculus	Few neurons and fibers	C
10.	Posterior nucleus of thalamus	Few small, scattered neurons; majority of fibers	S-2, S-3
11.	Medial leminiscus	Most fibers in cross section	S-2
12.	Medial geniculate body	Few scattered small neurons and fibers	S-3
13.	Substantia nigra	Few neurons and fibers	C
14.	Lateral geniculate	Few neurons of dorsal, more of ventral nucleus; scattered fibers	S-3
15.	Lateral habenular nucleus	Majority of neurons, scattered fibers	C
16.	Medial habenular nucleus	Few neurons and fibers	C
17.	Supraoptic nucleus	Globular neurons and fibers	C
18.	Anterior nucleus of thalamus	Numerous neurons and fibers of dorsomedial portion	C
19.	Medial nucleus of thalamus	Occasional neurons and fibers	C
20.	Nucleus reticularis of thalamus	Occasional small neurons, numerous scattered fibers	C
21.	Mammilo-thalamic tract	Most fibers in cross section	C
22.	Fimbria	Occasional neurons and fibers	C
23.	Dorsal hippocampal commissure	Occasional neurons and fibers	C
24.	Neocortex	Rare neurons and scattered fibers	Misc.

1 M = motoneuron; S = sensory neuron; C = correlation center; Misc. = miscellaneous. (The figures 1, 2 and 3 stand for primary, secondary and tertiary, respectively.)

The patients were all 'newcomers', hence reasonably free of 'contaminations', yet in need of immediate care. Noteworthy is that, apart from the initial differences (pre EST) as well as the terminal ones (post EST 30 days) in this series of investigations, the two groups displayed identical EEG changes at EST × 10 stage.

Autonomic Responses as Determined by Mecholyl Test Longitudinally in Conjunction with EST

The Mecholyl test as a means of evaluating the autonomic balance at the hypothalamic level (*Gellhorn,* 1957, 1967) was performed in three groups of schizophrenics. Transverse or single testings in three groups, whereby 'defect' schizophrenia is separated only in terms of the length of illness and of personality deterioration – hence it is nothing else but 'chronic' typical schizophrenia – resulted in some discrepancies as to the percentage of S, N, and P. However, a

Table III. Cases of paraphrenic group *('affektvoll')* (follow-up period: 5.6 years on average) (*Fukuda et al.,* 1972)

Case	Sex and age on first admission, years	Age of onset (by anamnesis), years	Psychological test and finding	Course (time period till remission), months	Outcome
T.H.	♀ 30	25–28	Rorschach: social deviation, aggressivity	3–6	Third admission; paranoid hallucinatory-delusional systematization
M.N.	♀ 33	20–25	MPI E 32 extravert, stable MPI N 15 extravert, stable MPI L 22 extravert, stable	ca. 18	Two admissions since
Y.S.	♀ 47	20–30 (?)	YG: E-type (emotionally stable, 'active')	ca. 30	Full remission
T.O.	♂ 39	36	MPI E 30 extravert, stable MPI N 5 extravert, stable MPI L 15 extravert, stable	ca. 10	Social remission
K.N.	♀ 36	22	MPI E 20 extravert, stable MPI N 3 extravert, stable MPI L 11 extravert, stable	ca. 1	No recovery, discharged

With the exception of case Y.S., all other four cases were college graduates, possessing above average IQ.

clearer discrepancy between the typical and atypical is demonstrable when the testings were pursued in a serial manner in relation to EST.

These findings seem, apart from stressing the importance of such consecutive testings, as criticizing hasty conclusions drawn hitherto from single testing with Mecholyl, especially in schizophrenic research, to argue the likelihood of differences in terms of underlying pathophysiology in two groups of schizophrenia; needless to add that changes in the autonomic reactivity or affectivity took place relatively well in parallel with clinical improvements. Moreover, much controversy surrounding EEG slowings versus EST (*Kalinowsky and Hoch,* 1952, 1961; *Solms,* 1963), in addition to the value of Mecholyl test in schizophrenics, appear reconcilable by virtue of these findings (*Fukuda and Matsuda,* 1969).

Further investigations currently under way are partly focused on the paraphrenia group. By applying the classification systems, both of *Leonhard* (1965) and of *Mitsuda* (1967), this particular group are now being devided into *'affektvoll'* and *'affektarm'* groups. So far, investigation has remained clinical and at EEG level, to uncover emotionally charged *('katathym')* systematization of paranoid hallucinations and delusions, as the cases show abnormal rhythmicity or rhythm formation of EEG, more or less similar to ' "Paren"-"Aidio"-rhythmie' (*Fukuda et al.,* 1972; see also: *Huber,* 1974). One of the rationales in picking up this particular group is to refine, if at all possible, the atypical group of schizophrenics, so as to be readily accessible to biological analysis of one type or another, leading ultimately to better understanding of pathophysiology of schizophrenia. Nevertheless, a total of over 30 cases of paraphrenia assembled to date seems to be dividable into *'affektvoll'* and *'affektarm'* groups; not only clinically, but also with the aid of EEG (*Hirota et al.,* in preparation). Table III should provide some profiles of the *'affektvoll'* group.

Correlative Synthesis

Thus far, no mention has been made of 'biological rhythm', somewhat on purpose, inasmuch as there is not even a single rhythm which remains uninfluenced under stressful conditions. A wide spectrum of 'biological rhythm', the best known being 'biological clock' (*Richter,* 1967), and/or 'circadian rhythm' (*Halberg,* 1961), can be listed up, chronologically throughout the lifetime from global down to molecular level, and then even in the order of milli- or micro-sec (see: *Chance,* 1967). Clinically, the most well defined and studied in this respect is the classical periodic catatonia (*Gjessing,* 1938) or periodic psychoses (*Hatotani,* 1974; *Jenner,* 1968), deserving at least the merit of irrefutableness of the 'samples' or 'materials'. Even with such a relatively 'pure' group of psychosis, all the findings reported so far seem to imply presumed disposition or constitution rendering possible such psychotic state to manifest at periods.

Table IV. Animal experiment

Stress (pre-loading)	Stress (overloading)	Parameters of investigation	Description of the conditions induced or changes observed	Behavioral peculiarity
ES	ES	EEG Vital stain	None significant ? (Neurons look as though suffering from cold)	'Memory loss', overly alert ('jumpy')
Enforced (running) activity	ES	EEG Vital stain AChE	Slowings ? Cortex ↑: brainstem ↓, etc.	Model 'atypical psychosis' (*Fukuda,* 1961, 1967, 1971)
Exposure to cold	ES	EEG Vital stain	Slowings ?	'Memory loss', overly alert
Hypophysectomy	ES	EEG Vital stain	Slowings ?	'Poor' looking in every aspect
Drugs (i.m.) Acetylcholine	ES	EEG Vital stain	Higher amplitude, fast activity None significant	Initial salivation, etc., later 'normalization' of EEG and behavior
Drugs (i.m.) Adrenaline	ES	EEG Vital stain	Blocking of slow wave None significant	No overt changes, later somewhat hyperactive
	Drugs (i.p.) Reserpine	Vital stain AChE	None significant None significant	'Autistic', motor retardation, etc.
	Drugs (i.p.) Methamphetamine	Vital stain AChE	None significant None significant	'Autistic' hyperactive or hypoactive
	Drugs (i.p.) 6-AN	Vital stain AChE	Gross morphologic changes None significant	'Autistic', hyperactive or hypoactive. Monoplegia or paraglegia

Table V. Clinical investigation

Diagnosis	Predisposing factor	Precipitation ('stress')	Parameters of investigation	Findings before treatment	Findings after treatment	Outcome
Schizophrenia, Typical	(Genetically predetermined) Underreactivity of the autonomic function (*Fukuda and Matsuda*, 1969)	Unknown ('intra'-personal?)	EEG Mecholyl test	(ES) Normal 'S' response	(ES) No significant change No significant change	Tendency to 'chronifization' with or without 'defect'
Schizophrenia, Atypical	Overreactivity of the autonomic function (*Fukuda and Matsuda*, 1969)	Psychophysio-logical	EEG Mecholyl test	(ES) A variety of abnormal EEG	(ES) Increase of slowing and amplitude	Full recovery without 'defect'
Schizophrenia, Intermediate	Overreactivity of the autonomic function (*Fukuda and Matsuda*, 1969)	Psychophysio-logical (plus frequent relapses)	EEG Mecholyl test	'P' response dominant	'N' response	Remission with 'defect'
Schizophrenia, Paraphrenic *'Affektvoll'*	Overreactivity of 'emotional circuit' (?)	Interpersonal + (?)	EEG Mecholyl test	(ES) Hypersynchronous, rhythmic sharp α-burst associated with frequency lability 'S' response	(ES) Some slowings 'S' response (unchanged)	Tendency to hallucinatory delusional systematization associated with deep emotional charge; some 'defect' may become evident
Schizophrenia, Paraphrenic *'Affektarm'*	Underreactivity of 'emotional circuit' (?)	('Intra'-personal?)	EEG	Some slow θ- or δ-'parenrhythmic' (drugs?)	No significant change (drugs?)	Tendency to 'chronifization' with 'defect'
Atypical psychoses	Endocrine, dien-cephalic, epileptoid lability and/or vulnerability	A diversity of 'overloadings'	EEG Mecholyl test	Varies greatly; 'S' or 'P' depending on the underlying factors	Usually consider-able slowings, 'N' response	Full recovery

With the schizophrenics, the situation has been complicated considerably by at least several other factors which involve diagnostic criteria, the hospitalization turmoils, etc. With the progress in sophistication of research methods, as aided by the rapid accumulation of data in basic areas of the brain physiology and chemistry in recent years, possible sources of errors as well as scrutinized research strategy are only beginning to shed light towards more sober directions (see: *Kety,* 1973). It is precisely here that adequate animal models are badly needed. While the identity of such models, claimed to be 'atypical psychosis' (*Fukuda,* 1962, 1971) and 'schizophrenia' (*Ellinwood,* 1974; *Machiyama et al.,* 1974; *Utena,* 1974), for example, remains to be verified, the productive value of these models has to be accredited with due attention. Interestingly enough, these animal models are 'preloaded' one way or the other which may be considered as 'sensitized' or 'stressed'.

Therefore, it is not without reason to conceive of the 'stress', a wholesale concept, in at least two steps which may differ from each other, either qualitatively or quantitatively, or both, as seems more likely. At present, physiological parameters or evidence to substantiate such an issue are fatally lacking and are indeed called for from what has been mentioned above.

With such an issue taken into account, the aforementioned data are summarized in two tables (tab. IV and V). Table V contains necessarily some speculations in part, which are yet to be validated or invalidated. Striking analogy does exist, however, between the animal and the schizophrenic illnesses if the 'stress' is divided and/or graded into two stages.

It will be remembered to this end that what we call stress in everyday life or clinical practice is the minor and predominantly emotional one, distressing or exciting or combined, which calls for continuous adaptation in order to maintain the homeostasis in good shape (*Smelik,* 1970). To promote the achievement of 'emotional homeostasis', 'one of the human goals' (*Haldane,* 1967), in schizophrenics may not be a mere fantasy if sufficient information becomes available from the physiological side. Conceptual, as well as various biological factors along this line, as documented recently, seem to argue in favor of such a view (*Huber,* 1974; *Itil,* 1974; *Pauleikhoff,* 1974; *Shagass,* 1974; *Tatetsu,* 1974; *Zerbin-Rüdin,* 1974; *Mitsuda and Fukuda,* 1974).

References

Bernard, C.: Leçons sur les propriétés physiologiques et les altérations pathologiques des liquides de l'organisme (Bailliers, Paris 1859).

Cannon, W.B.: The wisdom of body (Norton, New York 1939).

Chance, B.: Discussion; in *Richter* Sleep and altered states of consciousness. Res. Publ. Ass. nerv. ment. Dis. *45:* 29 (1967).

Ellinwood, E.H., jr.: Relationship of the amphetamine model psychosis to schizophrenia; in *Mitsuda and Fukuda* Biological mechanisms of schizophrenia and schizophrenia-like psychoses (Igaku-shoin, Tokyo, in press, 1974).

Fukuda, T.: A re-evaluation of the intra-arterial methylene blue (method of Ehrlich) in studies of the central nervous system. Arch. Jap. Chirur., Kyoto *26:* 505–514 (1957).

Fukuda, T.: Further study on the use of Ehrlich's method in studies of the nervous system with special reference to degenerating neurons. Arch. Jap. Chirur., Kyoto *27:* 861–879 (1958).

Fukuda, T.: Application of the methyl-green-pyronin (Brachet's test) onto the thiocholine method for neuronal acetylcholinesterase. Bull. Osaka med. Sch. *5:* 46–52 (1959).

Fukuda, T.: Cytological localization of acetylcholinesterase in the nerve cells (in Japanese). Adv. Neurol. Sci., Tokyo *4:* 622–624 (1960).

Fukuda, T.: Further studies of the nervous system by use of the modified thiocholine method. I. Effect *in vivo* of 5-fluoro-orotic acid on the synthesis of acetylcholinesterase of the autonomic ganglia. II. Distribution and cytological characteristics of cellular acetylcholinesterase in the central nervous system (in Japanese). Adv. Neurol. Sci., Tokyo *6:* 529–536 (1962).

Fukuda, T.: The place of intra-arterial methylene blue vital stain and histochemistry of acetylcholinesterase in experimental neurology and psychiatry. Folia psychiat. neurol. jap. *25:* 79–91 (1971).

Fukuda, T.: Transmission of schizophrenia as viewed from some pathophysiological studies on intrafamilial psychotics. Symposium on Transmission of Schizophrenia, 5th World Congr. of Psychiatry, Mexico 1971. Excerpta med. (in press, 1974).

Fukuda, T. and Funasaka, O.: Factor analysis of the slow waves induced by electroconvulsive shock. Further studies of EEG findings under various experimental conditions in adult albino rats. Bull. Osaka med. Sch. *7:* 130–140 (1961).

Fukuda, T. and Koelle, G.B.: The cytological localization of intracellular neuronal acetylcholinesterase. J. biophys. biochem. Cytol. *5:* 433–440 (1959).

Fukuda, T. and Matsuda, Y.: Comparative characteristics of slow wave EEG, autonomic function and clinical picture in typical and atypical schizophrenia during and following electroconvulsive shock treatment. Int. Pharmacopsychiat. *3:* 13–41 (1969).

Fukuda, T.; Matsuda, Y.; Ukawa, S.; Fujiiye, Y.; Yoshida, Y., and Kihara, T.: Effect of electroconvulsive shock and enforced activity on both cholinesterase in the rat brain (in Japanese). Adv. Neurol. Sci., Tokyo *10:* 233–238 (1965).

Fukuda, T; Stern, J.A., and Ulett, G.A.: EEG studies on the effects of electroconvulsive shock, experimental stress and subcutaneous injections of atropine on adult albino rats. J. Neuropsychiat. *1:* 11–16 (1959).

Fukuda, T.; Yamada, T.; Otsuka, F.; Imamichi, H.; Matsuda, Y.; Miyazaki, S.; Mimura, S.; Hirota, Y.; Uohashi, T.; Kuroda, Y.; Sugiyama, A.; Aoki, Y.; Miyazaki, T.; Ohiwa, T.; Shimizu, T., and Konno, H.: EEG findings in patients of so-called 'paraphrenia'. I. Cases of 'affektvolle Paraphrenie' *(K. Leonhard)* (in Japanese). Clin. EEG, Osaka *7:* 409–413 (1972).

Gellhorn, E.: Autonomic imbalance and the hypothalamus (University of Minnesota Press, Minnesota 1957).

Gellhorn, E.: Autonomic-somatic integrations (University of Minnesota Press, Minnesota 1967).

Gellhorn, E. and Kiely, W.F.: Autonomic nervous system in psychiatric disorder; in *Mendels* Biological psychiatry, pp. 235–261 (Wiley, New York 1973).

Gjessing, R.: Disturbances of somatic functions in catatonia with a periodic course, and their compensation. J. ment. Sci. *84:* 608–621 (1938).

Halberg, F.: Circadian systems. Report of the 39th Ross Conference on Pediatric Research, p. 57 (Ross Laboratories, Columbia, Ohio, June 1961).

Haldane, J.B.S.: Biological possibilities for the human species in the next ten thousand years; in *Wostenholme* Man and his future. Ciba Found. Vol. (paperback edition), pp. 337–361 (Churchill, London 1967).

Harris, G.W.: Neural control of the pituitary gland (Arnold, London 1955).

Hatotani, N.: Endocrinological studies on periodic psychoses; in *Mitsuda and Fukuda* Biological mechanisms of schizophrenia and schizophrenia-like psychoses (Igaku-shoin, Tokyo, in press, 1974).

Hess, W.R.: Das Zwischenhirn (Schwabe, Basel 1949).

Huber, G.: Verlauf und Dauerprognose schizophrener Erkrankungen; in *Mitsuda and Fukuda* Biological mechanisms of schizophrenia and schizophrenia-like psychoses (Igaku-shoin, Tokyo, in press, 1974).

Itil, T.M.: Computer-analysed EEG in schizophrenia with neuroleptic drugs; in *Mitsuda and Fukuda* Biological mechanisms of schizophrenia and schizophrenia-like psychoses (Igaku-shoin, Tokyo, in press, 1974).

Jasper, H.H.: Neurochemical mediators of specific and nonspecific cortical activation; in *Evans and Mulholland* Attention in neurophysiology, pp. 377–395 (Butterworths, London 1969).

Jenner, F.A.: Periodic psychoses in the light of biological rhythm research. Int. Rev. Neurobiol. *11:* 129–169 (1968).

Kalinowsky, L.B. and Hoch, P.H.: Shock treatments, psychosurgery and other somatic treatments in psychiatry (Grune and Stratton, New York 1952).

Kalinowsky, L.B. and Hoch, P.H.: Somatic treatments in psychiatry (Grune and Stratton, New York 1961).

Kety, S.S.: Problems in biological research in psychiatry; in *Mendels* Biological psychiatry, pp. 15–34 (Wiley, New York 1973).

Koelle, G.B.: The histochemical localization of cholinesterase in the central nervous system of the rat. J. comp. Neurol. *100:* 211–235 (1954).

Koelle, G.B.: Current concepts of synaptic structure and function. Ann. N.Y. Acad. Sci. *183:* 5–20 (1971).

Leonhard, K.: Die defektschizophrenen Krankheitsbilder (Thieme, Leipzig 1936).

Leonhard, K.: Aufteilung der endogenen Psychosen (Akademie Verlag, Berlin 1965).

Machiyama, Y.; Hsu, S.C.; Utena, H.; Katagiri, M., and Hirota, A.: Aberrant social behavior in animals induced by the chronic methamphetamine administration as a model for schizophrenia; in *Mitsuda and Fukuda* Biological mechanisms of schizophrenia and schizophrenia-like psychoses (Igaku-shoin, Tokyo, in press, 1974).

Mitsuda, H.: A clinico-genetic study of schizophrenia (published originally in Japanese in 1942); in *Mitsuda* Clinical genetics in psychiatry. Problems in nosological classification, pp. 49–90 (Igaku-shoin, Tokyo 1967).

Mitsuda, H. and Fukuda, T.: Biological mechanisms of schizophrenia and schizophrenia-like psychoses (Igaku-shoin, Tokyo, in press, 1974).

Pauleikhoff, B.: Zur psychopathologischen Klassifikation der atypischen Psychosen; in *Mitsuda and Fukuda* Biological mechanisms of schizophrenia and schizophrenia-like psychoses (Igaku-shoin, Tokyo, in press, 1974).

Richter, C.P.: Sleep and activity: their relation to the 24-hour clock; in Sleep and altered states of consciousness. Res. Publ. Ass. nerv. ment. Dis. *45:* 8–29 (1967).

Schneider, H. und Coper, H.: Morphologische Befunde am Zentralnervensystem der Ratte nach Vergiftung mit Antimetaboliten des Nicotinamids (6-Aminonicotinsäureamid und 3-Acetylpyridin) und einem Chinolinderivat (5-Nitro-8-Hydroxychinolin). Arch. Psychiat. Z. Neurol. *211:* 138–154 (1968).

Selye, H.: The story of the adaptation syndrome (Acta, Montreal 1953).

Shagass, C.: Evoked potential studies in schizophrenia; in *Mitsuda and Fukuda* Biological mechanisms of schizophrenia and schizophrenia-like psychoses (Igaku-shoin, Tokyo, in press, 1974).

Shute, C.C.D.: Distribution of cholinesterase and cholinergic pathways; in *Martini et al.* The hypothalamus, pp. 167–179 (Academic Press, New York 1970).

Smelik, P.G.: Integrated hypothalamic responses to stress; in *Martini et al.* The hypothalamus, pp. 491–498 (Academic Press, New York 1970).

Solms, H.: Die Krampfbehandlung; in Psychiatrie der Gegenwart, vol. 1/2, pp. 415–494 (Springer, Berlin 1963).

Tatetsu, S.: On histologic findings in schizophrenia and schizophrenic state. Panel discussion; in *Mitsuda and Fukuda* Biological mechanisms of schizophrenia and schizophrenia-like psychoses (Igaku-shoin, Tokyo, in press, 1974).

Ulett, G.A. and Johnson, M.: The effect of atropine and scopolamine upon electroencephalographic changes induced by electroshock. J. EEG clin. Neurophys. *9:* 217–224 (1957).

Utena, H.: On relapse liability; schizophrenia, amphetamine psychosis and an animal model. Panel discussion; in *Mitsuda and Fukuda* Biological mechanisms of schizophrenia and schizophrenia-like psychoses (Igaku-shoin, Tokyo, in press, 1974).

Zerbin-Rüdin, E.: Genetic aspects of schizophrenia (a survey); in *Mitsuda and Fukuda* Biological mechanisms of schizophrenia and schizophrenia-like psychoses (Igaku-shoin, Tokyo, in press, 1974).

Author's address: Prof. *T. Fukuda,* Neuropsychiatric Policlinic and Division of Mental Hygiene, Doshisha University, *Kyoto* (Japan)

Psychoneuroendocrinology. Workshop Conf. Int. Soc. Psychoneuroendocrinology, Mieken 1973, pp. 102–112 (Karger, Basel 1974)

Dynamics of the Regulation of CRF Content in the Rat Hypothalamus under Resting and Stressful Conditions[1]

T. Hiroshige, K. Abe, M. Kaneko, S. Wada and K. Fujieda[2]
Department of Physiology, Hokkaido University School of Medicine, Sapporo

Introduction

The major patterns of regulation of ACTH secretion are generally classified into three basic types; basal circadian rhythm, stress-induced activation and negative feedback inhibition. These basic patterns of regulation are known to be closely related under various conditions of enhanced ACTH secretion (*Ganong,* 1963; *Kendall,* 1971). On the other hand, ACTH release is controlled by hypothalamic neurohumoral principle or corticotropin-releasing factor (CRF). Thus, an analysis of dynamic changes of CRF activity is essential for the better understanding of mechanisms by which ACTH secretion is modulated by the central nervous system (CNS).

In spite of its importance, relatively few studies have been made on the CRF dynamics. This is mainly because of the lack of suitable CRF assay system. In addition, an extraordinary instability of CRF preparations may be another contributing factor. Furthermore, the CRF activity in the hypothalamus of animals appears to vary easily, depending upon the conditions under which the experiments are performed. Thus, it is considerably difficult to follow the CRF dynamics under conventional uncontrolled conditions.

Fortunately, we have been able to follow dynamic changes of hypothalamic CRF content by means of our intrapituitary injection for CRF assay (*Hiroshige et al.,* 1968b; *Hiroshige,* 1973). In this article, special emphasis will be laid on correlating specific aspect of CRF dynamics with basic patterns of ACTH secre-

1 This work was supported by grant No. 68804 from the Japanese Ministry of Education.
2 The authors are grateful to Prof. *S. Itoh* for his interest and encouragement throughout these studies.

tion. In addition, some recent findings on the CRF-like potency of several synthetic peptides will be mentioned, in the hope that this sort of information may shed some light on the chemical nature of CRF.

Corticotropin-Releasing Potency of Synthetic Peptides Related to Vasopressin

Since the hypothalamus is known to contain several biogenic amines and peptide hormones, it was thought essential to examine their inherent potential to release ACTH when placed directly into the anterior pituitary gland. Results obtained were reported previously (*Hiroshige et al.*, 1968a; *Hiroshige and Abe*, 1973). A survey of tested materials showed that substances such as norepinephrine, epinephrine, L-dopa, dopamine, acetylcholine, serotonin, putrescine and bradykinin did not cause an ACTH release at doses ranging from 1 to 100 ng per pituitary gland, whereas other substances such as vasopressin, angiotensin II and spermidine caused a significant release of endogenous ACTH at doses between 10 and 100 ng per gland. Among these, bradykinin and angiotensin II are of special interest, because both hormones have been postulated to be involved in the neuroendocrine mechanism in the control of ACTH secretion. *Redgate* (1968) and *Redgate et al.* (1973) indicated that, in addition to the CNS-CRF-ACTH path, two hormonal substances appear to participate in control of ACTH release, that is: (1) an angiotensin to CNS-CRF-ACTH path and (2) a bradykinin to sympathetic ganglionic neuron to ACTH cell path. Earlier, *Schally et al.* (1965) reported that intravenous injection of angiotensin II (1 μg) caused a significant increase in plasma corticosterone level of assay rats pretreated with Monase, dexamethasone, and morphine. *Osborne et al.* (1971) found that injected tritiated angiotensin II accumulated specifically into the pituitary gland in anesthetized rats. Thus, our findings with intrapituitary injection of angiotensin II are consistent with these results. However, *Gann* (1969) and *Gann and Cryer* (1973) found that infusion of angiotensin II (1–2 μg/min) led to increased secretion of cortisol in dogs with isolated median eminence and pituitary, but not in dogs with isolated anterior pituitary. *De Wied et al.* (1969) examined the ACTH-releasing activity of several substances of central nervous system origin. They found angiotensin II to be consistently potent in various *in vivo* assay systems, but in neither hypothalamic lesioned rats nor *in vitro* assay systems. Thus, our positive results with angiotensin II should be interpreted with caution, because this hormone was effective only when a large dose (75 ng) was applied.

As to the chemical nature of CRF, it should be recalled that the CRF activity of posterior pituitary extracts was abolished totally by pretreatment with thioglycollate, whereas that of median eminence extracts was abolished only partially (*Hiroshige*, 1971). *Chan et al.* (1970) and *Mulder et al.* (1970)

reported similar results. These results, therefore, suggest that the hypothalamic CRF may be a mixture of, at least, two components: one is related structurally to vasopressin, or possibly to β-CRF, and the other is not directly related to vasopressin, or possibly to α-CRF. In this connection, of special interest is the recent report by *Saffran et al.* (1972). These authors reported that pressinoic acid, a synthetic hexapeptide that corresponds to the ring of vasopressin, exhibited a CRF activity in *in vitro* assay in doses of 3–30 ng per ml. We also examined this peptide for its ACTH-releasing activity by our intrapituitary injection technique. Results obtained are shown in figure 1. Pressinoic acid was synthetized and supplied by Dr. *S. Sakakibara,* Protein Research Foundation, Osaka, Japan. It is of special interest to note that a marked activity was observed in doses around 1 ng per pituitary gland.

In order to confirm the inherent potential of the acid to release ACTH, we assayed the activity by use of another *in vivo* assay system (*Arimura et al.,* 1967). As shown in table I, pressinoic acid caused a significant release of endoge-

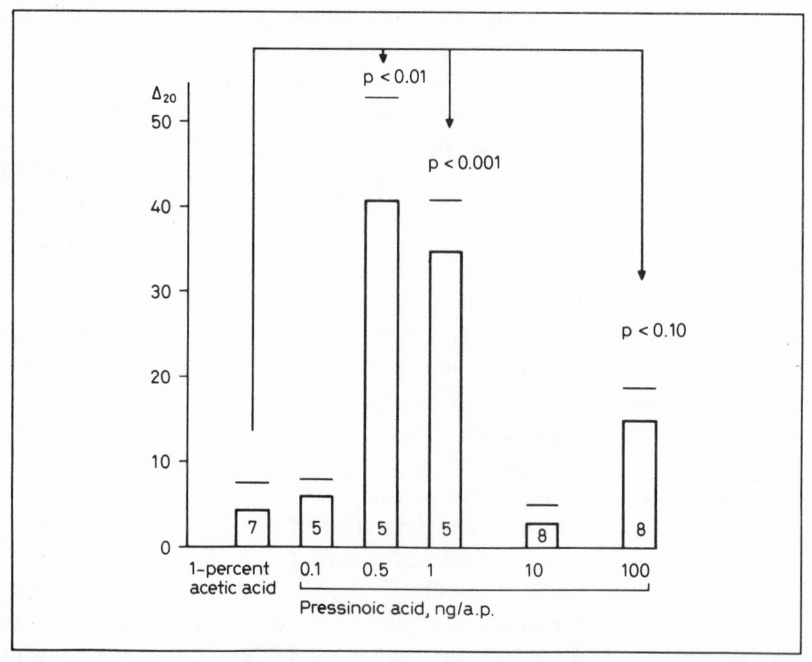

Fig. 1. ACTH-releasing activity of synthetic pressinoic acid as assayed by intrapituitary injection technique. Pressinoic acid was supplied either in powder or in ampoules. They were diluted by 1-percent acetic acid solution before injection. Numbers in the columns indicate number of determinations. Bars on the top denote standard errors of the mean. P values in the figure were obtained by Student's t-test.

nous ACTH at a dose of 50 μg (injected via the jugular vein), though the magnitude of response was small. Since *Arimura et al.* (1967) stated that in their assay system 1.5 μg of β-CRF was effective, this dose (50 μg) of pressinoic acid in our hands was not impressive at all. However, assuming that approximately a ten-thousandth of the cardiac output will reach the anterior pituitary gland in the rat, this would mean about 5 ng or less of 50 μg. This value is, in fact, comparable to the effective amount of the acid when tested by our intrapituitary injection technique.

Table I. ACTH-releasing potency of synthetic pressinoic acid. The activity was tested by the assay system of *Arimura et al.* (1967). P value was obtained by Student's t-test

Dose	No. of rats	Plasma corticosterone 15 min after i.v. injection, μg/100 ml plasma	
Control, 1-percent acetic acid	5	3.0 ± 0.4[1]	
Pressinoic acid			
5 μg	4	2.9 ± 0.4	
50 μg	4	10.2 ± 2.2	$p < 0.01$ (vs. control)

1 Mean ± SE.

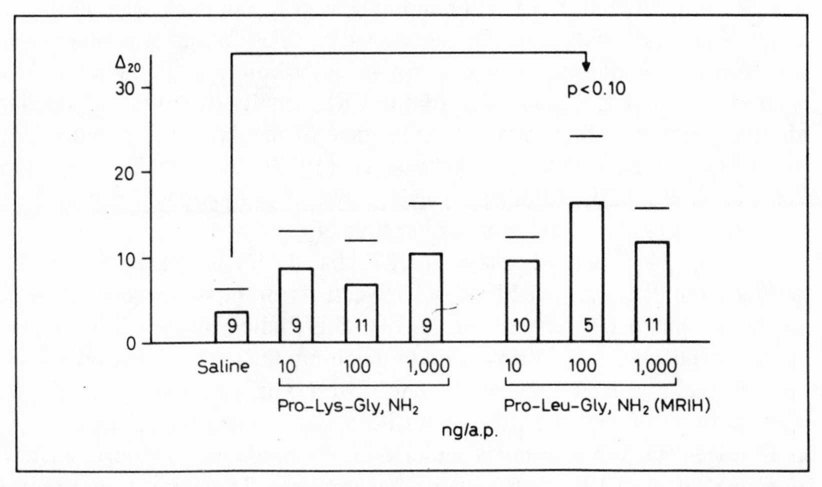

Fig. 2. CRF-like activity of synthetic tripeptides. Doses of the injected peptides are given on abscissa in terms of ng per anterior pituitary (ng/a.p.). For other details, see the legend to figure 1.

Considering the possibility that tripeptide moiety of vasopressin may play a role in the regulation of ACTH release, as in the case of oxytocin moiety in the MSH release (see *Marks,* this volume), we examined ACTH-releasing potency of synthetic tripeptides. Results are shown in figure 2. It appears that H-Pro-Leu-Gly (NH_2) (MSH-release inhibiting hormone) exhibited some activity at 100 ng per gland. However, neither of the tripeptides elicited an ACTH release comparable in magnitude to that of pressinoic acid.

These results indicate that pressinoic acid is a substance with the most potent CRF-like activity when tested with our intrapituitary injection technique. Thus, it is tempting to correlate this substance to so-called β-CRF. The final solution should await isolation and identification of pressinoic acid in the rat hypothalamus and at the same time we should validate functional significance of this substance under physiological conditions.

Phase Shift of Circadian Rhythm of CRF Activity

The pattern of circadian rhythm of hypothalamic CRF content in the rat is still controversial. We reported previously (*Hiroshige et al.,* 1969) that the CRF content in the male rat hypothalamus is lower in the morning and gradually increases toward the peak value in the afternoon. On the other hand, *David-Nelson and Brodish* (1969) observed a precipitous fall of CRF content 2 h prior to the onset of the darkness, even though these authors also observed a gradual rise of the CRF content from the morning level to the peak value in the afternoon. Still more controversial is the change of CRF rhythm after bilateral adrenalectomy. Nevertheless, it was shown in our laboratory (*Hiroshige and Sakakura,* 1971) that the peak value of the CRF activity in the rat hypothalamus showed a phase shift toward the earlier time of the day. Subsequently, similar observations were made by *Takebe et al.* (1972). In addition, these authors found that the CRF periodicity persists even after hypophysectomy, showing another phase shift toward more earlier time of the day.

During the course of study on CRF changes in the female rat, we unexpectedly found a high basal level of CRF activity in the morning and a low level in the afternoon (*Hiroshige et al.,* 1973). This finding thus renders the picture more complicated. In addition, after ovariectomy the peak value showed a phase shift toward the later time of the day. Figure 3 is a composite illustration of these patterns of circadian rhythm of CRF activity in rats. The phase shift of the peak value after ovariectomy is reminiscent of a similar shift, though in a reverse direction, of the CRF rhythm after adrenalectomy. Therefore, it is tempting to speculate that these phase shifts, though delicately modulated, appear to have something in common in their manifestation.

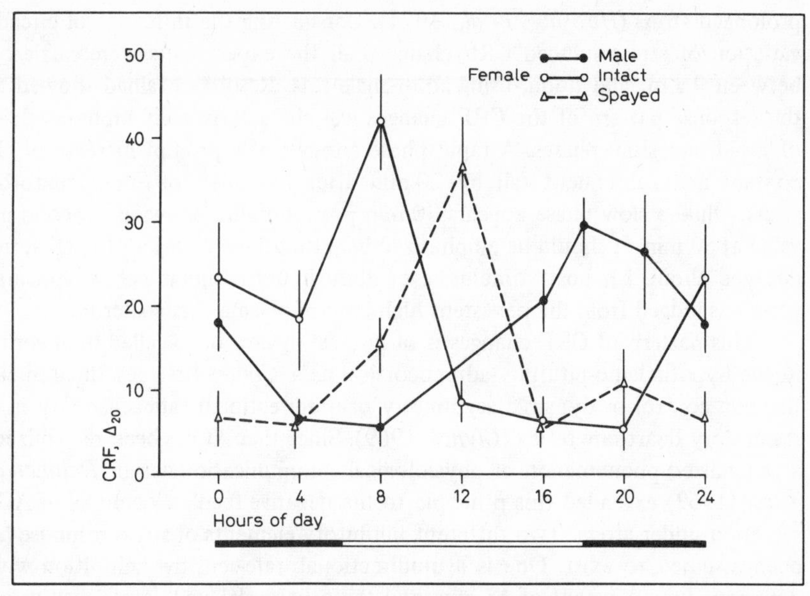

Fig. 3. Composite illustration of various patterns of circadian rhythm of hypothalamic CRF content in rats. ● = Male; ○ = female intact; △ = female spayed. Figure 6 in *Hiroshige et al.* (1973).

In this connection, of particular interest are recent findings by *Sakakura and Brodish* on the phase shift of CRF rhythm in the rat (personal communication). They found that male rats under chronic mild stress show a phase shift of CRF rhythm toward the earlier time of the day. These findings thus raise an important question. Is the sex difference observed genuine or is it explicable as due to sensitivity difference to environmental stress? In our subsequent study (*Hiroshige and Wada-Okada*, 1973), it was shown in well-gentled female rats, that female type of CRF rhythm is particularly manifest during proestrous and estrous stages of the estrous cycle. Thus, it appears essential to define the internal environmental conditions or hormonal status against the external conditions or environmental stress under which the animals were raised and handled.

CRF Dynamics in Relation to Negative Feedback Inhibition

Since we are confronted with a variety of stresses, a detailed analysis of the effects of these factors on the organism is not easy to achieve simultaneously. However, we can, to begin with, examine the effect of a typical noxious stimulus on the hypothalamic CRF content under somewhat specified conditions. This sort of study was performed in our laboratory, using ether laparotomy as a prolonged stress (*Hiroshige et al.*, 1971). Considering the influence of circadian variation of stress-induced CRF changes, all the experiments were carried out between 9 a.m. and noon, using adult male rats. Results obtained showed that the response pattern of the CRF changes was characteristically biphasic, that is, of rapid and slow phases. A rapid phase consists of a prompt increase of CRF content and subsequent fall by 20 min after the onset of ether laparotomy stress, while a slow phase appears 40 min post stimulus, showing a second peak value at 80 min. It should be emphasized here that these changes of CRF activity decayed about 2 h post stimulus, even though the stimulus *per se* appears to persist as judged from the persistent high level of plasma corticosterone.

This pattern of CRF changes is of interest in view of so-called 'rein control' in the hypothalamo-pituitary-adrenocortical axis. *Clynes* first described in 1958 the rein control or dynamic asymmetry of unidirectional rate sensitivity in the respiratory heart rate reflex (*Clynes,* 1969). Since then, it has been recognized as a generalized phenomenon of physiological communication design. *Dallman and Yates* (1969) extended this principle to the negative feedback control of ACTH secretion under stress. Two different inhibitory elements of stress response have been assumed to exist. One is a unidirectional, rate-sensitive inhibition with a dominant time constant of 15 min, and the other a delayed, level-sensitive inhibition with a time constant of about 2 h (*Yates and Maran,* 1973). *Espiner et al.* (1972) provided supporting evidence for this postulate. These investigators showed that the adrenal secretory response in conscious sheep to the stress of immobilization and venipuncture is biphasic. They interpreted that the postulated rate-sensitive feedback element detects the initial, rapid rise in adrenocortical response, and immediately inhibits the system, producing the early fall (within 30 min) toward basal levels. The rate-sensitive signal is then abolished and the system turns on again, to be inhibited later (1–5 h after the onset of stress) by the delayed, level-sensitive feedback element.

Since it is known that the hypothalamic CRF content is reduced by negative feedback inhibition, the whole pattern of the CRF changes in our rats can be explained by the postulate of *Dallman and Yates* (1969). That is to say, the rate-sensitive inhibitory element may account for the first peak of CRF activity, and the delayed, level-sensitive element for the second peak. The consistency appears to be surprisingly good, since we observed these CRF changes under prolonged stress in 1969 to 1970 without having noticed the elegant work of *Dallman and Yates* (1969).

In spite of this consistency, however, we observed also the CRF changes that are apparently not consistent with their theory. Figure 4 shows the results of the experiment in which an attempt was made to examine the ability of the hypothalamus to respond to a second stimulus that was superimposed on the first stimulus or ether laparotomy stress (*Hiroshige et al.*, 1971). As a second stimulus, unilateral leg break was used without anesthesia. When the animals

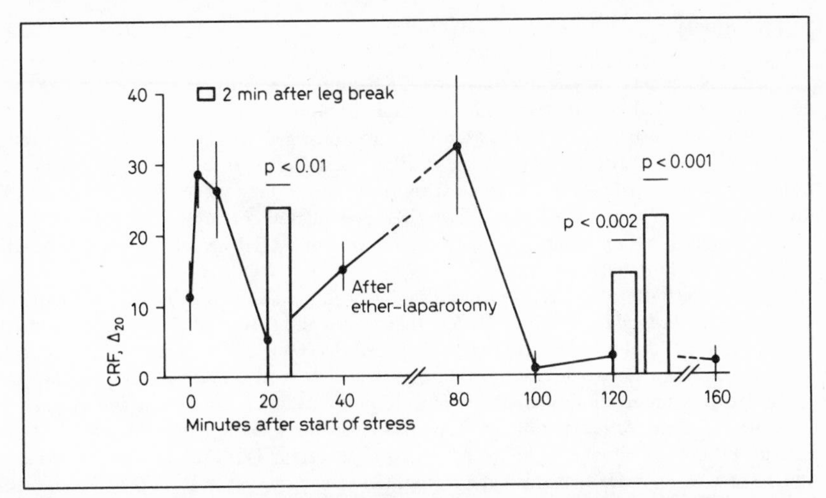

Fig. 4. Changes in the hypothalamic CRF content in adult male rats following a second stimulus applied at different time intervals from the first stimulus. Responses to the second stimulus (leg break) are shown as columns with standard errors on the top. Figure 7 in *Hiroshige* (1973).

were exposed to ether laparotomy stress and 2 h elapsed, no CRF activity was detected. Apparently, this was not due to the exhaustion of CRF neurons, because a second stimulus (leg-break stress) was found to be effective in inducing a rapid increment of the CRF content. Consequently, it appears that the hypothalamic CRF is ready to respond to a new stimulus, even though the CRF content has been depleted due to continuous drive by a previous stimulus. It is of interest to note that even in the midst of CRF fluctuations or at 20 min after the first stimulus, the second stimulus could effectively evoke a significant increase in the CRF content. Thus, it appears that neither of the postulated inhibitory elements worked effectively when a second stimulus was superimposed, as far as our data of CRF changes are concerned.

Quite recently, *Dallman and Jones* (1973) re-examined the problem, estimating plasma corticosterone and ACTH levels in intact and adrenalectomized rats under repeated stress. They found that the second response was equal to or greater in magnitude than the first. They concluded that the first stress elicits a prolonged period of hyper-responsiveness in either CNS or anterior pituitary components of the adrenocortical axis. Our previous findings on CRF changes under stress do indicate that this kind of hyper-responsiveness resides, at least, in the CNS, thus resulting in modulation of CRF dynamics under stress. How these changes in the responsiveness of the CNS under stress may affect the basal circadian rhythm, especially in relation to the phase shift, remains to be clarified.

Summary

Dynamic changes of the hypothalamic CRF content in adult rats were examined under resting and stressful conditions by means of our intrapituitary injection for CRF assay. In this article, two aspects are emphasized. One is concerned with phase shifts of circadian rhythm of CRF activity, especially in relation to sex difference in the pattern of CRF variation. Since it is possible that environmental stress may affect the pattern of CRF rhythm, the other topic is concerned with CRF dynamics under stress. This topic inevitably led us to another important aspect of the control of ACTH secretion, that is, negative feedback inhibition under stress.

The basic pattern of CRF dynamics under prolonged stress can well be explained by the theory of *Dallman and Yates* (1969) that claims the existence of a fast, rate-sensitive and a delayed, level-sensitive feedback element. However, CRF changes under repeated stress are not explained adequately by this theory. Recently, *Dallman* and her associates confirmed non-suppressibility of the second response after the first stress and claimed the existence of a prolonged period of hyper-responsiveness under stress in either CNS or anterior pituitary components of the adrenocortical system. Our findings on CRF dynamics under stress do indicate that this hyper-responsiveness resides, at least, in the CNS, resulting in the modulation of CRF dynamics under stress.

Finally, ACTH-releasing potency of synthetic pressinoic acid and other tripeptides was examined by the intrapituitary injection technique. It was found that pressinoic acid caused a marked ACTH release at doses lower than 1 ng per pituitary gland. This finding encourages us to pursue this specific substance for the future in the rat hypothalamus.

References

Arimura, A.; Saito, T., and Schally, A.V.: Assays for corticotropin-releasing factor (CRF) using rats treated with morphine, chlorpromazine, dexamethasone and Nembutal. Endocrinology, Springfield *81:* 235–245 (1967).

Chan, L.T.; Schaal, S.M., and Saffran, M.: The rat median eminence as a source of CRF; in *Meites* Hypophysiotropic hormones of the hypothalamus: assay and chemistry, pp. 253–259 (Williams & Wilkins, Baltimore 1970).

Clynes, M.: Cybernetic implications of rein control in perceptual and conceptual organization. Ann. N.Y. Acad. Sci. *156:* 629–670 (1969).

Dallman, M.F. and Jones, M.T.: Corticosteroid feedback control of stress-induced ACTH secretion; in *Brodish and Redgate* Brain-pituitary-adrenal interrelationships, pp. 176–196 (Karger, Basel 1973).

Dallman, M.F. and Yates, F.E.: Dynamic asymmetries in the corticosteroid feedback path and distribution-metabolism-binding elements of the adrenocortical system. Ann. N.Y. Acad. Sci. *156:* 696–721 (1969).

David-Nelson, M.A. and Brodish, A.: Evidence for a diurnal rhythm of corticotropin-releasing factor (CRF) in the hypothalamus. Endocrinology, Springfield *85:* 861–866 (1969).

Espiner, E.A.; Hart, D.S., and Beaven, D.W.: Cortisol secretion during acute stress and response to dexamethasone in sheep with adrenal transplants. Endocrinology, Springfield *90:* 1510–1514 (1972).

Gann, D.S.: Parameters of the stimulus initiating the adrenocortical response to hemorrhage. Ann. N.Y. Acad. Sci. *156:* 740–755 (1969).

Gann, D.S. and Cryer, G.L.: Feedback control of ACTH secretion by cortisol; in ***Brodish and Redgate*** Brain-pituitary-adrenal interrelationships, pp. 197–223 (Karger, Basel 1973).

Ganong, W.F.: The central nervous system and the synthesis and release of adrenocorticotropic hormone; in *Nalbandov* Advances in neuroendocrinology, pp. 92–149 (Univ. Illinois Press, Urbana 1963).

Hiroshige, T.: Role of vasopressin in the regulation of ACTH secretion: studies with intrapituitary injection technique. Med. J. Osaka Univ. *21:* 161–180 (1971).

Hiroshige, T.: CRF assay by intrapituitary injection through the parapharyngeal approach and its physiological validation; in ***Brodish and Redgate*** Brain-pituitary-adrenal interrelationships, pp. 57–78 (Karger, Basel 1973).

Hiroshige, T. and Abe, K.: Role of brain biogenic amines in the regulation of ACTH secretion; in *Yagi and Yoshida* Neuroendocrine control, pp. 205–228 (Univ. Tokyo Press, Tokyo 1973).

Hiroshige, T. and Sakakura, M.: Circadian rhythm of corticotropin-releasing activity in the hypothalamus of normal and adrenalectomized rats. Neuroendocrinology *7:* 25–36 (1971).

Hiroshige, T. and Wada-Okada, S.: Diurnal changes of hypothalamic content of corticotropin-releasing activity in female rats at various stages of the estrous cycle. Neuroendocrinology *12:* 316–319 (1973).

Hiroshige, T.; Sakakura, M., and Itoh, S.: Diurnal variation of corticotropin-releasing activity in the rat hypothalamus. Endocrinol. jap. *16:* 465–467 (1969).

Hiroshige, T.; Sato, T., and Abe, K.: Dynamic changes in the hypothalamic content of corticotropin-releasing factor following noxious stimuli: delayed response in early neonates in comparison with biphasic response in adult rats. Endocrinology, Springfield *89:* 1287–1294 (1971).

Hiroshige, T.; Abe, K.; Wada, S., and Kaneko, M.: Sex difference in circadian periodicity of CRF activity in the rat hypothalamus. Neuroendocrinology *11:* 306–320 (1973).

Hiroshige, T.; Kunita, H.; Ogura, C., and Itoh, S.: Effects on ACTH release of intrapituitary injections of posterior pituitary hormones and several amines in the hypothalamus. Jap. J. Physiol. *18:* 609–619 (1968a).

Hiroshige, T.; Kunita, H.; Yoshimura, K., and Itoh, S.: An assay method for corticotropin-releasing activity by intrapituitary microinjection in the rat. Jap. J. Physiol. *18:* 179–189 (1968b).

Kendall, J.W.: Feedback control of adrenocorticotropic hormone secretion; in ***Martini and Ganong*** Frontiers in neuroendocrinology 1971, pp. 177–207 (Oxford Univ. Press, New York 1971).

Mulder, A.H.; Geuze, J.J., and Wied, D. de: Studies on the subcellular localization of corticotropin releasing factor (CRF) and vasopressin in the median eminence of the rat. Endocrinology, Springfield *87:* 61–79 (1970).

Osborne, M.J.; Porter, N.; d'Auric, G.A.; Epstein, A.N.; Woral, M., and Meyer, P.: Metabolism of tritiated angiotensin II in anesthetized rats. Pflügers Arch. ges. Physiol. *326:* 101–114 (1971).

Redgate, E.S.: Role of the baroreceptor reflexes and vasoactive polypeptides in the corticotropin release evoked by hypotension. Endocrinology, Springfield *82:* 704–720 (1968).

Redgate, E.S.; Fahringer, E.E., and Szechtman, H.: Effects of the nervous system on pituitary adrenal activity; in ***Brodish and Redgate*** Brain-pituitary-adrenal interrelationships, pp. 152–175 (Karger, Basel 1973).

Saffran, M.; Pearlmutter, A.F.; Rapino, E., and Upton, G.V.: Pressinoic acid: a peptide with potent corticotrophin-releasing activity. Biochem. biophys. Res. Commun. *49:* 748–751 (1972).

Schally, A.V.; Carter, W.H.; Hearn, I.C., and Bowers, C.Y.: Determination of CRF activity in rats treated with Monase, dexamethasone, and morphine. Amer. J. Physiol. *209:* 1169–1174 (1965).

Takebe, K.; Sakakura, M., and Mashimo, K.: Continuance of diurnal rhythmicity of CRF activity in hypophysectomized rats. Endocrinology, Springfield *90:* 1515–1520 (1972).

Wied, D. de; Witter, A.; Versteeg, D.H.G., and Mulder, A.H.: Release of ACTH by substances of central nervous system origin. Endocrinology, Springfield *85:* 561–569 (1969).

Yates, F.E. and Maran, J.W.: Stimulation and inhibition of adrenocorticotropin (ACTH) release; in *Astwood and Greep* Handbook of physiology, Volume *Sawyer and Knobil* Hypothalamo-hypophysial system: subsection, the adenohypophysis. Amer. Physiol. Soc., Washington, D.C. (in press, 1974).

Author's address: Dr. *T. Hiroshige,* Department of Physiology, Hokkaido University School of Medicine, *Sapporo 060* (Japan)

Regulation of Hypothalamo-Pituitary Function

Psychoneuroendocrinology. Workshop Conf. Int. Soc. Psychoneuroendocrinology, Mieken 1973, pp. 114–121 (Karger, Basel 1974)

Hypothalamo-Hypophyseal Control of Ovulation[1]

M. Suzuki and K. Takahashi

Department of Obstetrics and Gynecology, Tohoku University School of Medicine, Sendai

Introduction

Ovulation is one of the most important phenomena during the reproductive process in the mammalian female. In this series of investigations, the female rabbit was mainly used. The female rabbit belongs to the class of induced ovulators as opposed to the much larger class of spontaneous ovulators. This physiological property has been utilized repeatedly in experiments dealing with the mechanism or control of ovulation (5).

Ovulation is caused by a complicated mechanism and effected by various factors, for example biological, chemical or physical stimulations in the female mammals. Ovulation is also effected by factors of season, time, species, age and the method of breeding, and induced by any stimulus which activates one of hypothalamus, hypophysis or ovary. However, the number of ova ovulated and the process of ovulation are different according to the site of stimulation. In this review, the authors wish to clarify some of the physiological phenomena and existence of some rules of ovulations in the female rabbits, mostly resulting from our own data.

Ovulation Caused by Hypothalamic Stimulation

Site of Action of Cupric Ovulation

A series of experiments to elucidate the mechanism as well as the site of action of the copper-induced ovulation has been carried out by the author and his co-workers, to understand better the mechanism of copper-induced ovulation.

1 This review was aided by grant No. M72.123 from the Population Council, New York.

Induction of ovulation by means of intravenously administered cupric gluconate was tested in nonpregnant, pseudopregnant and pregnant rabbits. In nonpregnant, estrogen-pretreated rabbits, injection of more than 3–4 mg/kg of the compound induced ovulation in all the animals. Several reports deal with the time of ovulation following mating, gonadotropin administration and other stimuli. According to our experiments, ovulation in these rabbits was not induced after 9 h but occurred between 10 and 13 h postinjection of the copper compound (5).

In pregnant and pseudopregnant animals, ovulation could not be induced by the injection of cupric gluconate, indicating a central rather than a peripheral action of this compound and suggesting inhibition of the copper-activated ovulation mechanism by humoral agents characteristic of pregnancy (5).

The fertilizability of copper-ovulated rabbit ova following artificial insemination was investigated in a series of does. Laparotomies performed at various stages of pregnancy indicated normal fertilization rates and normal embryonic development. The young, delivered at term, were likewise normal. Data were also presented concerning carbonic anhydrase activity of corpora lutea from copper-ovulated does, showing no difference between corpora lutea following mating and copper-induced ovulation (6).

Although high doses of estrogen have been found to inhibit the ovulatory response to stimulation, optimum dose of estrogen pretreatment, or 'priming', has also been found to facilitate the induction of ovulation by injection of copper salts into the anestrous rabbit. The optimal pretreatment of these animals proved to be 20 μg of 17β-estradiol benzoate daily for 4 days (7).

When copper salt was injected intravenously into the adult female rabbit, copper ions bound rapidly with albumin and the concentration of copper fell in a short time. Cupric albuminate could induce ovulation in the female rabbit. It may be said that cupric albuminate in blood-flow transports copper to the hypothalamus to induce ovulation (9).

The pituitary contents of FSH, LH and prolactin, as well as concentrations of FSH and LH in plasma, were biologically estimated before and after the injection with cupric salts in the female rabbit (13) and also in the female rat (12). Results of the above assay and of radioimmunoassay (3) at our laboratory show release of LH in the copper-injected rabbit and release of prolactin in the pseudopregnant rat.

Some kind of copper compounds are not effective to induce ovulation, but cupric sulfate, cupric chloride, cupric proline, cupric threonine, cupric gluconate, etc., are effective (10). *D*-penicillamine administration reduces effects as well as side effects of copper compounds (14).

For closer investigation of the site of action, a minute dose of copper salts was implanted stereotaxically into the brain of the rat (11) and the rabbit (1). Ovulation was then noted in the animals with implants in the vicinity to the posterior median eminence of the hypothalamus, but nothing was observed of those with implants in the other areas (anterior pituitary, etc.).

A method of incubating the anterior pituitary with or without the hypothalamus of the rat *in vitro* was used to study the neurohumoral regulation of the hypothalamus on the pituitary. Results suggest that the action of copper-salt caused a significant increase of LH and FSH under the addition of hypothalamic extract (15).

On the other hand, crude hypothalamic extract was injected into the pituitary of the rabbit *in vivo* and the increase of the rate of ovulation was observed. Injection of a crude hypothalamic extract, prepared in the presence of copper, into the pituitary caused a pronounced increase in the rate of ovulation (16).

The authors have injected a solution of copper sulfate intravenously into adult female Japanese mongrel rabbits and also infused the solution directly into the posterior median eminence (PME) of the hypothalamus. The 50-percent effective dose of copper sulfate required for infusion into the PME is one four-thousandth of the 50-percent effective dose required by intravenous injection (8).

Thus, results of these investigations, performed in different points of view, suggest conclusively that the site of action of copper salts is in the hypothalamus vicinity to the median eminence.

Ovulatory Phenomena Caused by Copper Compounds

Stimulants which act directly on the hypothalamus in the female rabbit such as electrical stimulation, sexual mating behavior, injection with a solution of cupric compounds, etc., have been reported elsewhere. Studies were performed by us on copper-induced ovulation in female adult rabbits. Ovulation rate (ratio of ovulated rabbits to all does given copper solution) increases in proportion to doses of copper which are administered intravenously or stereotaxically into the hypothalamus (8).

Superovulation such as that which occurs in gonadotropin-induced ovulation was not noted in copper-induced ovulation. The average number of ova discharged by rabbits subjected to copper-induced ovulation did not increase with doses of copper salts greater than the minimally effective dose (7, 8).

Ovulation Caused by Pituitary Stimulation

Values of LH, FSH and Prolactin in Administration with LH-RH

The female rat of Wistar strain weighing 200–300 g, 3 weeks after castration, was primed with estrogen and progestin. The animal was infused stereo-

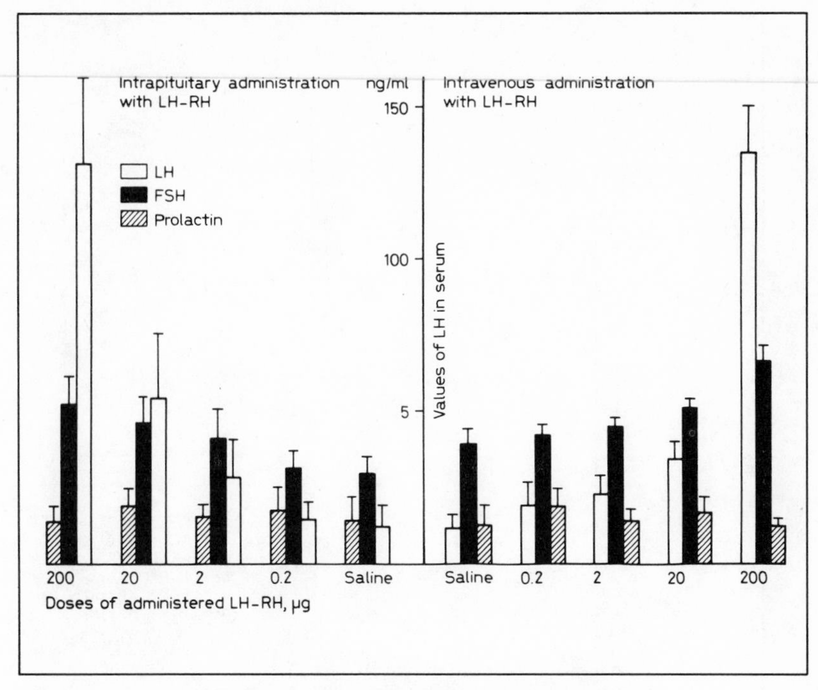

Fig. 1. Values of LH, FSH and prolactin in administration with LH-RH.

taxically with 0.5 μl of LH-RH solution into the anterior pituitary or injected with 50 μl intravenously. The doses administered were 0.02 ng, 0.2 ng, 2 ng and 20 ng of LH-RH in each group. Blood was collected by decapitation. LH, FSH or prolactin in serum were determined by radioimmunoassay (4).

Intrapituitary infusion of more than 0.2 ng of LH-RH raised the value of serum LH significantly. Intravenous injection of more than 2 ng of LH-RH caused a significant increase of serum LH. FSH values were also increased significantly with more than 2 ng of LH-RH intrapituitary. Intravenous administration of 20 ng of LH-RH caused a significant elevation of serum FSH. However, prolactin value in serum could not be found in any significant variation (fig. 1) (4).

Ovulatory Phenomena Caused by LH-RH

Various doses of LH-RH, namely 5, 10, 100, 200 and 1,000 μg per animal, were injected intravenously into the rabbit. Observation of ovulation by laparotomy was carried out before and 24, 48 and 96 h after the injection. The number of ovulated follicles showed the maximum at the administration of 100 μg of

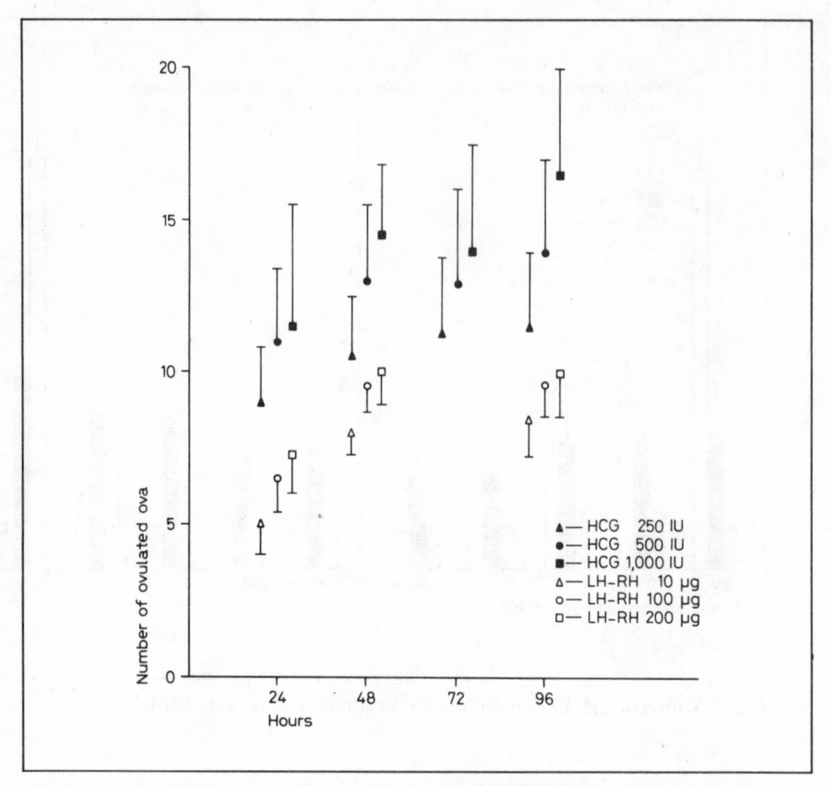

Fig. 2. Relationship between the number of ovulated ova and time course after administration of various doses of LH-RH and HCG in female rabbits.

LH-RH after 48 h, even if higher doses of LH-RH were given, and was slightly more than the number of ova by mating or natural ovulation. It was observed that there was an upper limit as to the number of ovulated follicles and that superovulation of rabbit could not be induced by the injection with LH-RH (fig. 2) (3).

Ovulation Caused by Ovarian Stimulation

Ovulatory Phenomena Caused by FSH and LH

It is a well-known phenomenon that superovulation in the doe is easily induced by the administration of high doses of exogenous FSH for a few days followed by administration of exogenous LH (2). Estrogen pretreatment affects promptly the induction of superovulation by gonadotropin in rabbits, but the maximum number of superovulated ova is almost limited (17). The number of superovulated ova increases in proportion to doses of injected gonadotropins, but does not exceed the maximum number which is specific for species of animals (4).

Ovulatory Phemomena Caused by LH

It is also known that administration with LH only is difficult to produce superovulation (3). High doses of HCG were injected to the female rabbit and observation was made up to 96 h. The number of ovulated ova increased slightly with time.

Propositional Laws about Ovulation

Ovulation in the adult female rabbit is easily induced by stimulation on either the hypothalamus, the hypophysis or the ovary. However, different kinds of ovulatory stimulants act on different kinds of organs, namely the hypothalamus, hypophysis or ovary. Such physiological phenomena of ovulation as already explained are summarized as follows:

Ratio of Ovulated Rabbits

The ratio of ovulated rabbits to all rabbits given ovulatory stimulation (abbreviated to ratio of ovulated rabbits) is in parallel with doses of an ovulatory stimulant, if the dose of the ovulatory stimulant is limited within a certain range.

Minimum doses of ovulatory stimulant do not induce ovulation in the rabbit. Optimal doses of the stimulant cause ovulation. However, less ovulatory doses induce ovulation in some of the stimulated rabbits, and more doses cause ovulation in all of them. In such phenomena, the ratio of ovulated rabbits increases in parallel with augmentation of stimulation. This phenomenon may be observed in some mammalia other than rabbits.

Number of Ovulated Follicles

When observation is made only on the ovulated female rabbit among the stimulated does, the following phenomenon is found.

1. Stimulation on the Ovary

When an ovulation-inducing stimulant, which is within a certain range in dose and which acts on the ovarian level, is given to the female rabbit, the average number of ovulated follicles per animal increases in proportion to doses of the administered ovulatory stimulant.

This phenomenon is observable in animals of gonadotropin administration. If the dose of gonadotropin is less, the rabbit does not ovulate. According to the increment of gonadotropin doses, not only the average number of ovulated rabbits increases but also the average number of ovulated follicles per ovulated rabbit as well. Superovulation is induced only when the animals are administered with both FSH and LH in optimal ratio and doses under well-timed conditions. Finally, a number of superovulated follicles reaches the maximum and does not increase even if high doses of gonadotropin are given.

2. Stimulation on the Pituitary

The pituitary secrets gonadotropin, as humoral stimulation, as seen in the ovary. The number of ovulated ova increases in parallel with increment of doses of the pituitary stimulant, as in ovulation caused by ovarian stimulation, when doses of the stimulant are within the limited range. A maximum number of ovulated ova induced by administration with exogenous LH-RH does not come up to superovulation, even if it exceeds the number of ovulated ova in the natural ovulation.

It may be said that there is the natural upper limit in the capacity of biosynthesis, storage and surge of gonadotropin from the pituitary.

3. Stimulation on the Hypothalamus

Mating, electrical stimulation of the brain, injection of copper compounds, etc., belong to the hypothalamic stimulants. The average number of ovulated follicles per ovulated rabbit used was constant, when the ovulation was caused by hypothalamic stimulation.

The above phenomena were illustrated in the rabbit and may be available in some of the other mammalia. Using these ovulatory phenomena, it may be possible to clarify the site of action of unknown ovulatory stimulants, experimentally.

Summary

Laws of ovulation in the adult female rabbit were proposed anew by the authors on the basis of data of a series of their investigations on reproductive physiology of ovulation. The propositional laws may be available in some of the other mammalia, and also be useful as an experimental procedure in reproductive biology.

References

1 *Hiroi, M.; Sugita, S., and Suzuki, M.:* Ovulation induced by implantation of cupric sulfate into the brain of the rabbit. Endocrinology, Springfield *77:* 963–967 (1965).

2 *Maurer, R.R.; Hunt, W.L., and Foote, R.H.:* Repeated superovulation following administration of exogenous gonadotropins in Dutch-belted rabbits. J. Reprod. Fertil. *15:* 93–102 (1968).

3 *Nakagawa, K. and Suzuki, M.* (in press, 1973).

4 *Sasamori, H. and Suzuki, M.* (in press, 1973).

5 *Suzuki, M. and Bialy, G.:* Studies on copper-induced ovulation in the rabbit. Endocrinology, Springfield *74:* 780–783 (1964).

6 *Suzuki, M. and Bialy, G.:* Fertilizability of copper-ovulated rabbit ova. Endocrinology, Springfield *75:* 288–289 (1964).

7 *Suzuki, M.; Watanabe, S., and Hoshii, M.:* Effect of estrogen on copper-induced ovulation in the rabbit. Endocrinology, Springfield *76:* 1205–1207 (1965).

8 *Suzuki, M.; Hiroi, M., and Sugita, S.:* Ovulation in rabbits following intravenous and intracerebral administration of copper sulphate. Acta endocrin., Kbh. *50:* 512–516 (1965).

9 *Suzuki, M.; Hirabayashi, T., and Hiroi, M.:* Mode of copper transportation in copper-induced ovulation. Acta med. biol., Niigata *16:* 27–33 (1968).

10 *Suzuki, M. and Hoshii, M.:* Studies of various copper compounds on the induction of ovulation and toxicity. J. jap. obstet. gynaec. Soc. *15:* 52–54 (1968).

11 *Suzuki, M.; Kato, T., and Hiroi, M.:* Effect of hypothalamic implantation of copper-salt on the estrous cycle in the rat. Acta med. biol., Niigata *15:* 221–225 (1968).

12 *Suzuki, M.; Hiroi, M.; Nagamatsu, M.; Seike, H.; Ueno, T.; Hojyo, T.; Otake, S.; Takahashi, T.; Nunokawa, O.; Terajima, T., and Tanemoto, Y.:* Copper-induced pseudopregnancy in the rat. Acta med. biol., Niigata *17:* 243–250 (1969).

13 *Suzuki, M.; Murai, A.; Hiroi, M.; Nagamatsu, M.; Seike, H.; Ueno, T.; Hojyo, T.; Otake, S.; Takahashi, T.; Nunokawa, O.; Terajima, T., and Tanemoto, T.:* Effect of copper-salt on the gonadotropin secretion in the rabbit. Acta obstet. gynaec. jap. *17:* 94–98 (1970).

14 *Suzuki, M.; Terajima, T., and Hiroi, M.:* Effects of *D*-penicillamine on the induction of pseudopregnancy by copper in rats and rabbits. Endocrinologia jap. *17:* 241–246 (1970).

15 *Suzuki, M.; Hiroi, M., and Takahashi, T.:* Mechanism of copper-induced ovulation by the incubation experiment of the anterior pituitary gland. Endocrinologia jap. *19:* 1–9 (1972).

16 *Suzuki, M.; Tanemoto, Y., and Takahashi, K.:* The effect of copper salts on ovulation, especially on hypothalamic ovulatory hormone releasing factor. Tohoku J. exp. med. *108:* 9–18 (1972).

17 *Ishijima, Y.; Ito, M.; Hirabayashi, T., and Sakuma, Y.:* Effect of the estrogen pretreatment on the induced superovulation by PMS in rabbits. II. Experiments on the speed of ova descending the genital tract and cleavage-rate of ova. Jap. J. Animal Reprod. *13:* 71–76 (1967).

Authors' address: Dr. *M. Suzuki* and Dr. *K. Takahashi,* Department of Obstetrics and Gynecology, Tohoku University School of Medicine, Seiryo-machi-l-ban-l-go, *Sendai 980* (Japan)

Psychoneuroendocrinology. Workshop Conf. Int. Soc. Psychoneuroendocrinology, Mieken 1973, pp. 122–127 (Karger, Basel 1974)

Steroidal Regulation of Hypothalamo-Pituitary Function in Relation to Ovulating Hormone Release in Rats

Fumihiko Kobayashi[1]

Shionogi Research Laboratory, Shionogi & Co. Ltd., Osaka

Introduction

It has been demonstrated by many investigators that ovarian steroids given exogenously exert their effects on ovulating hormone (OH) release by inhibiting or by facilitating action according to the timing of administration during 4- or 5-day estrous cycles (3, 6, 10, 15). Especially, the effect of integrated amounts of estrogen secreted from the afternoon of diestrus II until the morning of proestrus is responsible for central nervous system (CNS) mechanism(s) to induce OH release (11, 12). Progesterone is also known to stimulate OH release under various experimental conditions (6). It has been demonstrated that the secretion of this steroid does increase physiologically at around the time of OH release in proestrous afternoon (19). In the experiments presented in this paper, we attempted to elucidate the stimulatory action of progesterone on the CNS mechanism(s) regulating OH release in proestrous rats by various methods.

Experiments were carried out by using adult female Wistar rats weighing approximately 200 g. They were housed in an air-conditioned room (25 ± 1 °C, 50–60 % humidity) illuminated 12 h a day from 08.00 to 20.00 and maintained with ordinary laboratory chow and water *ad libitum.* Vaginal smearing was carried out daily and the rats showing at least three regular 4-day estrous cycles were used.

1 The author gratefully acknowledges Dr. *T. Miyake* for his advice and discussion, and Mr. *K. Hara* and Mr. *K. Sasaki* for their excellent technical assistance.

Facilitation by Progesterone of OH Release

Subcutaneous Effect of Progesterone in Normal and in Sodium Pentobarbital-Blocked Proestrous Rats

To determine whether the timing of OH release is affected by progesterone, the rats given 10 mg of the steroid on the morning of proestrus were timed hypophysectomized on the afternoon of the same day and autopsied on the following morning to inspect ovulation (13). Ovulation did not occur at all in control rats when hypophysectomized before 16.00 proestrus, but did occur in some rats operated at 17.00 and 18.00, and full ovulation was observed in the rats hypophysectomized at 19.00 or later. When progesterone was injected subcutaneously at any time on the morning of proestrus, ovulation was confirmed on estrus in nearly all the rats hypophysectomized at 15.00 proestrus, and all the rats ovulated when operated at 16.00 or 17.00. The timing of actual ovulation on the early morning of estrus was also advanced in the rats given progesterone on the morning of proestrus. These results indicate that progesterone, when given on the morning of proestrus, advances by 2 or 3 h not only the timing of OH release on the afternoon of proestrus, but also accelerates the subsequent ovulation occurring the next morning without changing the interval between OH release and ovulation. A similar advancement by progesterone of OH release (21) as well as preovulatory progesterone secretion (20) has been reported.

In order to elucidate additionally the facilitatory action of progesterone on OH release, the steroid (5 mg/rat) was injected to proestrous rats which had been treated with the minimal dose of sodium pentobarbital (PB) (25 mg/kg, i.p.) to inhibit spontaneous OH release (16). In contrast to control rats given PB alone (20–30 %), incidence of ovulation (number of rats ovulating against total number) markedly increased to attain about 80 % when progesterone was injected simultaneously with PB, or until 21.00 proestrus, 4 h after PB injection. Injection of progesterone to the PB-blocked rats at 22.00 or later, failed to show any increase in the incidence of ovulation.

It is speculated from these results, that the intrinsic CNS mechanism(s) to induce OH release starts to activate to a certain extent about 3 h before the timing of spontaneous OH release and lasts for about 6 h on the afternoon of proestrus.

Intrahypothalamic Implantation of Progesterone on OH Release in Proestrous Rats

Crystalline progesterone tapped into stainless steel tubing was implanted bilaterally in the anterior hypothalamic area (AHA), the medial basal hypothalamus (MBH), or the anterior pituitary before 11.30 on the morning of proestrus. All the rats were anesthetized with PB (40 mg/kg, i.p.) at 16.00 on the same day and were inspected for ovulation on the following morning (13). Ovulation was only observed in the rats having progesterone implantation in the median eminence-arcuate region of the hypothalamus, but not in the ventromedial nucleus. Implantation of the steroid in the AHA or in the anterior pituitary was almost ineffective. Implantation of crystalline cholesterol into these areas, as a control, was quite ineffective. These results suggest that the median eminence-arcuate region of the hypothalamus may be a site of action of progesterone in facilitating OH release in proestrous rats. Similar results were recently demonstrated by *Van Rees* (18). However, it has also been suggested that progesterone facilitates ovulation by acting on the medial preoptic area (1) or at other CNS sites more distant from the MBH (9).

Hypothalamic Deafferentation on Progesterone-Induced OH Release in Proestrous Rats

Further study was made to analyze the facilitatory mechanism of progesterone on OH release in proestrous rats by using deafferentation technique of *Halász and Pupp* (8). Neural connections to the MBH or to the preoptic-suprachiasmatic region of the hypothalamus were transected on the morning of proestrus with a curved knife or with an L-shaped knife. After the recovery from the surgical anesthesia, the rats were injected with progesterone (1 mg/rat) and anesthetized again with PB (35 mg/kg, i.p.) given at 16.15 proestrus and the occurrence of ovulation was observed on the following morning (14). The responses of sham-operated controls to progesterone and PB were the same as those of intact rats. As well as the sham-operated controls, posterior cut at the level of the mammillary body did not block the progesterone-induced ovulation. However, anterior cut at the level of retrochiasmatic region completely blocked progesterone-induced ovulation. The same blockade of progesterone-induced ovulation was also observed in the rats given more rostral and broad dorsal (roof-cut) deafferentations. These results are similar to those reported previously (9) and suggest that the extrahypothalamic location in the higher CNS may be responsible for the induction of OH release by progesterone. However, this does not agree with the result that progesterone implantation in the MBH advances OH release in proestrous rats (5, 13, 18).

It is assumed from these results that the action mechanism of progesterone to facilitate OH release in proestrous rats may be due to the sensitization of the MBH for afferent ovulatory impulses coming from the higher CNS *via* the anterior region of the hypothalamus. *Döcke and Dörner* (5) and *Van Rees* (18) demonstrated in preoptic lesioned or in retrochiasmatic deafferented rats that progesterone pretreatment resulted in the decrease in the threshold of effective electrical stimulation given on the MBH to induce gonadotropin release.

Adrenal Participation on OH Release

Although many studies clearly demonstrate the facilitatory effect of progesterone on OH release in proestrous rats, it is much more difficult to ascribe a physiological role to this steroid than to estrogen. Indeed, ovarian progesterone secretion does increase prior to ovulation, but many data indicate that ovarian progesterone secretion starts after, but not before, the beginning of OH release (2, 19). This suggests that progesterone cannot be the signal for the onset of OH release, even if it can cause OH release under experimental conditions. However, *Barraclough et al.* (2) have demonstrated that peripheral plasma concentrations of progesterone significantly elevate before OH release occurs. This finding is substantiated by the observations of *Feder et al.* (7). Both of these groups have proposed a possibility that gradual increase in peripheral plasma concentrations of progesterone occurring before OH release may be due to adrenal secretion and plays a role in inducing OH release. The adrenal has been known to be a source of progesterone in rat and the circadian secretion of adrenocorticoid is greatest on the day of proestrus just before the onset of OH release (4).

To analyze the adrenal participation on OH release, the effect of adrenalectomy on the timing of OH release was examined. Rats were adrenalectomized on the morning of proestrus and were timed hypophysectomized on the afternoon of the same day beginning at 16.00. They were autopsied on the afternoon of the next day to inspect ovulation. The results are presented in table I. Sham-operated controls did not ovulate at all when hypophysectomized at

Table I. Effect of adrenalectomy on the timing of ovulating hormone release in proestrous rats

Operation[1], 10.00–11.00 on proestrus	Hour of hypophysectomy[1]	Number of rats ovulating / total number of rats	Average number of ova per ovulating rat
Sham-adrenalectomy	16.00	0/5	0
	19.00	6/6	12.2 ± 1.3[2]
	22.00	5/5	10.6 ± 1.0
Adrenalectomy	16.00	0/5	0
	19.00	4/9	7.3 ± 2.3
	21.00	6/12	8.8 ± 1.8
	22.00	4/4	13.0 ± 2.1

1 Operations were performed under methylhexabital sodium anesthesia.
2 Mean ± SE.

16.00 proestrus, but did ovulate by the hypophysectomy performed at 19.00 or later. On the contrary, the rats adrenalectomized on the morning of proestrus did not show full ovulation when hypophysectomized at 19.00 or 21.00 proestrus. Full ovulation in these adrenalectomized rats was observed when hypophysectomy was postponed until 22.00 proestrus, 3 h later than from sham-operated controls. This clearly indicates that the timing of OH release is delayed in some rats 3 h longer than in intact rats. A similar delay by adrenalectomy of OH release and mating behavior has been demonstrated (7, 17). These results suggest that progesterone secreted from the adrenal gland may have some role to synchronize the timing of OH release in proestrous rats.

Summary and Comment

These experiments clearly demonstrate that progesterone has an activity to facilitate OH release in proestrous rats. Since progesterone injection given on diestrus II or before 02.00 proestrus completely inhibits spontaneous ovulation (10), it is considered that the appearance of facilitatory action of progesterone on OH release depends on the pre-existing estrogen level in the circulating blood. A possibility that progesterone directly induces the secretion of hypothalamic neurohumor(s) from the MBH into the portal vessels is denied, because the facilitation by progesterone of OH release does not take place in the rats deafferented at the retrochiasmatic or more rostral levels. It is likely, therefore, that, even if the median eminence-arcuate region of the hypothalamus is the primary site of progesterone action in facilitating OH release, the facilitation by this steroid may be due to the increase in the responsiveness of the MBH to ovulatory stimulus coming from the higher CNS. The results that acute adrenalectomy performed on the morning of proestrus postpones the time period of OH release, suggest an important role of this gland for the onset of OH release. Progesterone secreted by the adrenals, though not essential for OH release to occur, may account for the synchronization of the timing of OH release in proestrous rats. Preovulatory secretion of progesterone from the ovary is the consequence of OH release, rather than the cause, and may be responsible for the prolongation and heightening of the CNS ovulating activity, for the increase in the responsiveness of the MBH to ovulatory stimulus, and for full sexual receptivity in the hours before ovulation.

References

1 *Barraclough, C.A.; Yrarrazaval, S., and Hatton, R.:* A possible hypothalamic site of action of progesterone in the facilitation of ovulation in the rat. Endocrinology, Springfield *75:* 838–845 (1964).

2 *Barraclough, C.A.; Collu, R.; Massa, R., and Martini, L.:* Temporal interrelationships between plasma LH, ovarian secretion rates and peripheral plasma progestin concentrations in the rat: effects of nembutal and exogenous gonadotrophins. Endocrinology, Springfield *88:* 1437–1447 (1971).

3 *Brown-Grant, K.:* The induction of ovulation by ovarian steroids in the adult rat. J. Endocrin. *43:* 553–562 (1969).

4 *Critchlow, V.; Liebelt, R.A.; Bar-Sela, M.; Mountcastle, W., and Lipscomb, H.S.:* Sex difference in resting pituitary-adrenal function in the rat. Amer. J. Physiol. *205:* 807–815 (1963).
5 *Döcke, F. and Dörner, G.:* A possible mechanism by which progesterone facilitates ovulation in the rat. Neuroendocrinology *4:* 139–149 (1969).
6 *Everett, J.W.:* Central neural control of reproductive functions of the adenohypophysis. Physiol. Rev. *44:* 373–431 (1964).
7 *Feder, H.H.; Brown-Grant, K., and Corker, C.S.:* Preovulatory progesterone, the adrenal cortex and the 'critical period' for luteinizing hormone release in rats. J. Endocrin. *50:* 29–39 (1971).
8 *Halász, B. and Pupp, L.:* Hormone secretion of the anterior pituitary gland after physical interruption of all nervous pathways to the hypophysiotrophic area. Endocrinology, Springfield *77:* 553–562 (1965).
9 *Kaasjager, W.A.; Woodbury, D.M.; Van Dieten, J.A.M., and Van Rees, G.P.:* The role played by the preoptic region and the hypothalamus in spontaneous ovulation and ovulation induced by progesterone. Neuroendocrinology *7:* 54–64 (1971).
10 *Kobayashi, F.; Hara, K., and Miyake, T.:* Effects of steroids on the release of luteinizing hormone in the rat. Endocrin. jap. *16:* 251–260 (1969a).
11 *Kobayashi, F.; Hara, K., and Miyake, T.:* Causal relationship between luteinizing hormone release and estrogen secretion in the rat. Endocrin. jap. *16:* 261–267 (1969b).
12 *Kobayashi, F.; Hara, K., and Miyake, T.:* Further studies on the causal relationship between the secretion of estrogen and the release of luteinizing hormone in the rat. Endocrin. jap. *16:* 501–506 (1969c).
13 *Kobayashi, F.; Hara, K., and Miyake, T.:* Facilitation of luteinizing hormone release by progesterone in proestrous rats. Endocrin. jap. *17:* 149–155 (1970).
14 *Kobayashi, F. and Miyake, T.:* Acute effect of hypothalamic deafferentation on progesterone-induced ovulating hormone release in the rats. Endocrin. jap. *18:* 395–401 (1971a).
15 *Kobayashi, F.; Hara, K., and Miyake, T.:* Induction of delayed or advanced ovulation by estrogen in 4-day cyclic rat. Endocrin. jap. *18:* 389–394 (1971b).
16 *Kobayashi, F.; Hara, K., and Miyake, T.:* Facilitation by progesterone of ovulating hormone release in sodium pentobarbital-blocked proestrous rats. Endocrin. jap. *20:* 175–180 (1973).
17 *Nequin, L.G. and Schwartz, N.B.:* Adrenal participation in the timing of mating and LH release in the cyclic rat. Endocrinology, Springfield *88:* 325–331 (1971).
18 *Rees, G.P. van:* Control of ovulation by the anterior pituitary gland. Progr. Brain Res., vol. 38, pp. 193–210 (Elsevier, Amsterdam 1972).
19 *Uchida, K.; Kadowaki, M., and Miyake, T.:* Ovarian secretion of progesterone and 20α-hydroxy-pregn-4-en-3-one during rat estrous cycle in chronological relation to pituitary release of luteinizing hormone. Endocrin. jap. *16:* 227–237 (1969a).
20 *Uchida, K.; Kadowaki, M., and Miyake, T:* Effect of exogenous progesterone on the preovulatory progesterone secretion in the rat. Endocrin. jap. *16:* 485–491 (1969b)
21 *Zeilmaker, G.H.:* The biphasic effect of progesterone on ovulation in the rat. Acta endocrin., Kbh. *51:* 461–468 (1966).

Author's address: Dr. *F. Kobayashi,* Shionogi Research Laboratory, Shionogi & Co. Ltd., Fukushima-ku, *Osaka 553* (Japan)

Psychoneuroendocrinology. Workshop Conf. Int. Soc. Psychoneuroendocrinology, Mieken 1973, pp. 128–135 (Karger, Basel 1974)

Prolactin Secretion in Normal and Neonatally Estrogenized Persistent-Diestrous Rats at Advanced Ages[1]

S. Kawashima, T. Asai and K. Wakabayashi

Zoological Institute, Faculty of Science, University of Tokyo, Tokyo, and Institute of Endocrinology, Gunma University, Maebashi

Introduction

According to *Bloch* (6) estrous phase was prolonged in female rats of the Wistar strain at about 10 months of age and they showed persistent estrus at 17 months. *Aschheim* (2) and *Aschheim and Pasteels* (3) have pointed out that in such animals the secretion of gonadotropins by the hypophysis was impaired, while the secretion of prolactin was enhanced. *Clemens and Meites* (7) reached a similar conclusion by the measurement of the hypophyseal LH, FSH and prolactin contents in old Sprague-Dawley rats showing spontaneous persistent estrus. The hypophyseal prolactin levels in 21-month-old rats were two to three times more than those in young cycling females (7). The primary determinant for the induction of persistent-estrous (PE) state at advanced ages appears to be estrogen secreted from the ovaries, since ovariectomy could abolish the establishment of the senile change (3) and, on the contrary, the addition of small amounts of estrogen to adult female rats brought about early induction of persistent estrus (12). In relation to these findings, the present study was undertaken to show if there was an elevation in prolactin secretion in old persistent-diestrous (PD) rats induced by neonatal estrogen administration as well as in old ovariectomized female rats. The results were briefly reported elsewhere (13).

Materials and Methods

Normal and Persistent-Diestrous Rats

The animals used in this study were male and female rats of the Wistar strain maintained in the Zoological Institute, University of Tokyo. When female rats reached the age of

1 Supported in part by a grant-in-aid of Fundamental Scientific Research from the Japanese Ministry of Education (No. 854183).

about 11 months, daily vaginal smears were recorded. If a female rat was in persistent estrus for 1 month or more, it was used for experiment. Eighteen- or 24-month-old females which showed prolonged estrus or prolonged diestrus and which had shown persistent estrus at 12 months of age were also used. Cycling females at 3 months of age served as young controls. Some other normal female rats were ovariectomized at 3 months of age and sacrificed 9 months later. PD rats were secured by neonatal administration of estrone in daily injections to female rats for 30 successive days commencing from the day of birth, i.e., 50 μg daily in 0.02 ml of sesame oil for the first 10 days, 100 μg daily in 0.04 ml of oil for the middle 10 days and 200 μg daily in 0.08 ml of oil for the last 10 days. After the attainment of puberty, the animals were continuously in the state of diestrus. Their ovaries contained small follicles, neither large follicles nor corpora lutea being found. Normal male rats as well as PD rats were killed by decapitation at either 3, 12, 18 or 24 months of age.

Morphological Studies of Prolactin Cells

Both normal and PD rats of varying ages were killed by decapitation and the hypophysis was quickly taken out. For the cytological study the hypophysis was fixed in Bouin-Holland-Sublimate for 2 h after having been placed in a mixture of 3-percent $K_2Cr_2O_7$, saturated $HgCl_2$ solution and neutral formalin (8:1:1) (22) for 1 h. Frontal sections, 5 μ thick, were cut in paraffin and stained according to *Herlant*'s tetrachrome technique (10). Erythrosinophils in the hypophyses were counted in 15–25 oil immersion fields in each of three sections at three comparable levels.

Radioimmunoassay of Prolactin

Prolactin concentrations in the serum and the pars distalis of hypophysis in the experimental animals were determined with the NIAMDD-radioimmunoassay kit from Rat Pituitary Hormone Distribution Program, National Institute of Arthritis, Metabolism and Digestive Diseases, National Institute of Health, Bethesda, Md., which consisted of NIAMDD-rat prolactin-I-1, NIAMDD-anti-rat prolactin serum-S-1 and NIAMDD-rat prolactin-RP-1 (11 IU/mg in mouse deciduoma assay), according to the procedure described in the instruction paper with a slight modification. The results were expressed as μIU or mIU of NIAMDD-rat prolactin-RP-1.

Blood samples were collected from decapitated animals between 14.00 and 17.00 (photoperiod: 06.00–20.00) and the pars distalis of hypophysis was simultaneously taken out and weighed on a torsion balance. The hypophysis was homogenized in 0.05 *M* phosphate buffer at pH 7.5 followed by freezing and thawing. Serum and supernatant of the hypophyseal homogenate were kept frozen at −20 °C until radioimmunoassay. Serum and hypophyseal samples from rats bearing spontaneous mammary gland tumors or pituitary tumors were discarded.

Results

Microscopical Observations

In tetrachrome preparations of the hypophysis, prolactin cells were identified by the affinity of the cells to erythrosin. Erythrosinophils markedly increased in percent of total pituitary cells in 1- or 2-year-old normal rats; however, the affinity to the dye was very weak in many cells. Prolactin cells predominantly distributed in the central portion in each half of the pars distalis. In contrast, in estrogenized PD rats, there was no significant increase in number of prolactin cells. The cells only slightly increased when the PD animals reached the age of 2 years. In nine of 11 normal rats at 1 year of age and four of seven animals at 2 years of age, the mammary glands contained hyperplastic milk-containing alveoli. Duct-alveolar system in PD rats was poorly developed.

Prolactin Concentrations in the Serum and the Hypophysis

Serum Prolactin

In a preliminary study, serum prolactin levels during the estrous cycle of the rat at 3 months of age were determined. Significant increase in prolactin concentrations was noted on the afternoon of proestrus, representing a 3–4-fold rise over base-line values on the second day of diestrus (pooled value of proestrus at 14.00–18.00:659 ± 57 μIU, n = 11, and diestrus-2 at 10.00–14.00:139 ± 86, n = 5; the difference is statistically significant).

In old female rats, either at the age of 12 or 18 months, mean serum prolactin values were higher than values at any stage of the estrous cycle of young rats. However, the variances calculated from prolactin titers in old rats were heterogenous to the variance in young rats. State of mammary glands was studied in order to find the source of great variances in old rats. In seven of 14 rats at 12 or 18 months of age, mammary glands were well-developed, containing patches of milk-containing alveoli. Serum prolactin levels of these rats were more

Table I. Serum prolactin levels in normal (N) and persistent-diestrous (PD) rats

Group	Age, months	Prolactin μIU/ml, mean ± SE
N♀	12 and 18 (without milk)	674 ± 113 (7)
N♀	12 and 18 (with milk)[1]	4,666 ± 640 (4)
N ♀[2]	12	343 ± 34 (5)
PD ♀	3	357 ± 88 (4)
PD ♀	18	364 ± 62 (4)
N ♂	3	179 ± 12 (5)
N ♂	18	247 ± 19 (5)

1 Two of 12-month-old and one of 18-month-old rats were omitted from the mean value, since serum levels exceeded 7,680 μIU/ml.

2 Ovariectomized at 3 months of age. The numbers of rats are indicated in parentheses.

than 3,000 μIU/ml. The levels of the remaining seven rats were between 300 and 1,200 μIU/ml. Since there was no difference between 12-month and 18-month groups, they were pooled and the pooled data were divided into two subgroups of rats according to the state of mammary gland (F = 36.3 with 6 and 6 df).

The serum prolactin value of normal old rats with milk was much higher than that of young rats at proestrus (pooled proestrus value of the young versus old PE rats; $p < 0.01$), and even the value of old rats without milk was approximately at proestrous level. On the other hand, serum of PD rats contained less prolactin at 18 months of age than normal old females with milk ($p < 0.01$, table I), There was no significant difference between 3- and 18-month-old PD animals. Regardless of age, prolactin levels of all PD rats exist within the normal range of young female rats. The levels of serum prolactin of normal male rats at 3 months of age were lower than the levels of PD rats and were approximately the base-line values of young females on the second day of diestrus. There was slight but significant difference in the serum prolactin levels between 3- and 18-month-old male rats ($0.01 < p < 0.02$). When normal female rats were ovariectomized at 3 months of age and the blood was collected by decapitation 9 months later, the serum prolactin level was not elevated to the control level.

Hypophyseal Prolactin

The weight of the hypophysis, prolactin concentration and total content are presented in table II. There was a marked increase in weight of the pars distalis of the hypophysis in normal females at advanced ages. Old PD rats and old male rats also showed some increase in hypophyseal weight ($p < 0.01$ in either group);

Table II. Prolactin content in the pars distalis of hypophysis of normal (N) and persistent-diestrous (PD) rats

Group	Age, months	Weight, mg	Prolactin mIU	
			whole gland	per mg
N ♀ (diestrus-1)	3	5.7 ± 0.7 (4)[1]	145 ± 36	29.0 ± 10.0
N ♀ (without milk)	12 and 18	15.3 ± 1.6 (7)	701 ± 289	52.0 ± 23.0
N ♀ (with milk)	12 and 18	15.8 ± 0.7 (7)	1,886 ± 373	117.0 ± 22.0
N ♀[2]	12	10.9 ± 0.5 (5)	81 ± 9	7.5 ± 0.9
PD ♀	3	5.0 ± 0.4 (4)	110 ± 36	24.0 ± 9.0
PD ♀	18	8.6 ± 0.6 (6)	236 ± 92	30.0 ± 13.0
N ♂	3	7.9 ± 0.1 (5)	66 ± 5	8.3 ± 0.7
N♂	18	10.7 ± 0.4 (5)	88 ± 5	8.3 ± 0.4

1 Mean ± SE. The number of rats is indicated in parentheses.
2 Ovariectomized at 3 months of age.

however, their weights were less than that in normal females of the comparable age ($p < 0.01$).

Along with the increase in weight, the total prolactin content of the hypophysis significantly increased in old females bearing mammary glands with milk-containing alveoli ($p < 0.01$). Concentration was also elevated in these rats ($0.01 < p < 0.02$). Old female rats without milk also contained more prolactin than young females but the difference was statistically not significant. Old PD rats contained decidedly less prolactin than old normal rats with milk on the bases of concentration and total content.

There was no difference in prolactin concentration between young and old male rats, but on the whole gland basis 18-month-old male rats contained more prolactin than 3-month-old males ($0.01 < p < 0.02$).

Ovariectomy of normal females at 3 months of age resulted in the maintenance of prolactin content and concentration at a low level. The weight of hypophysis was also less than that of normal females of the same age ($p < 0.01$).

Discussion

The present results showing that serum and hypophyseal prolactin concentrations in normal female rats in spontaneous persistent-estrus have markedly increased at advanced ages are in good agreement with observations by *Aschheim* (2), *Aschheim and Pasteels* (3) and *Clemens and Meites* (7). From the observations that the content of hypothalamic prolactin-inhibiting factor (PIF) did not appreciably alter in old persistent-estrous (PE) rats, *Clemens and Meites* (7) concluded that the increase in the hypophyseal concentration of prolactin was due to estrogen secreted from the ovaries of PE rats, possibly acting directly on the hypophysis to stimulate prolactin secretion.

There was ample evidence for the stimulatory effects of estrogen on prolactin secretion (e.g. 8, 9, 19, 20). *Nicoll and Meites* (20) demonstrated that cultured anterior hypophysis actively produced prolactin for prolonged periods of time and that estrogen added to the medium stimulated the secretion by acting directly on the hypophyseal cells. *Pantić and Genbačev* (21) were of the same opinion that estrogen acted directly on the prolactin cells. They further reported that the cells in male rats were more readily influenced than those in females by single neonatal injection of estrogen. If adult female rats were given repeated injections of estrogen, pituitary tumor could be induced. *Hymer et al.* (11) and recently *Watari and Tsukagoshi* (26) showed that prolactin cell of the rat hypophysis, following estrogen treatment for more than 3 months, was characterized by a well-developed endoplasmic reticulum, enlarged Golgi apparatus and a few secretory granules. The hypophyses of old female rats contained a great number of prolactin cells bearing similar ultrastructural characteristics as cells of estrogen-stimulated hypophysis, while in old PD rats prolactin cells did not show significant signs of hyperactivity (*Kawashima and Yamamoto,* unpublished observation). Although the prolactin content of the hypophysis in PD rats was lower than those in PE or normal rats, the potency of the hypophysis to secrete prolactin was little, if any, affected by neonatal steroid treatment, since injections of reserpine or placement of the organ under the renal capsule could enhance the elaboration of the hormone (14, 15, 23).

However, it is not decided whether the retardation of aging processes in PD rats in terms of prolactin secretion was due to decreased titer of estrogen and/or to decreased sensitivity to estrogen or to extra estrogen factors. Concerning the mechanism underlying early steroid treatment on prolactin cell function, experiments are now in progress to study if there are long-lasting direct effects of estrogen on the hypophyseal cells under culture conditions.

In relation to the effects of neonatal steroid treatment on prolactin secretion as adult, *Kikuyama* (15) reported that corpora lutea induced by exogenous gonadotropins were not spontaneously functional in androgenized PE rats. *Barraclough and Fajer* (4) also reported that serum prolactin was very low in androgenized rats on the basis of progesterone production by gonadotropin-induced corpora lutea. Hypophyseal prolactin content in androgenized females was less than that in normal females (16). On the contrary, in PE rats of the Holtzmann strain induced by neonatal androgenization, *Mallampati and Johnson* (17) recently found a very high concentration of serum prolactin level measured by radioimmunoassay. There may be some strain difference in response to early steroid treatment. *Zeilmaker et al.* (27) stated that in neonatally androgenized Osborne-Mendel rats the neural mechanism to stimulate prolactin secretion was impaired; however, in Charles River CD rats the similar treatment of androgen presented evidence of elevated prolactin secretion in company with persistent-estrus and milk production. *Neill* (19) also stated that base-line prolactin level was higher in gonadectomized males or ovariectomized androgen-sterilized females than in untreated gonadectomized females, and that mammary glands were enlarged and contained milk in androgen-sterilized PE rats. From these results he suggested a possibility that neonatal androgen acted to inhibit the organization of the anterior hypothalamic 'surge center', which is normally functioning in females during the estrous cycle (e.g. 1, 18, 24, 25) and simultaneously suppressed the PIF-producting function of the medial hypothalamus. *Blake et al.* (5) recently reported in PD rats, secured by complete deafferentation of the medial basal hypothalamus, that not only 'proestrous surge' was blocked, but also at all times serum concentration of the hormone was kept as low as the diestrous level, and that these changes were ascribable to alterations in circulating estrogen.

The present findings and those of previous workers, indicate that the blood titer of estrogen, which is more in old normal females than PD rats secured by neonatal estrogenization, or male rats, or old females ovariectomized during early adult life, is playing a key role in enhancing the aging processes of the hypothalamus in terms of prolactin secretion.

Summary

Prolactin secretion in normal and neonatally estrogenized persistent-diestrous (PD) rats at advanced ages was investigated. Prolactin cells increased in number and in percent of total hypophyseal cells in normal female rats at the age of 1 or 2 years compared to those of young rats, but the increase was not observed in PD rats. Radioimmunoassay of prolactin showed that the concentrations in the serum, as well as the hypophysis, were less in old PD rats or in male rats, than those in old normal females. Together with the result that the increase in prolactin concentrations in old females was nullified by ovariectomy at 3 months of age, it is concluded that the blood titer of estrogen, which had been more in old normal females than the other groups of rats, may be playing an important role in enhancing the prolactin secretion at advanced ages.

References

1 *Amenomori, Y.; Chen, C.L., and Meites, J.:* Serum prolactin levels in rats during different reproductive states. Endocrinology, Springfield *86:* 506–510 (1970).

2 *Aschheim, P.:* Œstrus permanent et prolactine. C.R. Acad. Sci. *255:* 3053–3055 (1962).

3 *Aschheim, P. et Pasteels, J.L.:* Etude histophysiologique de la sécrétion de prolactine chez les rates séniles. C.R. Acad. Sci. *257:* 1373–1375 (1963).

4 *Barraclough, C.A. and Fajer, A.B.:* Progestin secretion by gonadotropin-induced corpora lutea in ovaries of androgen-sterilized rats. Proc. Soc. exp. Biol. Med. *722:* 781–785 (1968).

5 *Blake, C.A.; Weiner, R.I., and Sawyer, C.H.:* Pituitary prolactin secretion in female rats made persistently estrous or diestrous by hypothalamic deafferentation. Endocrinology, Springfield *90:* 862–866 (1972).

6 *Bloch, S.:* Untersuchungen über das funktionelle Altern tierischer Genitalorgane. Gynaecologia *144:* 313–316 (1957).

7 *Clemens, J.A. and Meites, J.:* Neuroendocrine status of old constant-estrous rats. Neuroendocrinology *7:* 249–256 (1971).

8 *Clemens, J.A.; Sharr, C.J.; Tandy, W.A., and Roush, M.E.:* Effects of hypothalamic stimulation on prolactin secretion in steroid treated rats. Endocrinology, Springfield *89:* 1317–1322 (1971).

9 *Hayashi, S.:* Prolactin content of the anterior hypophysis of neonatally estrogenized ovariectomized adult rats following postpuberal estrogen injections. Annot. Zool. Japon *43:* 88–92 (1970).

10 *Herlant, M.:* Etude critique de deux techniques nouvelles destinées à mettre en évidence les différentes catégories cellulaires présentes dans la glande pituitaire. Bull. Micr. appl. *10:* 37–44 (1960).

11 *Hymer, W.C.; McShan, W.H., and Christiansen, R.G.:* Electron microscopic studies of anterior pituitary glands from lactating and oestrogen treated rats. Endocrinology, Springfield *69:* 81–90 (1961).

12 *Kawashima, S.:* Influence of continued injections of sex steroids on the estrous cycle in the adult rat. Annot. Zool. Japon *33:* 226–233 (1960).
13 *Kawashima, S.:* Prolactin secretion in normal and neonatally estrogenized persistent-diestrous rats at advanced ages. U.S.–Japan seminar on long-term effects of perinatal hormone administration, Tokyo 1972.
14 *Kawashima, S. and Takewaki, K.:* Basophils and erythrosinophils in the anterior hypophysis of steroid-sterilized female rats. Annot. Zool. Japon *39:* 23–29 (1966).
15 *Kikuyama, S.:* Secretion of luteotrophic hormone by the anterior hypophysis in persistent-estrous and -diestrous rats. J. Fac. Sci. Univ. Tokyo, Sec. IV *10:* 231–242 (1963).
16 *Kurcz, M.; Kovacks, K.; Tiboldi, T., and Orosz, A.:* Effect of androgenization on adenohypophyseal prolactin content in rats. Acta endocrin., Kbh. *54:* 663–667 (1970).
17 *Mallampati, R.S. and Johnson, D.C.:* Serum and pituitary prolactin, LH and FSH in androgenized female and normal male rats treated with various doses of estradiol benzoate. Neuroendocrinology *11:* 46–56 (1973).
18 *Neill, J.D.:* Effects of 'stress' on serum prolactin and luteinizing hormone levels during the estrous cycle of the rat. Endocrinology, Springfield *87:* 1192–1197 (1970).
19 *Neill, J.D.:* Sexual differences in the hypothalamic regulation of prolactin secretion. Endocrinology, Springfield *90:* 1154–1159 (1972).
20 *Nicoll, C.S. and Meites, J.:* Estrogen stimulation of prolactin production by rat adenohypophysis *in vitro.* Endocrinology, Springfield *70:* 272–277 (1962).
21 *Pantić, V. and Genbačev, O.:* Pituitaries of rats neonatally treated with oestrogen. I. Luteotropic and somatotropic cells and hormones content. Z. Zellforsch. *126:* 41–52 (1972).
22 *Pasteels, J.L.:* Recherches morphologiques et expérimentales sur la sécrétion de prolactine. Arch. Biol. *74:* 439–553 (1963).
23 *Takewaki, K.:* Secretion of luteotropin in persistent-diestrous rats. Endocrin. jap. *11:* 1–8 (1964).
24 *Taya, K. and Igarashi, M.:* Changes in FSH, LH and prolactin secretion during estrous cycle in rats. Endocrin. jap. *20:* 199–205 (1973).
25 *Uchida, K.; Kadowaki, M.; Miyake, T., and Wakabayashi, K.:* Effects of exogenous progesterone on the ovarian progestin secretion and plasma LH and prolactin levels in cyclic rats. Endocrin. jap. *19:* 323–333 (1972).
26 *Watari, T. and Tsukagoshi, N.:* Electron microscopic observations on the estrogen-induced pituitary tumor. Gunma Symp. Endocr. *6:* 297–313 (1969).
27 *Zeilmaker, G.H.; Everett, J.W.; Redmond, W.C., and Quinn, D.L.:* Defective control of prolactin secretion in androgen-sterilized (TP) female rats. Acta endocrin., Kbh. *119:* 109 (1967).

Authors' addresses: Dr. *S. Kawashima,* Zoological Institute, Faculty of Science, University of Tokyo, *Tokyo 113;* Dr. *T. Asai* and Dr. *K. Wakabayashi,* Institute of Endocrinology, Gunma University, *Maebashi 371* (Japan)

Psychoneuroendocrinology. Workshop Conf. Int. Soc. Psychoneuroendocrinology, Mieken 1973, pp. 136–143 (Karger, Basel 1974)

Possible Role of Pituitary Gonadotropin in the Regulation of Neonatal Testicular Activity and Sexual Differentiation of the Brain in the Rat[1]

Y. Arai and K. Serisawa

Department of Anatomy, Juntendo University School of Medicine, Tokyo

Introduction

Differentiation of postpubertal secretory pattern of gonadotropin (GTH) and sexual behavior normally occurs sometime between the fourth and sixth day of life in the male rat. Removal of the testes during the first few neonatal days produces 'feminine males' which are capable of secreting GTH in a cyclic female pattern and which can display feminine sexual behavior (lordosis). It is now believed that internal secretions of the infantile testes are responsible for normal masculine neuroendocrine development and that the specific hormone involved is an androgen (*Barraclough,* 1967; *Gorski,* 1971, *Arai,* 1973). However, control mechanisms of this critical activity of the testes are still not conclusive. Recently, the existence of pituitary testicular feedback regulation has been suggested by the compensatory testicular hypertrophy that occurred within 3 days of hemiorchidectomy performed on the day of birth (*Yaginuma et al.,* 1969). *Goldman and Gorski* (1971) found that serum GTH levels were low in the neonatal male rat but increased rapidly after castration, and that this increase in turn was inhibited by exogenous sex steroids. In the present study, as one step to elucidate a possible role of pituitary gland in the sexual differentiation of the brain, it was attempted to determine whether or not exogenous GTH or LH-releasing hormone (LH-RH) could stimulate the neonatal testes and advance the masculinization of GTH regulation.

Materials and Methods

Newborn male Wistar rats which were raised in our vivarium were treated with GTH or synthetic LH-RH twice daily from day 1 (= the day of birth) to day 2. Subcutaneous injections of ovine LH (NIH-LH-S18), ovine FSH (NIH-FSH-S9), HCG (Gonatropin,

1 This work was aided by research grants from the Japanese Ministry of Education.

Teikoku Hormone Mfg. Co. Ltd., Tokyo) or PMSG (Serotropin, Teikoku Hormone Mfg. Co. Ltd., Tokyo) were given in 0.04 ml saline. Synthetic LH-RH (kindly supplied by Chugai Pharmaceutical Co. Ltd., Tokyo) was administered intraperitoneally. Orchidectomy was performed under cold anesthesia on day 3. Removed testes were weighed, fixed in Boiun's fluid. Sections cut at 10 μ in paraffin were stained with H-E. A limited number of testes removed from LH or HCG-treated and control rats were sectioned at 15 μ with a cryostat (temperature $< -20\,°C$) for histochemical demonstration of 3β-hydroxysteroid dehydrogenase (3β-HSD). Sections were treated by the method of *Levy et al.* (1959). At approximately 30 days of age, half of an ovary from a littermate was transplanted subcutaneously on the ventral body wall. At 60 days of age, these rats were autopsied. The histological structure of ovarian grafts was used to monitor the pattern of GTH secretion. The presence of numerous corpora lutea (CL) of normal size and healthy appearance was considered evidence of the female pattern of cyclic release of GTH. The absence of CL in a polyfollicular ovarian graft was considered to be evidence of the masculine pattern of GTH secretion.

Results

Effects of Exogenous GTHs on Neonatal Testes and Masculinization of GTH Regulation

Table I presents the changes in body and testicular weights following GTH administration from days 1 to 2. There was no significant difference in body weight among the GTH-treated rats and saline controls when measured upon orchidectomy on day 3. Testes of the rats treated with GTH (groups 3–5) were significantly heavier than those of the controls. Histological study showed that

Table I. Body and testicular weights (± SE) on day 3 rats injected with GTH for the first 2 days of life

Group	Treatment	No. of rats	Body weight, g	Testicular weight (left and right), mg/100 g body weight
1	Saline (0.04 ml × 2/day) from days 1–2	21	7.87 ± 0.16	81.61 ± 1.49
2	40 μg LH (20 μg × 2/day) from days 1–2	14	7.75 ± 0.24	103.07 ± 3.31
3	40 μg FSH (20 μg × 2/day) from days 1–2	18	7.26 ± 0.23	96.07 ± 10.50
4	40 IU HCG (20 IU × 2/day) from days 1–2	12	7.88 ± 0.15	98.92 ± 3.45
5	40 IU PMSG (20 IU × 2/day) from days 1–2	15	7.68 ± 1.45	99.45 ± 0.71

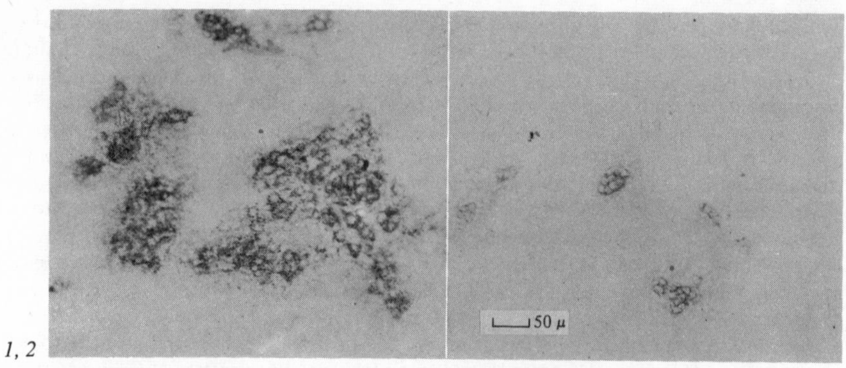

1, 2

Fig. 1. 3β-HSD activity of testes of 3-day-old rat treated with 40 μg LH from days 1–2. Note intense formazan deposition in the interstitial cells.

Fig. 2. 3β-HSD activity of testes of 3-day-old rat treated with saline alone.

Table II. Effects of neonatal treatment with GTH on differentiation of the postpubertal pattern of GTH secretion in the male rat

Group	Treatment	No. of rats	Description of ovarian graft at autopsy		Males with masculine pattern of GTH secretion	
			with CL	polyfollicular[1]	No.	%
1	Saline (0.04 ml × 2/day) from days 1–2	16	14	2	2/16	12.5
2	40 μg LH (20 μg × 2/day) from days 1–2	10	3	7	7/10	70.0
3	40 μg FSH (20 μg × 2/day) from days 1–2	11	9	2	2/11	18.2
4	40 IU HCG (20 IU × 2/day) from days 1–2	8	0	8	8/8	100.0
5	40 IU PMSG (20 IU × 2/day) from days 1–2	10	7	3	3/10	30.0
6	Castration on day 1 and 40 μg LH from days 1–2	5	5	0	0/5	0.0
7	Castration on day 1 and 40 IU HCG from days 1–2	4	4	0	0/4	0.0

1 Since it has been reported that polyfollicular ovaries, which are indicative of masculine pattern of GTH release, may show local luteinization or occasional CL formation (*Harris and Levine*, 1965; *Gorski and Wagner*, 1965), this type of polyfollicular ovaries was included in the masculine type of ovaries in the present study.

the hyperplasia and hypertrophy of the interstitial cells were noticeable in the testes of rats injected with 40 μg of LH or 40 IU of HCG (groups 2 and 4). Strong enzyme activity for 3β-HSD was demonstrated in these testes after incubation with dehydroepiandrosterone as the substrate. This has been considered helpful in determining the site of steroid hormone formation (*Levy et al.*, 1959; *Baillie et al.*, 1966). Intense deposition of formazan which filled the hypertrophied cytoplasm of the interstitial cells was seen (fig. 1), while the reaction was weak in the control testes (fig. 2). Interstitial cell hypertrophy was also recognized in the testes of rats treated with 40 IU of PMSG (group 5), but was less in degree. However, treatment with 40 μg FSH increased testicular weight, but it was ineffective in stimulating the interstitial cells in group 3 rats.

Effects of neonatal treatment with GTHs on the postpubertal pattern of GTH secretion are summarized in table II. Fourteen of 16 rats treated with saline alone and castrated on day 3 became 'feminine males' (group 1). Their ovarian grafts contained numerous CL and an irregular number of developing follicles. Only two of group 1 rats (12.5 %) had a polyfollicular ovarian graft. However, in the male rats pretreated with 40 μg of LH or 40 IU of HCG for the first 2 days of life (groups 2 and 4), neonatal castration on day 3 failed to induce 'feminine males'. Ovarian grafts recovered from seven of ten (70 %) rats treated with 40 μg of LH or all rats treated with 40 IU HCG were of the masculine type. In contrast, pretreatment with 40 μg FSH did not alter the incidence of 'feminine males' significantly (group 3), compared with saline controls. Although PMSG stimulated interstitial cells of neonatal testes as described above, it was ineffective in masculinizing GTH secretory pattern (group 5). In the rats which were castrated on the day of birth and treated with 40 μg of LH or 40 IU of HCG from that day to day 2 (groups 6 and 7), none of the rats was masculinized.

Effect of Low Dosages of LH or HCG on Masculinization of GTH Regulation

The effects of varying amounts of LH or HCG administered from day 1 to day 2 are summarized in tables III and IV. At 20 μg LH, 71.4 % (10/14) of the injected rats were masculinized. This percentage did not drop at the 1-μg level of LH. However, with 0.1 μg, only 30 % (3/10) of the treated rats were masculinized.

A similar situation was obtained with HCG (table IV). When the dosage of HCG was decreased from 10 IU to 0.5 IU, the percentage of masculinized rats still showed high levels (10 IU, 81.8 %; 0.5 IU, 83.3 %). At 0.1 IU, 50 % (8/16) of the injected rats were of the masculine type. Further decrease was obtained with 0.01 IU. Only 20 % of rats were masculinized.

Table III. Effect of neonatal treatment with varying dosages of LH on differentiation of GTH regulation in the male rat

Treatment	Testicular weight (day 3 rat testes), mg/100 g body weight	No. of rats	Description of ovarian graft at autopsy		Males with masculine pattern of GTH secretion	
			with CL	poly-follicular	No.	%
Saline from days 1–2	81.61 ± 1.49 (21)[1]	16	14	2	2/16	12.5
20.0 μg LH from days 1–2	89.77 ± 2.45 (13)	14	4	10	10/14	71.4
10.0 μg LH from days 1–2	93.40 ± 2.04 (15)	11	4	7	7/11	63.6
5.0 μg LH from days 1–2	86.15 ± 2.01 (8)	5	2	3	3/5	60.0
1.0 μg LH from days 1–2	85.74 ± 1.38 (15)	11	3	8	8/11	72.7
0.1 μg LH from days 1–2	82.57 ± 2.18 (13)	10	7	3	3/10	30.0

1 Number of rats whose testes were weighed on castration on day 3.

From these data, the minimal effective dose which is capable of masculinizing GTH regulation in a majority of rats will exist between 1 and 0.1 μg for LH and around 0.1 IU for HCG. This assumption is well supported by significant stimulation of testicular weights when measured upon orchidectomy on day 3. Testes of the rats receiving 1–20 μg LH or 0.1–10.0 IU HCG were significantly heavier than those of saline controls (tables III and IV).

Effect of Synthetic LH-RH on Differentiation of Postpubertal GTH Regulation

A different approach to clarify the possible regulation of neonatal testicular activity by pituitary gland was made by injecting a hypophysiotropic hormone, LH-RH. The results are shown in table V. Treatment with 1 or 0.1 μg synthetic LH-RH caused a significant increase in the weight of day 3 rat testes upon their removal. But the testicular weight was not significantly different from that of saline controls when 0.01 μg LH-RH was administered. At autopsy on day 60, 83.3 % (5/6) of rats injected with saline alone intraperitoneally (group A) became 'feminine males'. This incidence of 'feminine males' is almost compa-

Table IV. Effect of varying doses of HCG given neonatally on differentiation of the postpubertal secretory pattern of GTH in the male rat

Treatment	Testicular weight (day 3 rat testes), mg/100 g body weight	No. of rats	Description of ovarian graft at autopsy		Male with masculine pattern of GTH secretion	
			with CL	poly-follicular	No.	%
Saline from days 1–2	81.61 ± 1.49 (21)	16	14	2	2/16	12.5
10.0 IU HCG from days 1–2	93.02 ± 2.95 (13)	11	2	9	9/11	81.8
5.0 IU HCG from days 1–2	91.40 ± 1.57 (19)	8	3	5	5/8	62.5
1.0 IU HCG from days 1–2	94.11 ± 2.58 (12)	8	2	6	6/8	75.0
0.5 IU HCG from days 1–2	90.31 ± 1.68 (16)	12	2	10	10/12	83.3
0.1 IU HCG from days 1–2	87.12 ± 2.12 (20)	16	8	8	8/16	50.0
0.01 IU HCG from days 1–2	83.24 ± 1.86 (8)	5	4	1	1/5	20.0

rable with that of controls receiving subcutaneous saline injections. However, treatment of newborn males with LH-RH effectively reduced the incidence of 'feminine males'. At 1 μg, 100 % (8/8) of the injected rats were masculinized (group B). With 0.1 μg, 60 % were still masculinized (group C). However, 0.01 μg of LH–RH was not so effective (group D) as the treatments with higher doses, 40 % of rats were of the masculine type. Of the rats castrated before LH-RH injection was started on day 1, all rats became 'feminine males' (group E).

Discussion

The results of the present study clearly indicate that exogenous LH or HCG can accelerate the masculinizing process of the neonatal male rat hypothalamus. Since the treatment with LH or HCG was ineffective in the rats castrated on day 1, the presence of the testes seems to be essential for the initiation of the masculinization of GTH regulatory mechanisms. The adrenal cortex may not be primarily involved. Recently, it has been reported that testicular tissues from newborn rats convert approximately half of progesterone to testosterone *in vitro* (*Ficher and Steinberger,* 1971). *Resko et al.* (1968) demonstrated that testos-

Table V. Effect of synthetic LH-RH on differentiation of postpubertal GTH regulation in the male rat

Group	Treatment	Testicular weight (day 3 rat testes), mg/100 g body weight	Description of ovarian graft at autopsy		Males with masculine pattern of GTH secretion	
			with CL	polyfollicular	No.	%
A	Saline (IP)[1] from days 1–2	80.43 ± 1.68 (7)[2]	5	1	1/6	16.7
B	1.0 μg LH-RH (IP) from days 1–2	97.96 ± 3.22 (10)	0	8	8/8	100.0
C	0.1 μg LH-RH (IP) from days 1–2	85.83 ± 1.70 (20)	6	9	9/15	60.0
D	0.01 μg LH-RH (IP) from days 1–2	82.50 ± 1.13 (12)	6	4	4/10	40.0
E	Castration on day 1 and 1.0 μg LH-RH from days 1–2	–	5	0	0/5	0.0

1 Intraperitoneal injections were given in 0.04 ml saline twice daily.
2 The number of rats whose testes were weighed on castration on day 3.

terone actually exists not only in the testes but also in the blood of neonatal male rats. Although androgen levels of the testes and the blood were not measured in the present study, increased production and secretion of androgen, which was responsible for the initiation of the masculinization of the hypothalamus, would be assumed in the rats receiving LH or HCG neonatally. Histochemical evidence suggesting an increase in steroid biosynthesis in 3-day-old rat testes pretreated with LH or HCG will support this conjecture. Failure of FSH to secure the masculine type of GTH release may be explained by the lack of the stimulatory effect on the interstitial cells of day 3 rat testes. However, it was not certain why PMSG failed to secure the masculine type, in spite of the interstitial cell stimulation.

Our experiments have demonstrated that LH-RH treatment, as well as GTH treatment, does stimulate neonatal testes and apparently advance the masculinization of GTH regulation in the male rat. Recently *Miyachi et al.* (1973) reported that infantile male rat pituitaries (from 8- to 26-day-old rats) were capable of secreting GTHs following intraperitoneal injection of LH-RH. It is possible that increased release of GTH (presumably LH) with consequent increased secretion of androgen occurs in our rats during the treatment with LH-RH. Obviously, further quantitative study is required to establish this point.

However, it can be assumed that pituitary secretory activity of neonatal male rats is dependent on the hypothalamic output of releasing hormones. Thus, the testicular function which is responsible for masculine neuroendocrine development will not be autonomous, but it will be regulated by this hypothalamo-pituitary axis.

Acknowledgements

The authors are greatly indebted to Dr. *R.W. Bates*, NIAMD, Bethesda, Md., USA for generous supplies of LH and FSH, and to Dr. *S. Suzuki*, Chugai Pharmaceutical Co. Ltd., Tokyo, Japan for his gift of the sample of synthetic LH-RH.

References

Arai, Y.: Sexual differentiation and development of the hypothalamus and steroid-induced sterility; in *Yagi and Yoshida* Neuroendocrine control, pp. 27–55 (Univ. Tokyo Press, Tokyo 1973).

Baillie, A.H.; Ferguson, M.M., and Hart, D.Mck.: Developments in steroid histochemistry (Academic Press, London 1966).

Barraclough, C.A.: Modifications in reproductive function after exposure to hormones during the prenatal and early postnatal period; in *Martini and Ganong* Neuroendocrinology II, pp. 61–99 (Academic Press, New York 1967).

Ficher, M. and Steiberger, E.: In vitro progesterone metabolism by rat testicular tissue at different stages of development. Acta endocrin., Kbh. *68:* 285–292 (1971).

Goldman, B.D. and Gorski, R.A.: Effects of gonadal steroids on LH and FSH in neonatal rats. Endocrinology, Springfield *89:* 112–115 (1971).

Gorski, R.A.: Gonadal hormones and the perinatal development of neuroendocrine function; in *Martini and Ganong* Frontiers in neuroendocrinology, 1971, pp. 237–290 (Oxford Univ. Press, New York 1971).

Gorski, R.A. and Wagner, J.W.: Gonadal activity and sexual differentiation of the hypothalamus. Endocrinology, Springfield *76:* 226–239 (1965).

Harris, G.W. and Levine, S.: Sexual differentiation of the brain and its experimental control. J. Physiol., Lond. *181:* 379–400 (1965).

Levy, H.; Dean, H.W., and Rubin, B.L.: Visualization of steroid-3β-ol-dehydrogenase activity in tissue of intact and hypophysectomized rats. Endocrinology, Springfield *65:* 932–943 (1959).

Miyachi, Y.; Nieschlag, E., and Lipsett, M.B.: The secretion of gonadotropins and testosterone by the neonatal male rat. Endocrinology, Springfield *92:* 1–5 (1973).

Resko, J.A.; Feder, H.H., and Goy, R.W.: Androgen concentrations in plasma and testes of developing rats. J. Endocrin. *40:* 485–491 (1968).

Yaginuma, K.; Matsuda, A.; Murasawa, Y.; Kobayashi, T., and Kobayashi, T.: Presence of hypothalamo-pituitary testicular axis in the early postnatal period. Endocrin. jap. *16:* 5–10 (1969).

Author's address: Dr. *Y. Arai*, Department of Anatomy, Juntendo University School of Medicine, 2-1-1 Hongo, *Tokyo 113* (Japan)

Psychoneuroendocrinology. Workshop Conf. Int. Soc. Psychoneuroendocrinology, Mieken 1973, pp. 144–152 (Karger, Basel 1974)

Sexual Differentiation of the Hippocampus: Effects of Immobilization Stress on Gonadotropin Secretion in the Rat

E. Terasawa and M. Kawakami[1]

Second Department of Physiology, Yokohama City University School of Medicine, Yokohama

Introduction

Reports have been accumulating that stressful stimuli influence not only the pituitary adrenal system, but also the pituitary ovarian system. Surgical stress can block ovulation on the day of diestrus (*Schwartz and McCormack,* 1972), and advance the time of LH release as well as the time of lordosis behavior on the morning of proestrus (*Nequin and Schwartz,* 1971). Stressful stimuli cause luteinization in the neonatally estrogenized persistent-estrous rat (*Takasugi,* 1956) and in the male rat with ovarian transplant (*Takewaki,* 1956; *Machida,* 1968, 1970), but not in the female rat (*Machida,* 1968). Surgical stress under ether anesthesia enhances the release of prolactin in the female rat but not in the male rat (*Neill,* 1970). These experiments also appear to indicate a sexual differentiation in the stress-induced gonadotropins and prolactin secretions. Since both pituitary-adrenal and pituitary-ovarian system relay on common sites in the brain, sexual differentiation of the stress-induced gonadotropin secretion may be due to the sexual differentiation of the brain.

Generally, the medial preoptic basal hypothalamus is considered to be the site of the brain responsible for the sterilization (*Barraclough,* 1967; *Gorski,* 1968). However, previous work in our laboratory has indicated that neonatal treatment with androgen in the female rat induces some impairment in the amygdala and the hippocampus responding to electrical stimulation rather than the medial preoptic area (*Kawakami and Terasawa,* 1972b). Since numerous data

1 The authors wish to express their thanks to the National Institute of Arthritis and Metabolic Diseases, Rat Pituitary Hormones Program, for kindly supplying the materials for radioimmunoassay.

have suggested the participation of the limbic forebrain structures, especially the hippocampus, in control of adrenocorticotropin secretion (*Kawakami et al.*, 1968, 1971; *Knigge,* 1961; *Knigge and Hays,* 1963), it is of interest to investigate the effects of hippocampal ablation on stress-induced gonadotropin secretion.

Effects of Immobilization Stress on Serum Corticosterone, and Serum and Pituitary Prolactin

Application of immobilization stress by fixing a rat in spinal position for 1 h, increased serum corticosterone in intact proestrous females, males, and gonadectomized estrogen-primed females and males, so that ACTH was apparently released by the stress. The magnitude of responses is greater in females than in males.

Immobilization stress increased serum prolactin in intact proestrous females, ovariectomized estrogen-primed females, and orchidectomized estrogen-primed males, but not in intact males. Thus, there were sexual differentiation in stress-induced increase of serum prolactin being due to the blood level of estrogen. This assumption was supported by *Neill* (1970) that the stress of ether laparotomy induced an increase in serum prolactin in females, but not in males, and effects of the stress-induced prolactin release was greatest on the day of proestrus and greater on the day of diestrus III in 5-day cycling rats. Pituitary prolactin was not changed by immobilization stress in any of the examined rats.

Effects of Immobilization Stress on Serum Gonadotropins

Application of immobilization stress did not induce any change in serum and pituitary LH, either in intact or ovariectomized estrogen-primed females, or in intact males. No effects on serum LH by the stress of ether laporatomy in intact females and males were reported (*Neill,* 1970). However, in orchidectomized estrogen-primed males, serum LH had a tendency to increase by immobilization stress. The stress increased serum LH significantly in three of five orchidectomized estrogen-primed males. Pituitary LH in the orchidectomized estrogen-primed males also increased by immobilization stress. Thus, the stressful stimuli enhanced synthesis and release of LH in the male rat under estrogen influence. This fact agrees with the studies by *Takewaki* (1956) and *Machida* (1968, 1970) that stressful stimuli, by making a subcutaneous pouch of croton oil, caused luteinization in the castrated male rats with ovarian transplantation, otherwise anovulatory polyfollicular characteristics and stressful stimuli caused luteinization in the neonatally estrogenized persistent-estrous rat (*Takasugi,*

1956). Therefore, the stress would be effective to stimulate the mechanism of LH release in the male type of animals under estrogen influence.

Application of the stress, increase serum and pituitary FSH in the female, but not in the intact male rats. In orchidectomized estrogen-primed males, the stress increased pituitary FSH, but not serum FSH. Therefore, the stress-induced increase of FSH may be influenced by the blood level of estrogen, as is prolactin.

Effects of Hippocampal Ablation on Stress-Induced Changes in Serum Corticosterone or Serum and Pituitary Prolactin

Dorsal fornix and hippocampus were ablated by suctioning 40 days prior to experimentation as shown in figure 1. Most of these female rats resumed their 4-day vaginal cycles eventually (*Terasawa and Kawakami,* 1973). The stress was applied on the day of proestrus in females. Hippocampal ablation resulted in an

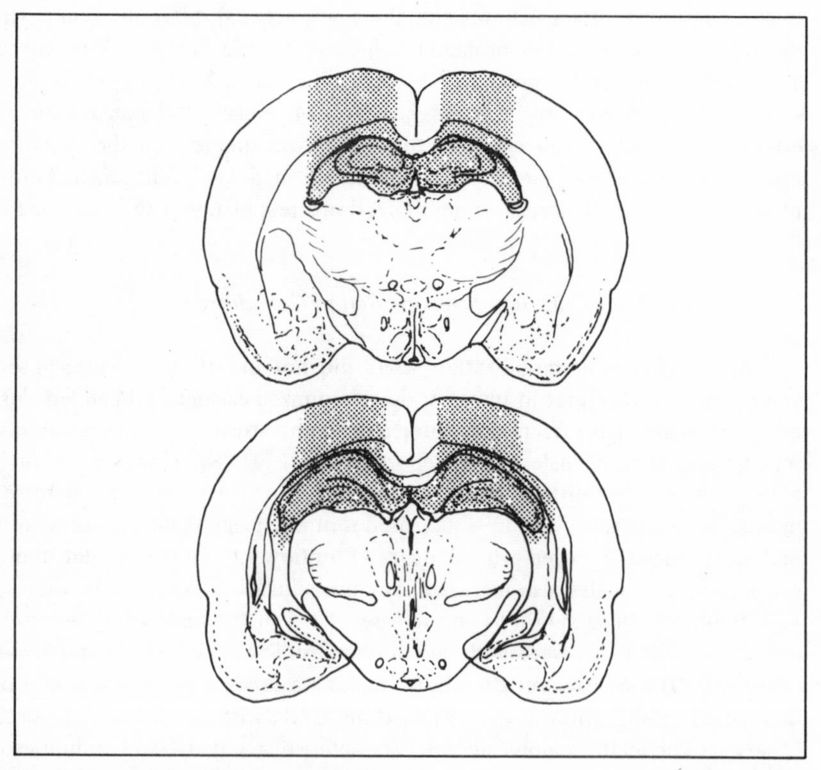

Fig. 1. Sites and size of hippocampal ablation.

increase of serum corticosterone in females and a decrease of serum corticosterone in orchidectomized estrogen-primed males. The stress increased serum corticosterone in these hippocampal ablated male and female rats, although the magnitude of increases was smaller in females and greater in males with hippocampal ablation, as compared to intact animals. From the results of ablation, the hippocampus appears to inhibit the basal secretion of ACTH in female rats, while it facilitates the secretion of ACTH in orchidectomized estrogen-primed males. Although effects or no effects of fornix ablation on diurnal fluctuation in plasma corticosterone in the male rat were reported (*Möberg et al.*, 1971; *Lengvari and Halász,* 1973), such a sexual differentiation by hippocampal ablation was not observed. In the present experiment, hippocampal ablation did neither inhibit nor facilitate stress-induced ACTH secretion in both sexes; in detail inspection, however, an increased rate of corticosterone to the stress was reduced by hippocampal ablation in females, while it was enhanced in males, inasmuch as the basal levels of corticosterone were changed by hippocampal ablation. Effects of hippocampal ablation on adrenocortical response to the stress were reported (*Knigge,* 1961; *Knigge and Hays,* 1963).

Ablation of the hippocampus did not induce any change in serum prolactin. Hippocampal ablation did not influence the stress-induced increase of serum prolactin in both females and orchidectomized estrogen-primed males. The stress increased serum prolactin in both hippocampal ablated proestrous females and orchidectomized estrogen-primed males the same as in intact controls. The stress increased pituitary prolactin in orchidectomized estrogen-primed males, but did not induce any change in hippocampal ablated proestrous females. Thus, the hippocampus is not involved in the mechanism of stress-induced prolactin release, while it seems to be involved in the mechanism of ACTH release, at least partially.

Effects of Hippocampal Ablation on Stress-Induced Serum and Pituitary Gonadotropins

Hippocampal ablation slightly increased serum LH in females and decreased serum LH in orchidectomized estrogen-primed males. Hippocampal ablation did not change in pituitary LH in females, but increased in orchidectomized estrogen-primed males. These data of proestrous females agree with previous observations (*Terasawa and Kawakami,* 1973).

Hippocampal ablation deprived the stress-induced LH increases in orchidectomized estrogen-primed males, in contrast to hippocampal ablation which increased LH in proestrous females under stressful conditions. Thus, hippocampal ablation enhanced not only basal secretion of LH, but also stress-induced release of LH in females, that is, the hippocampus may inhibit both basal secre-

Table I. Effects of immobilization stress on serum corticosterone, gonadotropins and prolactin, and pituitary gonadotropins and prolactin in female rats; all animals except for ovariectomized, were sacrificed on the day of proestrus

Animals	Number of rats	Weights of APG, mg	Pituitary concentration of			Serum concentration of			
			LH, μg/mg	FSH, μg/mg	prolactin, μg/mg	LH, ng/ml	FSH, ng/ml	prolactin, ng/ml	ACH, μg/dl
Intact control	8	10.4 ± 0.4[1]	0.80 ± 0.06	1.92 ± 0.15	3.60 ± 0.46	1.14 ± 0.05	276 ± 16	57.4 ± 6.5	24.9 ± 4.1
Immobilization stress	8	10.5 ± 0.5	0.86 ± 0.05	2.71 ± 0.19[2]	3.29 ± 0.46	1.85 ± 0.12	404± 30[2]	124.9 ± 13.0[2]	108.7 ± 8.0[2]
HPC ablated, control	8	11.8 ± 1.2	0.88 ± 0.11	2.91 ± 0.42	3.01 ± 0.72	1.22 ± 0.06	382 ± 26	61.4 ± 7.0	52.1 ± 9.4
HPC ablated, immobilization stress	8	10.8 ± 1.4	0.89 ± 0.03	2.59 ± 0.36	3.93 ± 0.42	9.00 ± 7.27	402 ± 58	141.7 ± 43.4[2]	122.3 ± 9.5[2]
Ov-X + E, control	6	10.7 ± 0.7	1.92 ± 0.18	25.09 ± 2.53	1.42 ± 0.21	3.34 ± 0.40	1,235 ± 92	37.6 ± 1.5	18.3 ± 2.8
Ov-X + E, stress	7	12.4 ± 1.5	1.25 ± 0.11[2]	19.96 ± 2.81	1.57 ± 0.15	2.89 ± 0.47	1,144 ± 86	109.6 ± 11.2[2]	90.1 ± 10.3[2]

1 Mean ± SE.

2 $p < 0.05$ versus control.

Amounts are expressed in terms of NIH-LH-S1, NIH-FSH-S1 and NIAMD-rat-prolactin-RP-1.

Abbreviations: APG = anterior pituitary gland; HPC = hippocampus (ablation of the hippocampus was performed about 40 days prior to experimentation); Ov-X + E = ovariectomized and estrogen-primed rats (animals were ovariectomized 3 weeks prior to, and received estrogen injection 48 h prior to experimentation).

Table II. Effects of immobilization stress on serum corticosterone, gonadotropins and prolactin, and pituitary gonadotropins and prolactin in male rats

Animals	Number of rats	Weights of APG, mg	Pituitary concentration of			Serum concentration of			
			LH, μg/mg	FSH, μg/mg	prolactin, μg/mg	LH, ng/ml	FSH, ng/ml	prolactin, ng/ml	ACH, μg/dl
Intact control	5	11.0 ± 1.7[1]	1.59 ± 0.26	20.4 ± 4.8	3.38 ± 0.52	1.17 ± 0.11	499 ± 117	48.7 ± 16.1	16.0 ± 0.5
Immobilization stress	5	10.4 ± 0.8	1.55 ± 0.14	23.0 ± 2.8	1.97 ± 0.90	0.80 ± 0.11	440 ± 68	57.3 ± 17.7	40.3 ± 2.8[2]
Orch-X + E, control	5	10.9 ± 0.7	2.48 ± 0.43	19.2 ± 3.2	2.45 ± 0.40	7.50 ± 1.04	1,971 ± 197	37.5 ± 4.6	27.3 ± 3.9
Orch-X + E, immobilization stress	5	9.4 ± 0.9	3.89 ± 0.13[2]	26.8 ± 2.4[2]	2.83 ± 0.39	24.90 ± 10.51	2,292 ± 807	192.6 ± 40.2[2]	67.0 ± 2.9[2]
HPC ablated Orch-X + E, control	5	9.7 ± 1.3	4.84 ± 0.59	32.1 ± 4.2	4.23 ± 0.16	3.21 ± 0.34	1,640 ± 351	36.2 ± 2.9	11.5 ± 1.5
HPC ablated Orch-X + E, stress	5	9.3 ± 0.7	5.30 ± 0.88	28.9 ± 2.1	2.98 ± 0.29[2]	3.18 ± 0.27	1,555 ± 267	162.7 ± 22.0[2]	53.8 ± 3.1[2]

1 Mean ± SE.
2 $p < 0.05$ versus control.
Amounts are expressed in terms of NIH-LH-S1, NIH-FSH-S1 and NIAMD-rat-prolactin-RP-1.
Abbreviations: APG = anterior pituitary gland; HPC = hippocampus (ablation of hippocampus was performed 40 days prior to experimentation); Orch-X + E = orchidectomized estrogen-primed rats (animals were orchidectomized 50 days prior to, and received 10 μg of estrogen injection 48 h prior to experimentation).

tion of LH and stress-induced release of LH in females. By contrast, in orchidectomized estrogen-primed males, hippocampal ablation decreased serum LH (so that it accumulated in the pituitary), and it deprived the stress-induced release of LH, i.e. the hippocampus in these rats may facilitate not only basal secretion of LH but also stress-induced release of LH. Opposite function of the hippocampus on LH release in female and male animals, such as neonatally androgenized or orchidectomized estrogen-primed rats were suggested as follows: Electrical stimulation of the hippocampus increased serum LH in androgenized (*Kawakami and Terasawa,* 1972b), estrogenized (*Kawakami et al.,* 1973a), and orchidectomized estrogen-primed rats (*Kawakami and Terasawa,* 1974), while stimulation did not alter the basal level of serum LH and rather inhibit its release in proestrous females (*Velasco and Taleisnik,* 1969; *Kawakami et al.,* 1972; *Kawakami et al.,* 1973b). Stimulation of the hippocampus in androgenized rats resulted in a few ovulating rats (*Terasawa,* 1971) and exerted weaker inhibitory influence on preoptic-induced ovulation (*Kawakami and Terasawa,* 1972b), while stimulation inhibits spontaneous ovulation as well as preoptic-induced ovulation (*Velasco and Teleisnik,* 1969; *Kawakami et al.,* 1973b). Therefore, it is assumed that neonatal treatment with androgen impairs inhibitory function of the hippocampus and alters its nature to rather stimulatory functions. In this regard, stress-induced LH release in orchidectomized estrogen-primed males may be due to the lack of inhibitory function of the hippocampus on LH release.

Hippocampal ablation increased serum and pituitary FSH in proestrous females (*Terasawa and Kawakami,* 1973, tables I and II) and increased pituitary FSH in orchidectomized estrogen-primed rats, so that the hippocampus is of an inhibitory nature in the control of FSH secretion. However, the stress-induced increase of serum or pituitary FSH in females and orchidectomized estrogen-primed males were abolished by hippocampal ablation, inasmuch as further increase of serum and pituitary FSH by the stress were not observed in both hippocampal ablated males and females. Participation of the hippocampus in control of FSH secretion was suggested, such as the stimulation of the dorsal hippocampus increased serum FSH on the day of estrus, while it decreased serum FSH on the day of diestrus II in cycling rats (*Kawakami et al.,* 1972), hippocampal stimulation of the prepuberal rats increased serum FSH (*Kawakami and Terasawa,* 1972a), and hippocampal stimulation with medial preoptic stimulation increased pituitary FSH throughout the estrous cycle (*Kawakami et al.,* 1973b). Therefore, the hippocampus may be involved in the stress-induced FSH secretion to some extent.

The stress-induced release of LH, FSH and prolactin are essentially controlled by the medial preoptic basal hypothalamus. However, the hippocampus may be involved in the increases of LH and FSH, caused by stressful stimuli, which induce different effects on both sexes. Thus, limbic structures modulate the secretion of gonadotropins under certain conditions.

Summary

Stress-induced gonadotropin secretion was investigated in both sexes in relation to limbic participation in its mechanisms. Application of the immobilization stress induced the release of serum corticosterone in all of intact females, males, ovariectomized females and orchidectomized estrogen-primed males. Serum prolactin was also increased by the stress in intact proestrous females, ovariectomized estrogen-primed females and orchidectomized estrogen-primed males, but not in intact males. Thus, stress-induced prolactin release appears to be influenced by estrogen. Hippocampal ablation did not affect the stress-induced release of corticosterone and prolactin.

Serum LH was neither increased nor decreased by the stress in intact proestrous females, ovariectomized estrogen-primed females, and males, but it was increased in orchidectomized estrogen-primed males. Hippocampal ablation, however, induced a facilitatory effect on LH release in stressed females and abolished stress-induced increase in LH in orchidectomized estrogen-primed males. Thus, the hippocampus in females appears to inhibit the stress-induced LH release, while the hippocampus in male animals under estrogen influences do not inhibit it.

The stress increased serum or pituitary FSH in intact proestrous females and orchidectomized estrogen-primed males, but this effect was suppressed by hippocampal ablation.

In conclusion, there was sexual differentiation in stress-induced gonadotropin secretions, and the hippocampus may be involved in its control mechanism.

References

Barraclough, C.A.: Modification in reproductive function after exposure to hormones during neonatal and early postnatal period; in *Martini and Ganong* Neuroendocrinology, vol. 2, pp. 61–99 (Academic Press, New York 1967).

Gorski, R.A.: The neural control of ovulation; in *Assali* Biology of gestation, vol. 1, pp. 1–66 (Academic Press, New York 1968).

Kawakami, M.; Kimura, F., and Seto, K.: Effects of electrical stimulation of the brain on serum and pituitary concentration of gonadotropin in the estrogenized rat. Endocrinol. jap. *20:* 59–66 (1973a).

Kawakami, M.; Kimura, F., and Wakabayashi, K.: Electrical stimulation of the hippocampus under the chronic preparation and changes of LH, FSH and prolactin levels in serum and pituitary. Endocrinol. jap. *19:* 85–96 (1972).

Kawakami, M.; Seto, K.; Kimura, F., and Yanase, M.: Difference in the buffer action between the limbic structures and the hypothalamus to the immobilization stress in rabbits; in *Ford* Influence of hormones on the nervous system, pp. 105–120 (Karger, Basel 1971).

Kawakami, M.; Seto, K.; Terasawa, E.; Yoshida, K.; Miyamoto, T.; Sekiguchi, M., and Hattori, Y.: Influence of electrical stimulation and lesion in limbic structures upon biosynthesis of adrenocorticoid in rabbit. Neuroendocrinology *3:* 337–348 (1968).

Kawakami, M. and Terasawa, E.: Electrical stimulation of the brain on gonadotropin secretion in the female prepuberal rat. Endocrin. jap. *19:* 335–347 (1972a).

Kawakami, M. and Terasawa, E.: A possible role of the hippocampus and the amygdala in the androgenized rat. Effect of electrical stimulation of the brain on gonadotropin secretion. Endocrin. jap. *19:* 349–358 (1972b).

Kawakami, M. and Terasawa, E.: Further studies on sexual differentiation of the brain. Response to electrical stimulation in gonadectomized estrogen-primed rats. Endocrin. jap. (in press, 1974).

Kawakami, M.; Terasawa, E.; Kimura, F., and Wakabayashi, K.: Modulating effect of limbic structures on gonadotropin release. Neuroendocrinology *12:* 1–16 (1973b).

Knigge, K.M.: Adrenocortical response to stress in rats with lesions in hippocampus and amygdala. Proc. Soc. exp. Biol. Med. *108:* 18–21 (1961).

Knigge, K.M. and Hays, M.: Evidence of inhibitory role of hippocampus in neural regulation of ACTH release. Proc. Soc. exp. Biol. Med. *114:* 67–69 (1963).

Kurata, H.: Methods of measurement of plasma 17-OHCS (cortisol) and its application; in Clin. exam. *14:* 857–864 (1970, in Japanese).

Lengvari, I. and Halász, B.: Evidence for a diurnal fluctuation in plasma corticosterone levels after fornix transection in the rat. Neuroendocrinology *11:* 191–196 (1973).

Machida, T.: Formation of *corpora lutea* in ovarian grafts in orchidectomized rats following application of croton oil and injection of progesterone. J. Fac. Sci., Univ. Tokyo, sec. 4 *11:* 429–436 (1968).

Machida, T.: Luteinization of ovarian transplants in gonadectomized male and female rats under stressful conditions and its relation to sexual differentiation of the hypothalamus. Endocrin. jap. *17:* 189–193 (1970).

Möberg, G.P.; Scapagnini, U.; de Groot, J., and Ganong, W.F.: Effect of sectioning the fornix on diurnal fluctuation in plasma corticosterone in the rat. Neuroendocrinology *7:* 11–15 (1971).

Neill, J.D.: Effect of 'stress' on serum prolactin and luteinizing hormone levels during the estrous cycle of the rat. Endocrinology, Springfield *87:* 1192–1197 (1970).

Nequin, L.G. and Schwartz, N.B.: Adrenal participation in the timing of mating and LH release in the cyclic rat. Endocrinology, Springfield *88:* 325–331 (1971).

Niswender, G.D.; Midgley, jr., A.R.; Monroe, S.C., and Reichert, jr., L.E.: Radioimmunoassay for rat luteinizing hormone with antiovine LH serum and ovine LH-^{131}I. Proc. Soc. exp. Biol. Med. *128:* 807–811 (1968).

Schwartz, N.B. and McCormack, C.E.: Reproduction. Gonadal function and its regulation. Annu. Rev. Physiol. *34:* 425–472 (1972).

Takasugi, N.: Untersuchungen über die hypophysäre, gonadotrope Aktivität der daueröstrischen und normalen Ratten unter Stress-Situationen. J. Fac. Sci., Univ. Tokyo, sec. 4 *7:* 605–625 (1956).

Takewaki, K.: Luteinization of follicles in ovarian grafts in castrated male rat following injections of progesterone or deoxycorticosterone acetate, or under stressful condition. Annot. Zool. Japon. *29:* 1–6 (1956).

Terasawa, E.: Changes in electrical activity in the androgenized rat and the anterior deafferented rat. Folia endocrin. jap. *46:* 1081–1096 (1971).

Terasawa, E. and Kawakami, M.: Effects of limbic forebrain ablation on pituitary gonadal function in the female rat. Endocrin. jap. *20:* 277–289 (1973).

Velasco, M.E. and Taleisnik, S.: Effect of hippocampal stimulation on the release of gonadotropins. Endocrinology, Springfield *85:* 1154–1159 (1969).

Authors' address: Dr. *E. Terasawa* and Prof. *M. Kawakami*, Second Department of Physiology, Yokohama City University School of Medicine, 2–33 Urafune-cho, Minami-ku, *Yokohama 232* (Japan)

Psychoneuroendocrinology. Workshop Conf. Int. Soc. Psychoneuroendocrinology, Mieken 1973, pp. 153–160 (Karger, Basel 1974)

Modulation of Responsiveness of the Hypothalamic Unit Activity to LH-RH, LH and FSH by Ovarian Hormones

M. Kawakami and Y. Sakuma[1]

Second Department of Physiology, Yokohama City University School of Medicine, Yokohama

Introduction

It is well established that the medial basal hypothalamic region (MBH) participates in the control of gonadotropin release of anterior pituitary, as an 'interface' between the neural and the endocrine system. There is general agreement that this system is modulated directly or indirectly by gonadotropin itself and ovarian hormones. For instance, electrophysiological investigations have revealed that estrogen injection facilitates single unit activity in the MBH of ovariectomized animals (*Kawakami and Saito,* 1967), whereas progesterone injection decreases single unit activity in the MBH (*Ramirez et al.*, 1967). These facilitatory and inhibitory effects of ovarian hormones on the neuronal activity of the MBH may appear as either direct or indirect action of these hormones on the MBH. In fact, some studies attributed the responses of the MBH neurons to systemic injection of LH in the estrogen-treated ovariectomized rat to the effect of the higher nervous structures which were considered to be sensitive to LH (*Terasawa et al.*, 1969; *Gallo et al.*, 1972). In the present experiment, we attempted to find the very response of the MBH neurons to LH and FSH as well as LH-RH, in order to define the effect of ovarian hormones on the neuronal activity more precisely, employing microiontophoretical method for the application of the hormones.

1 The authors are grateful to NIAMDD for their generous supply of LH and FSH.

Table I. Responses of the ARC neurons to LH-RH, LH and FSH

A. Normal cycling female rats							
(1) Diestrus I		LH [10]			FSH [10]		
		5[1] (50)	1[2] (10)	4[3] (40)	0[1] (0)	1[2] (10)	9[3] (90)
LH-RH	4[1] (31)	1	1	–	–	–	2
[13]	0[2] (0)	–	–	–	–	–	–
	9[3] (69)	4	–	4	–	1	7
(2) Proestrus		LH [7]			FSH [6]		
		4[1] (57)	1[2] (14)	2[3] (29)	3[1] (50)	1[2] (17)	2[3] (33)
LH-RH	5[1] (39)	2	1	1	1	–	1
[13]	3[2] (23)	1	–	–	–	1	–
	5[3] (39)	1	–	1	2	–	1
B. Ovariectomized (Ovx) rats							
(1) Ovx only		LH [8]			FSH [8]		
		7[1] (88)	0[2] (0)	1[3] (13)	5[1] (63)	0[2] (0)	3[3] (38)
LH-RH	7[1] (64)	6	–	–	4	–	2
[11]	0[2] (0)	–	–	–	–	–	–
	4[3] (36)	1	–	1	1	–	1
(2) Ovx + estrogen		LH [12]			FSH [10]		
		8[1] (67)	3[2] (25)	1[3] (8)	3[1] (30)	2[2] (20)	5[3] (50)
LH-RH	9[1] (45)	4	2	1	3	–	3
[20]	8[2] (40)	2	1	–	–	1	–
	3[3] (15)	2	–	–	–	1	2
(3) Ovx + progesterone		LH [11]			FSH [7]		
		4[1] (36)	0[2] (0)	7[3] (64)	1[1] (14)	0[2] (0)	6[3] (86)
LH-RH	8[1] (47)	2	–	3	1	–	3
[17]	0[2] (0)	–	–	–	–	–	–
	9[3] (53)	2	–	4	–	–	3

1 Facilitation.
2 Inhibition.
3 Indeterminate.
( ) = number, and [ ] = percentage of the tested neurons.

Materials and Methods

Female rats of Wistar strain (200–270 g) were kept in light controlled (light on from 05.00 to 19.00) quarters, and were given food and water ***ad libitum***. Vaginal smears were examined every morning and only the animals which showed 4-day cycles were used in the experiment. Some rats were ovariectomized and used 10–14 days later, with or without pretreatment with estrogen (10 μg of estradiol benzoate per day for two successive days prior to the experiment) or progesterone (5 mg of progesterone propionate given s.c. at 08.00 on the day of the experiment). Experiment was performed under light urethane anesthesia (1.2 g/kg body weight) which was administered i.p. at 09.00 on the day of the experiment. Animals were secured in a stereotaxic apparatus in the supine position, and ventral surface of the brain was exposed by parapharyngeal approach. A bipolar silver ball electrode for antidromic activation of the hypothalamic arcuate (ARC) neurons was placed on the junction of the pituitary stalk and the median eminence. Recording of the extracellular potentials and microiontophoretic application of the hormones were carried out by using five-barreled glass microelectrodes constructed from pirex tubes of 3 mm outer diameter. The microelectrode was inserted into the ARC stereotaxically according to the coordinates of the atlas of *Albe-Fessard et al.* (1966) from the exposed ventral surface of the brain. Recordings were made from the central barrel of the electrode which was filled with 4 *M* NaCl. The other barrels were filled with the solution of LH-RH (Chugai), 100 μg/ml, LH (Armour), 1 mg/ml, and FSH (Armour), 1 mg/ml, in the physiological saline solution. One of the barrels was filled with physiological saline for eliminating the current effects. Application of the hormones was performed on the ARC neurons which was activated antidromically by the stimulation of the median eminence with a train of single square pulses of 0.2–1.4 mA with a duration of 0.5 msec, by applying anodal currents up to 50 nA. Only the rats in proestrus which were anesthetized with urethane administered at 09.00 were used for the experiment on the neurons of medial preoptic area (MPO). The animals were placed in a stereotaxic apparatus in the prone position. The parietal surface of the cranial bone was removed and a bipolar side-by-side electrode was placed in the ARC region for the antidromic activation of the MPO neurons. The hormones were applied on the antidromically activated neurons, the same as the ARC neurons were. In both cases, the animals were sacrificed by cardiac perfusion of 10-percent formalin and the tip position was examined in the serial sections.

Results

The ARC neurons were activated antidromically by electrical stimulations of the median eminence with latencies of 2–20 msec. Mean of 97 responses observed was 5.4 ± 3.2 (SD) msec. Seventy-four of 97 antidromically activated neurons showed spontaneous discharges and such neurons were offered to the present experiment. LH-RH was applied on all of the 74 neurons, and LH and FSH were applied on 48 and 41 of them, respectively. Eventually, successive application of LH-RH, LH and FSH was performed on 39 identical neurons.

(a) LH-RH: As shown in table I, 44 of 74 neurons showed changes in the neuronal discharge to the application of LH-RH, of which 33 increased and 11 decreased the rate of the discharge. LH-RH responsive neurons were found chiefly in the ARC, but in some cases they ranged in the periventricular region of the medial basal hypothalamus and their distribution seems not to be restricted by the demarcation of the hypothalamic nuclei.

Dose-response relationship was observed between the discharge rate and the amplitude of the current flow applied for the iontophoresis (fig. 1). Responsive-

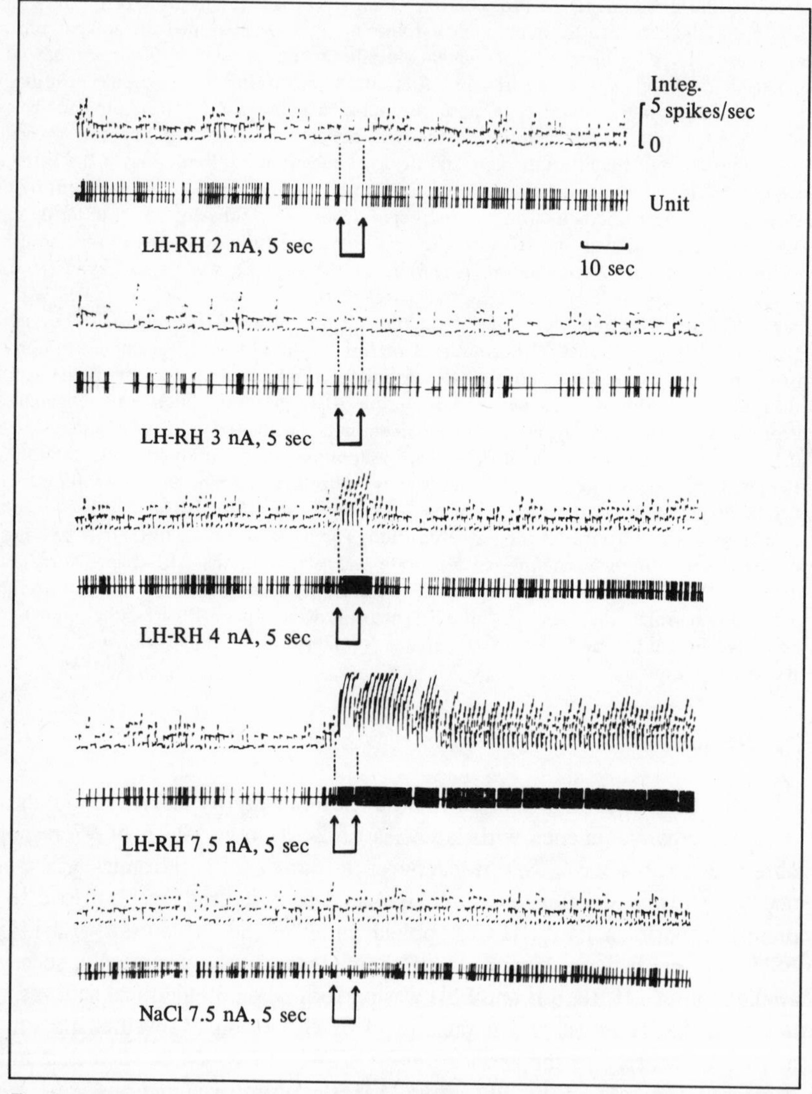

Fig. 1. Responses of the ARC neurons to LH-RH. The triggered pulses and the integrated number of the pulses during 1 sec are shown. *a* Ovariectomized and estrogen treated rat.

ness of the neurons to the iontophoretic infusion of LH-RH showed fluctuations according to the changes in the hormonal environments of the animal. The largest ratio of LH-RH responsive neurons was seen in ovariectomized and estrogen treated rats (85 %), whereas in rats in diestrus I, only 31 % of the examined

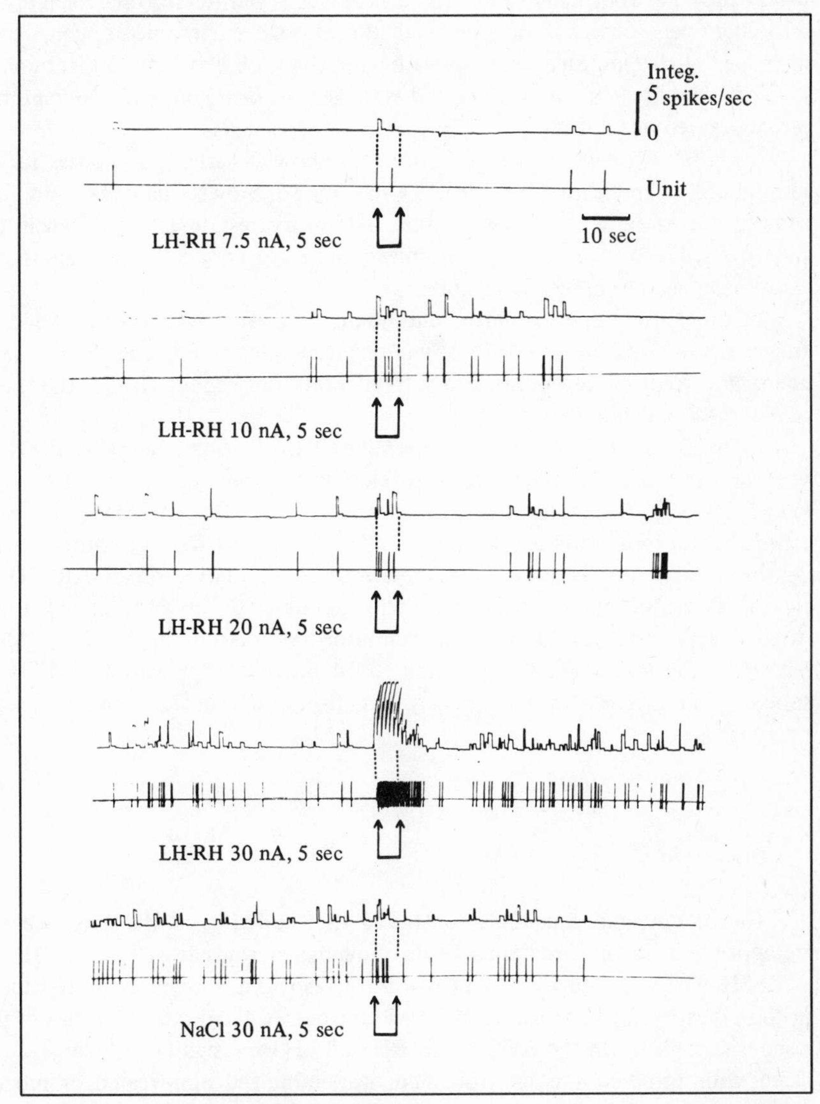

1b

b Ovariectomized and progesterone primed rat. Note the differences in the amplitude of the current flow required for eliciting the responses.

neurons showed response. Inhibitory responses of the neurons were seen exclusively in the proestrous and in the ovariectomized and estrogen treated rats.

(b) LH: Thirty-three (68 %) of 48 neurons showed responses to LH, of which five neurons decreased and 28 increased their rate of discharge to the application. The neurons which showed inhibitory responses were seen in the proestrous, diestrous I and ovariectomized and estrogen treated rats. Facilitatory responses were observed in rats in all the internal environments which were examined. LH-responsive neurons were frequently observed in ovariectomized and estrogen treated rats (92 %) and were less in ovariectomized and progesterone treated rats (36 %).

(c) FSH: Sixteen (38 %) of 41 neurons showed variable responses to the iontophoretic application of FSH. Twelve of 16 showed increases and four showed decreases in the discharge rate. Inhibitory responses were seen in the proestrous, diestrous I and ovariectomized and estrogen treated rats, and facilitatory responses were seen in all conditions.

In proestrous, ovariectomized and non-treated, and ovariectomized and estrogen treated rats, the majority of neurons responded to FSH, but in diestrous, and ovariectomized and progesterone treated rats, only about 10 % of the tested neurons responded in each group.

Neurons of the medial part of the MPO (0.25–0.75 mm lateral to the third ventricle) responded antidromically to electrical stimulation of the ARC. The latency ranged from 4 to 25 msec with a mean of 11.8 ± 5.9 (SD) msec in 76 responses. Iontophoretic applications of LH-RH and LH were performed on 45 neurons which were responsive to the electrical stimulation of the ARC. Only two of 45 showed increment in the discharge rate to the application of LH-RH, whereas application of LH exerted facilitatory effects on the 12 (27 %) MPO neurons. FSH was applied in addition to 14 neurons. Application of FSH resulted in an increase of the discharges in five, and a decrease in one, of 14 neurons.

Discussion

The present experiment demonstrated that activities of the ARC neurons are dependent on the circulating levels of ovarian steroids as well as LH-RH, LH and FSH. They seemed to be activated under estrogen dominance in the blood. In the proestrous, and ovariectomized and estrogen treated rats, the ratios of the responsive neurons in the ARC to LH-RH and LH were significantly large, compared with those in the diestrous and ovariectomized non-treated or progesterone treated rats. The responsiveness of the ARC neurons to LH-RH may be interpreted as a result of the feedback action of LH-RH. *Hyyppa et al.* (1971)

have proposed the possibility of 'ultrashort' feedback control of FSH-releasing factor on the central nervous system, observing an inhibitory effect of systemic application of hypothalamic extract which contained FSH-releasing factor exerted on the hypothalamic content of this releasing factor.

The possible 'short' feedback effect of LH on the central nervous system was reported earlier by *Sawyer and Kawakami* (1959), who observed an EEG after-reaction phenomenon, following copulation of the estrous rabbit. The investigations adopting the single unit recording in the hypothalamus was introduced to the study of the feedback action of the hormones by *Kawakami and Saito* (1967) and *Ramirez et al.* (1967). They administered LH intravenously and observed a decrease of the hypothalamic neuronal discharges, thus assuming the site for negative feedback of LH in the medial basal hypothalamus. However, the results of their experiments could not exclude an indirect effect which might be induced by other neural structures responsive to LH. In the present experiment, adopting the microiontophoretic method for the application of the hormones on the single neurons, we could find that although the majority of neurons responded facilitatory to LH, but only in the proestrous and ovariectomized and estrogen treated rats, some neurons responded inhibitory. Recently, *Gallo et al.* (1972) observed inhibitory and facilitatory responses in the multiunit activity of the ARC in the ovariectomized and estrogen treated rat following the systemic application of LH. They have attributed these responses to the changes in the activity of the ventral hippocampus which they supposed as a receptor site for LH. However, in the present study, we have elucidated that a direct application of LH on the surface of the ARC neurons was effective in eliciting the response. Therefore, the response in the ARC may be regarded as a direct effect of LH exerted on neuronal membranes of the ARC. *Whitmoyer et al.* (1973) reported the validity of microiontophoretic ejection of LH and FSH *in vitro*. Observations that long-term exposure of the ovariectomized rat to equine serum gonadotropin (PMS) administered systemically (*Szontágh and Uhlarik,* 1964) or to LH implanted in the median eminence in a cocoa-butter vehicle (*Corbin and Cohen,* 1966; *Dávid et al.*, 1966) lowers the pituitary content of LH, may provide fundamental evidences for the hypotheses on the ARC as a site of negative control of LH.

As the number and the ratio of the neurons which responded to FSH were relatively small when compared with those which responded to LH-RH and/or LH, it might be assumed that the ARC is not a major site for the feedback of FSH. This observation agrees with that of *Dufy et al.* (1973) that FSH was ineffective on the hypothalamic activity when administered intravenously.

The refractoriness of the MPO neurons to LH-RH seems to be somewhat interesting. Because this region of the brain was reported not to be included in the hypophysiotropic area (*Halász et al.*, 1962), our present results may imply the distribution of LH-RH responsive neurons in the hypophysiotropic area.

References

Albe-Fessard, D.; Stutinsky, F. et Libouban, S.: Atlas stéréotaxique du diencéphale du rat blanc, pp. 1–74 (Editions du CNRS, Paris 1966).

Corbin, A. and Cohen, A.I.: Effect of median eminence implants of LH on pituitary LH of female rats. Endocrinology, Springfield *78:* 41–46 (1966).

Dávid, M.A.; Frashini, F., and Martini, L.: Control of LH secretion: role of a short feedback mechanism. Endocrinology, Springfield *72:* 936–946 (1966).

Dufy, B.; Vincent, J.D.; Bensch, C., and Faure, J.M.A.: Effects of vaginal stimulation and luteinizing hormone on hypothalamic single units in the freely moving rabbit. Neuroendocrinology *11:* 119–129 (1973).

Gallo, R.V.; Johnson, J.H.; Kalra, S.P.; Whitmoyer, D.I., and Sawyer, C.H.: Effects of luteinizing hormone on multiple-unit activity in the rat hippocampus. Neuroendocrinology *9:* 149–157 (1972).

Halász, B.; Pupp, L., and Uhlarik, S.: Hypophysiotropic area in the hypothalamus. J. Endocrin. *25:* 147–154 (1962).

Hyyppa, M.; Motta, M., and Martini, L.: 'Ultrashort' feedback control of follicle-stimulating hormone-releasing factor. Neuroendocrinology *7:* 227–235 (1971).

Kawakami, M. and Saito, H.: Unit activity in the hypothalamus of the cat: effect of genital stimuli, luteinizing hormone and oxytocin. Jap. J. Physiol. *17:* 466–486 (1967).

Ramirez, V.D.; Komisaruk, B.R.; Whitmoyer, D.I., and Sawyer, C.H.: Effects of hormones and vaginal stimulation on the EEG and hypothalamic units in rats. Amer. J. Physiol. *212:* 1376–1384 (1967).

Sawyer, C.H. and Kawakami, M.: Characteristics of behavioral and electroencephalographic after-reactions to copulation and vaginal stimulation in the female rabbit. Endocrinology, Springfield *65:* 622–630 (1959).

Szontágh, F.E. and Uhlarik, S.: The possibility of a direct 'internal' feedback in the control of pituitary gonadotropin secretion. J. Endocrin. *29:* 203–204 (1964).

Terasawa, E.; Whitmoyer, D.I., and Sawyer, C.H.: Effect of luteinizing hormone on multiple-unit activity in the rat hypothalamus. Amer. J. Physiol. *217:* 1119–1126 (1969).

Whitmoyer, D.I.; Ramirez, V.D., and Sawyer, C.H.: In vitro validation of microiontophoresis of gonadotropins to study their effect on neural activity. Proc. Endocrine Soc. 54th Meet. Abstr. 177 (1973).

Authors' address: Prof. *M. Kawakami* and Dr. *Y. Sakuma,* Second Department of Physiology, Yokohama City University School of Medicine, 2–33 Urafune-cho, Minami-ku, *Yokohama 232* (Japan)

Psychoneuroendocrinology. Workshop Conf. Int. Soc. Psychoneuroendocrinology, Mieken 1973, pp. 161–169 (Karger, Basel 1974)

On the Action of Gonadotropin-Releasing Factor

K. Wakabayashi and B. Tamaoki

Institute of Endocrinology, Gunma University, Maebashi, and National Institute of Radiological Sciences, Chiba

The hypothesis of the humoral control of the anterior pituitary gland by hypothalamus (*Harris,* 1955) has been evidenced by many investigators. Among them, luteinizing hormone-releasing factor (LH-RF) was found to be a decapeptide with an amino acid sequence of pyroGlu-His-Trp-Ser-Tyr-Gly-Leu-Arg-Pro-Gly-NH_2 (*Matsuo et al.,* 1971; *Baba et al.,* 1971). In the present paper, the characteristic features of LH-RF action were studied with rat anterior pituitary glands *in vitro* through comparison with other agents which have *in vitro* LH-releasing activity.

Materials and Methods

Adult male rats of Wistar strain were employed. The anterior pituitaries were removed immediately after decapitation, and cut into halves. In some cases, these halves were used as pairs, the one served as controls and the others as experimentals. In other cases, these halves were randomized among 3–5 treatment groups (*Wakabayashi et al.,* 1972). The glands were preincubated in a Dubnoff-type metabolic incubator for 30 min at 37 °C under an atmosphere of 95-percent O_2 and 5-percent CO_2 with a constant shaking of 90 c/min in 2 ml of Krebs-Henseleit-glucose solution (KH-G) (*Wakabayashi et al.,* 1972). The glands were then blotted lightly, and weighed, and transferred to the incubation flask containing 2 ml of incubation medium consisting of KH-G and test materials. Test materials were firstly dissolved in 1-percent bovine serum albumin-phosphate buffered saline (0.01 *M* phosphate, 0.14 *M* NaCl, pH 7.5) and added to 19 volumes of KH-G. A low Ca^{++} medium was prepared by replacing $CaCl_2$ for NaCl, and a high K^+ medium was prepared by replacing a part of NaCl for KCl (*Wakabayashi et al.,* 1973). Incubation was performed for 3 h under the same condition to preincubation. After incubation, the medium was separated from the glands by decantation, and cooled in ice water, then centrifuged at 2,000 *g* for 15 min. The supernatant fluid was kept frozen until hormone assay.

Rat LH, FSH and prolactin were assayed by radioimmunoassay with the NIAMDD kits according to the method of *Monroe et al.* (1968) with minor modifications. The results were expressed as the equivalent of NIH-LH-S1 and NIH-FSH-S1 for LH and FSH, respectively, and as mIU for prolactin.

Synthetic LH-RF and a peptide pyroGlu-Tyr-Arg-Trp-NH_2 were prepared and provided by Chugai Pharmaceutical Co., Tokyo.

Glucose oxidation in the anterior pituitary glands was examined by incubating the glands in 2 ml of KH-G containing 2 μc of uniformly labeled ^{14}C-glucose for 3 h in a flask with a center well with the subsequent trapping of $^{14}CO_2$ (*Wakabayashi et al.*, 1972).

Results

Gonadotropin-Releasing Effect of Synthetic LH-RF in vitro

When the anterior pituitaries from male rats were incubated with varied concentrations of synthetic LH-RF, LH and FSH releases were dependent upon LH-RF concentration, whereas prolactin release was not influenced at all by LH-RF (fig. 1).

The minimum effective concentration of LH-RF to male rats anterior pituitaries seemed to be between 5–25 ng/ml (fig. 2). When theophylline was added to the incubation medium at a concentration of 5×10^{-3} *M*, 1 ng of LH-RF/ml

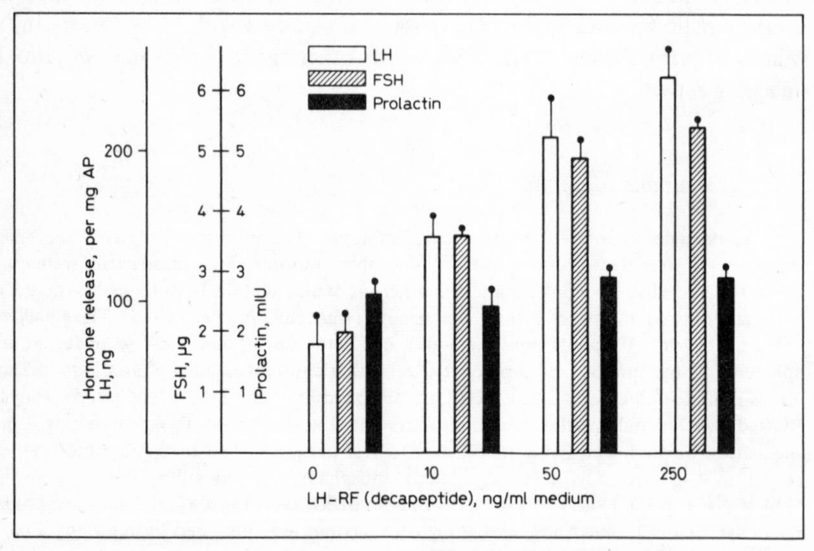

Fig. 1. In vitro effects of varied amount of LH-RF on LH, FSH and prolactin release from the anterior pituitaries of male rats. The hemipituitaries were randomized among four groups. A group consists of three flasks each containing three halves. Each column shows the mean with SE.

caused a highly significant increase in LH release, showing an enhancement of LH-releasing effect of LH-RF by this agent (fig. 2).

Various Chemical Agents Causing LH Release in vitro

Catecholamines and their chemically related compounds were examined for *in vitro* LH-releasing and prolactin-inhibiting activities by incubating the male anterior pituitaries with 5×10^{-4} *M* of these compounds. Table I shows the results which are expressed as percentages of the hormonal releases from the controls. Direct effects of these compounds on LH and prolactin were also examined to see whether the release was really inhibited or the immunochemical activity of the hormones was modified by the compound. For this purpose, two kinds of experiments were performed. In experiment 1, the aliquots of the supernatant fluid of rat anterior pituitary homogenates, which had been frozen-thawed and centrifuged, were incubated with the compounds for 3 h, and the recovery of the hormones was estimated. In experiment 2, the anterior pituitaries were preincubated for 3 h, and the preincubation media were pooled, and its aliquots were further incubated with the compounds. The recovery percentage of the hormones were then estimated.

Norepinephrine, isoproterenol, propranolol and chlorpromazine caused significant increase in LH release. Among them, norepinephrine caused about 50-

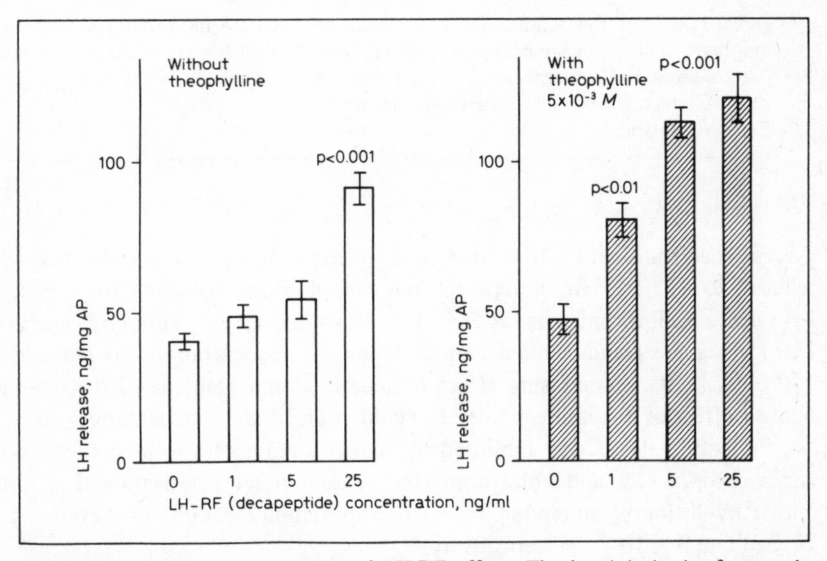

Fig. 2. Theophylline augmentation of LH-RF effect. The hemipituitaries from male rats were randomized among four groups. A group consists of three flasks each containing three halves. Each column shows the mean with SE. Significance of the difference versus control is shown as p value on the column.

Table I. In vitro effects of catecholamines and their related compounds

Compounds	Percent release		Direct effects of the compounds			
			exp. 1		exp. 2	
	LH	prolactin	LH	prolactin	LH	prolactin
Norepinephrine	144	36[3]	68	94	53	100
Isoproterenol	200[1]	76[2]	102	103	99	131
Phenethylamine	105	65[1]	105	105	107	108
Amphetamine	120	71[1]	97	102	106	130
Tyramine	105	58[2]	123	100	108	94
Dopamine	11[3]	22[2]	24	80	22	54
Propranolol	344[2]	57[3]	100	41	110	33
Chlorpromazine	3,557[3]	50[3]	111	51	113	50
Ephedrine	100	107	–	–	–	–
N-methyl ephedrine	88	105	–	–	–	–
Monoethanolamine	102	109	–	–	–	–
N-methyl EA	91	103	–	–	–	–
N,N-dimethyl EA	98	109	–	–	–	–

Concentration of the compounds: 5×10^{-4} *M*.
1 $p < 0.05$.
2 $p < 0.01$.
3 $p < 0.001$ versus control.
Direct effects of the compounds were examined by incubating anterior pituitary homogenates (exp. 1) or preincubation medium (exp. 2) with 5×10^{-4} *M* compounds for 3 h followed by hormone assay. The results were expressed as percentages of the controls which were incubated without the compounds.
EA = ethanolamine.

percent inactivation of LH at this concentration. When its concentration was reduced to 5×10^{-5} *M*, norepinephrine caused about 100-percent increase in LH release. Chlorpromazine, at 5×10^{-4} *M*, caused an extraordinary release of LH, leaving the glands almost empty. When the concentration was reduced to 1×10^{-4} *M*, its LH-releasing effect became moderate (table II). LH-release inhibiting effect of dopamine seemed to be due to its direct inactivation of LH.

Prolactin release was inhibited by many compounds. Among them, dopamine, propranolol and chlorpromazine caused direct inactivation. Prolactin-release inhibition by norepinephrine and isoproterenol was observed even at 5×10^{-6} *M* and 2×10^{-5} *M*, respectively.

Besides these compounds, high K^+ medium containing 6×10^{-2} *M* K^+, i.e. 10-fold concentration of K^+ usually present in KH-G and other physiological fluids, caused significant increase of LH release (table II). This high K^+ medium

Table II. Effects of stimulating agents and theophylline on LH release *in vitro*

Agents	No. of flasks	I, control	II, agent alone	III, theophylline alone	IV, agents and theophylline
Dibutyryl cAMP (1×10^{-3} *M*)	4	71.7 ± 8.7	67.6 ± 7.0	82.3 ± 11.6	114.3 ± 3.7[2,4]
High K^+ medium (6×10^{-2} *M*)	6	26.8 ± 4.1	89.0 ± 9.3[3]	43.6 ± 8.9	196.2 ± 14.8[3,5]
pGlu-Tyr-Arg-Trp-NH_2 (100 µg/ml)	4	46.5 ± 5.1	53.6 ± 2.2	43.9 ± 2.7	63.3 ± 1.6[1,4]
Chlor-promazine (1×10^{-4} *M*)	4	35.6 ± 4.6	108.5 ± 20.0[3]	38.0 ± 4.0	162.0 ± 16.8[3,4]
Propranolol (5×10^{-4} *M*)	4	55.1 ± 7.9	106.8 ± 5.6[2]	67.3 ± 5.7	98.2 ± 11.0[1]
Isoproterenol (5×10^{-4} *M*)	4	80.3 ± 13.4	159.2 ± 15.7[2]	113.2 ± 13.7	188.5 ± 37.6[1]

1 $p < 0.05$, versus control (I).
2 $p < 0.01$, versus control (I).
3 $p < 0.001$, versus control (I).
4 $p < 0.05$, versus agent alone (II).
5 $p < 0.001$, versus agent alone (II).
Mean ± SE, ng × NIH-LH-S1/mg anterior pituitary.

Table III. Effects of chlorpromazine and propranolol on LH release in low Ca^{++} medium

Agents	No. of flasks	I, control	II, agent alone	III, low Ca^{++} medium	IV, agent in low Ca^{++} medium
Chlor-promazine (1×10^{-4} *M*)	4	29.7 ± 3.9	208.5 ± 12.9[1]	41.3 ± 3.0	156.5 ± 7.2[1,2]
Propranolol (5×10^{-4} *M*)	4	36.1 ± 2.1	149.0 ± 6.8[1]	44.3 ± 2.7	207.1 ± 23.9[1,3]

1 $p < 0.001$, versus control (I).
2 $p < 0.05$, versus agent alone (II).
3 $p < 0.01$, versus agent alone (II).
Mean ± SE, ng × NIH-LH-S1/mg anterior pituitary.

increased prolactin and FSH release as well. Dibutyryl cyclic AMP, and a tetrapeptide pyroGlu-Tyr-Arg-Trp-NH_2, reported by *Bowers et al.* (1971), have very weak LH-releasing activity and we did not observe the effect at practical concentrations. Table II shows the effect of theophylline upon the LH-releasing actions of these compounds. Significant enhancement of the activity was caused by theophylline with dibutyryl cAMP, high K^+ medium, the tetrapeptide and chlorpromazine, whereas no significant enhancement was observed with propranolol and isoproterenol.

It was reported that LH-releasing effects of LH-RF and high K^+ medium were completely abolished in low Ca^{++} medium (*Samli and Geschwind,* 1968; *Wakabayashi et al.,* 1969). Table III shows that LH-releasing effect of chlorpromazine was only partly reduced in low Ca^{++} medium, and that the effect of propranolol was not reduced at all.

Glucose oxidation in the anterior pituitary was promoted, to a small extent, by hypothalamic extract, high K^+ medium and dibutyryl cAMP (*Wakabayashi et al.,* 1972). As was shown in table IV, chlorpromazine, isoproterenol and propranolol, at concentrations effective for stimulation of LH release, caused some increase in glucose oxidation in the anterior pituitary *in vitro.* However, more concentrated chlorpromazine which would cause an extraordinary LH release, inhibited the glucose oxidation significantly.

Table IV. Effects of chlorpromazine, isoproterenol and propranolol on glucose oxidation in the anterior pituitaries

Agents	No. of flasks	Glucose oxidation, $^{14}CO_2$ c.p.m./mg AP/3 h
Control	4	3,212 ± 87
Chlorpromazine 1 × 10^{-4} *M*	4	4,516 ± 123[2]
Isoproterenol 5 × 10^{-4} *M*	4	4,301 ± 294[1]
Propranolol 5 × 10^{-4} *M*	4	4,193 ± 113[2]
Control	4	2,603 ± 30
Chlorpromazine 5 × 10^{-5} *M*	4	3,416 ± 60[2]
Chlorpromazine 1 × 10^{-4} *M*	4	3,414 ± 42[2]
Chlorpromazine 2 × 10^{-4} *M*	4	2,192 ± 68[2]

1 $p < 0.05$, versus control.
2 $p < 0.001$, versus control.
Mean ± SE.
The anterior pituitaries were incubated with KH-G containing 2 μc of ^{14}C-glucose (U) in a flask with a center well filled with 95-percent O_2 and 5-percent CO_2. CO_2 formed was captured in NCS after the incubation.

Discussion

The augmentation of LH-RF action by theophylline (*Jutisz et al.*, 1970; *Wakabayashi et al.*, 1972, 1973), an inhibitor of phosphodiesterase (*Butcher and Sutherland,* 1962), and the increased adenyl cyclase activity and cAMP level caused by hypothalamic extract (*Zor et al.*, 1969, 1970; *Steiner et al.*, 1970) indicated a possible involvement of adenyl cyclase system in the mechanism of action of LH-RF like that of other hormones (*Sutherland and Robinson,* 1966; *Sutherland et al.*, 1968). Our experiments showed that the effect of synthetic LH-RF was also enhanced by theophylline.

Various kinds of substances were reported to influence the hormonal release from pituitaries *in vitro.* Norepinephrine and polyamines were known to increase FSH release *in vitro* (*Kamberi and McCann,* 1969), while prolactin release was inhibited by norepinephrine, epinephrine and tyramine (*MacLeod,* 1969). Our present observation with catecholamines and their related compounds showed that LH release was also influenced by these amines. If we compare the structures of these compounds, we can guess some structure-activity relationship. As to LH-release promotion, catechol group of norepinephrine seemed to be essential, and the loss of a hydroxyl group on the benzene ring causes the loss of the activity. However, because propranolol was active, a certain volume of the moiety might be necessary around the benzene ring. Direct inactivation effect upon LH may derive from the catechol and primary amine groups on both ends of the molecule. The hydroxyl group on the side chain seemed to be essential to LH-releasing activity. Modification of the amino group would not influence the activity. For prolactin-release inhibiting activity, the hydroxyl groups on the benzene ring are not necessary, the hydroxyl group on the side chain being not essential, either.

As a result of our comparative experiments employing theophylline and low Ca^{++} medium, we can classify these agents into at least two groups with different modes of action. The first group contains modes, the effect of which was enhanced by theophylline and Ca^{++}-dependent, including LH-RF, the tetrapeptide, 60 m*M* K^+, and chlorpromazine. The LH-releasing effect of the second group was not influenced by theophylline nor by low Ca^{++} medium, and propranolol, isoproterenol, and probably norepinephrine would be included. From their influence on glucose oxidation in the gland, these compounds possibly do not destroy the cells, and the hormonal release is not a non-physiological one like that caused by iodoacetate (*Wakabayashi et al.*, 1972).

The existence of this type of agent may indicate another type of LH-releasing mechanism than that via the adenyl cyclase system. Chlorpromazine, at a low concentration, may act like LH-RF, but at 5×10^{-4} *M*, it caused an abnormal release owing to the metabolic disturbances, and probably, to the death of the LH-secreting cells. This was suggested by the partial inhibition of its effect in low Ca^{++} medium, and by its inhibiting action on glucose oxidation in the gland.

Though 60 m*M* K^{+} was influenced by theophylline and low Ca^{++} medium, there is some possibility that it might work with another type of mechanism because of its nonspecificity and its additive effect to LH-RF (*Samli and Geschwind*, 1968, *Wakabayashi et al.*, 1969, 1973).

Summary

On the bases of enhancement by theophylline and inhibition in low Ca^{++} medium, which are observed with LH-releasing effect of synthetic LH-RF, various agents which stimulate LH-release *in vitro* were examined. LH-RF, pyroGlu-Try-Arg-Trp-NH_2, 60 m*M* K^{+}, and low concentration of chlorpromazine, may belong to a group whose effect was influenced by theophylline and low Ca^{++} medium, and adenyl cyclase system may be involved in their mechanism of action. On the other hand, isoproterenol, propranolol, and probably norepinephrine belong to a second group whose effect was independent from theophylline and low Ca^{++} medium, and seemed to increase the hormonal release through different mechanism.

Acknowledgement

The authors wish to express their thanks to Rat Pituitary Hormone Distribution Program, NIAMDD, NIH, Bethesda, Md., USA, for their kind supply of radioimmunoassay kits for rat gonadotropins.

References

Baba, Y.; Matsuo, H., and Schally, A.V.: Structure of the porcine LH- and FSH-releasing hormone, II. Confirmation of the proposed structure by conventional sequential analyses. Biochem. biophys. Res. Commun. *44:* 459–463 (1971).

Bowers, C.Y.; Chang, J.K.; Sievertsson, H.; Bogentoft, C.; Currie, B.L., and Folkers, K.: Activity of a new synthetic tetrapeptide in hypothalamic luteinizing and follicle stimulating releasing hormone assay system. Biochem. biophys. Res. Commun. *44:* 414–421 (1971).

Butcher, R.W. and Sutherland, E.W.: Adenosine 3',5'-phosphate in biological materials. I. Purification and properties of cyclic 3',5'-nucleotide phosphodiesterase and use of this enzyme to characterize adenosine 3',5'-phosphate in human urine. J. biol. Chem. *237:* 1244–1250 (1962).

Harris, G.W.: Neural control of the pituitary gland (Arnold, London 1955).

Jutisz, M.; Kerdelhue, B., and Berault, A.: Further studies on mechanism of action of luteinizing hormone releasing factor using *in vivo* and *in vitro* techniques; in *Rosenberg and Paulsen* Human testis, pp. 221–228 (Plenum Press, New York 1970).

Kamberi, I.A. and McCann, S.M.: Effect of biogenic amines, FSH-releasing factor (FRF) and other substances on the release of FSH by pituitaries incubated *in vitro.* Endocrinology, Springfield *85:* 815–824 (1969).

MacLeod, R.M.: Influence of norepinephrine and catecholamine-depleting agents on the synthesis and release of prolactin and growth hormone. Endocrinology, Springfield *85:* 916–923 (1969).

Matsuo, H.; Baba, Y.; Nair, R.M.G.; Arimura, A., and Schally, A.V.: Structure of the porcine LH- and FSH-releasing hormone. I. The proposed amino acid sequence. Biochem. biophys. Res. Commun. *43:* 1334–1339 (1971).

Monroe, S.E.; Parlow, A.F., and Midgley, A.R., jr.: Radioimmunoassay for rat luteinizing hormone. Endocrinology, Springfield *83:* 1004–1012 (1968).

Samli, M.H. and Geschwind, I.I.: Some effects of energy-transfer inhibitors and of Ca^{++}-free or K^{+}-enhanced media on the release of luteinizing hormone (LH) from the rat pituitary gland *in vitro.* Endocrinology, Springfield *82:* 225–231 (1968).

Steiner, A.L.; Peake, G.T.; Utiger, R.D.; Karl, I.E., and Kipnis, D.M.: Hypothalamic stimulation of growth hormone and thyrotropin release *in vitro* and pituitary 3′,5′-adenosine cyclic monophosphate. Endocrinology, Springfield *86:* 1354–1360 (1970).

Sutherland, E.W. and Robinson, G.A.: The role of cyclic-3′,5′-AMP in response to catecholamines and other hormones. Pharmacol. Rev. *18:* 145–161 (1966).

Sutherland, E.W.; Robinson, G.A., and Butcher, R.W.: Some aspects of the biological role of adenosine 3′,5′-monophosphate (cyclic AMP). Circulation *37:* 279–306 (1968).

Wakabayashi, K.; Kamberi, I.A., and McCann, S.M.: In vitro response of the rat pituitary to gonadotrophin-releasing factors and to ions. Endocrinology, Springfield *85:* 1046–1056 (1969).

Wakabayashi, K.; Antunes-Rodrigues, J.; Tamaoki, B., and McCann, S.M.: In vitro effect of hypothalamic extract and other stimulating agents on glucose oxidation and luteinizing hormone (LH) release from rat anterior pituitary glands. Endocrinology, Springfield *90:* 690–699 (1972).

Wakabayashi, K.; Date, Y., and Tamaoki, B.: On the mechanism of action of luteinizing hormone-releasing factor and prolactin release inhibiting factor. Endocrinology, Springfield *92:* 698–703 (1973).

Zor, U.; Kaneko, T.; Schneider, H.P.G.; McCann, S.M.; Lowe, I.P.; Bloom, G.; Borland, B., and Field, J.B.: Stimulation of anterior pituitary adenyl cyclase activity and adenosine 3′,5′-cyclic phosphate by hypothalamic extract and prostaglandin E_1. Proc. nat. Acad. Sci., Wash. *63:* 918–925 (1969).

Zor, U.; Kaneko, T.; Schneider, H.P.G.; McCann, S.M., and Field, J.B.: Further studies of stimulation of anterior pituitary cyclic adenosine 3′,5′-monophosphate formation by hypothalamic extract and prostaglandins. J. biol. Chem. *245:* 2883–2888 (1970).

Authors' addresses: Dr. *K. Wakabayashi,* Institute of Endocrinology, Gunma University, *Maebashi;* Dr. *B. Tamaoki,* National Institute of Radiological Sciences, *Chiba* (Japan)

Psychoneuroendocrinology. Workshop Conf. Int. Soc. Psychoneuroendocrinology, Mieken 1973, pp. 170–177 (Karger, Basel 1974)

A New Finding in Luteinizing Hormone-Releasing Mechanism[1]

M. Fukushima, K. Kushima and K. Okuyama

Department of Obstetrics and Gynecology, Akita University Medical School, Akita

It is well known that gonadotropin secretion in the pituitary is regulated by the hypothalamus (3, 4). Recently, *Baba, Matsuo and Schally* (1) have purified successfully the luteinizing hormone-releasing factor (LH-RF) from pig hypothalami, proven it to be a polypeptide and shown its amino acid sequence to be (pyro) Glu- His- Tyr- Gly- Leu- Arg- Pro- Gly- NH_2. They reported that this decapeptide possessed a high follicle-stimulating hormone-releasing factor (FSH-RF) activity as well as LH-RF activity (6).

However, it is difficult to explain why there is a marked difference in the response or pattern of secretion between FSH and LH. There might be separate mechanisms at the pituitary level by which the gonadotropins produced by hypothalamic LH-RF are selectively released.

Samli and Geshwind (5) and *Schneider and McCann* (7) are among those who suggest a separate mechanism for regulating the synthesis and the release of LH.

We have previously shown in rats that LH stored in the anterior pituitary was released in sufficient amounts into the blood by a factor different from LH-RF, and named this the *LH-trigger factor (LH-TF)* (2).

Extraction of LH-TF from Bovine Pituitaries

The procedures used for the extraction of LH-TF from frozen pituitaries and hypothalami of bovine origin were: lyophilization, defatting, extraction with 2 *N* acetic acid, and gel filtration on basic ion exchange Sephadex, DEAE Sephadex A-25.

In the chromatography of the extract, the active fraction was eluted after big fractions by 0.2 *M* pyridin acetate buffer, pH 3.85 ± 0.05. This biologically

1 This study was supported in part by the Japanese Education Ministry, grant No. 767090.

active fraction appeared as a single spot with the positive ninhydrin reaction on a thin-layer chromatogram. This spot was not seen in the bovine hypothalami.

Bioassay for LH-TF

In an *in vitro* experiment, adult male Wistar-Imamichi or Sprague-Dawley strain rats were used as donors. Donor's pituitaries were incubated with the extract or synthetic LH-RF (Mochida) in the 200 mg% glucose Krebs-Ringer phosphate buffer, pH 7.4 for 2 h. LH content was measured separately in the anterior pituitary (hormones remaining in the tissue) and in the medium (hormone released), comparing the two groups by ovarian ascorbic acid depletion method.

After addition of LH-TF to the medium, as shown by the dotted line in figure 1, the LH content in the pituitary was rapidly decreased for 2 h, and then gradually decreased thereafter. The LH in the medium showed a sharp peak after 15 min, then decreased nearly to the control.

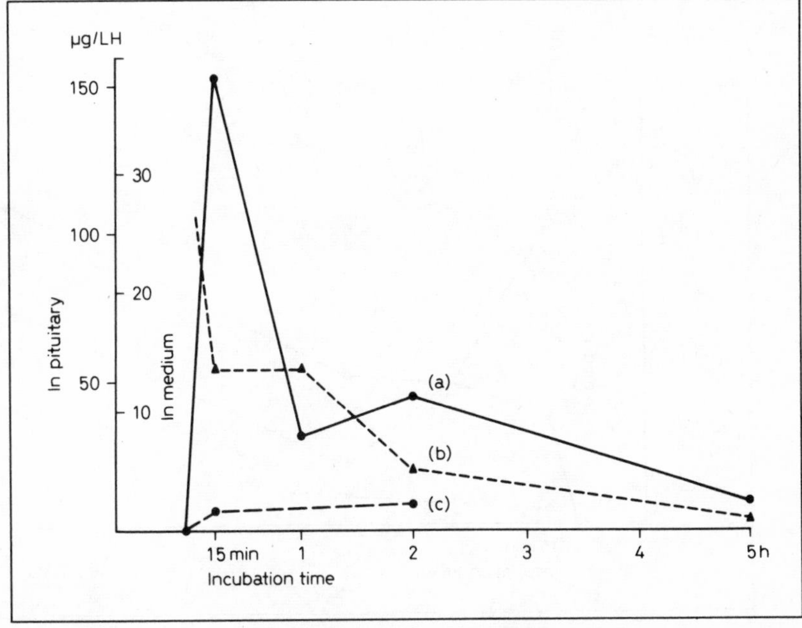

Fig. 1. Changes of LH content in pituitary and medium, incubated with extracted LH-TF. (a) LH in medium – LH-TF group; (b) LH in pituitary – LH-TF group; (c) LH in medium – control.

With the addition of synthetic LH-RF, the LH content was also decreased in the pituitary and increased in the medium during 15 min to 2 h, but afterwards LH was gradually increased, both in the pituitary and in the medium, suggesting LH synthesis by LH-releasing factor (fig. 1 and 2).

In an *in vivo* experiment, the pituitaries from some donors who were given a preliminary intravenous injection of the extract, or synthetic LH-RF, were homogenized and treated with trypsin and heated (100 °C, 10 min) to destroy LH and LH-RF decapeptides in the tissues. After centrifugation, the supernatant was used as a medium to incubate with the untreated anterior pituitaries of other donors. After incubation, the medium was removed and LH was assayed, for comparison with the control treated in the same process after intravenous saline injection.

As shown in figure 3, 15 min after intravenous injection of LH-trigger factor, LH-TF activity appeared both in the pituitary and hypothalamus, although the pituitary content was much higher than the hypothalamic one.

In the case of synthetic LH-releasing factor, 30 min after intravenous injection, LH-trigger factor activity appeared and consequently LH was released into

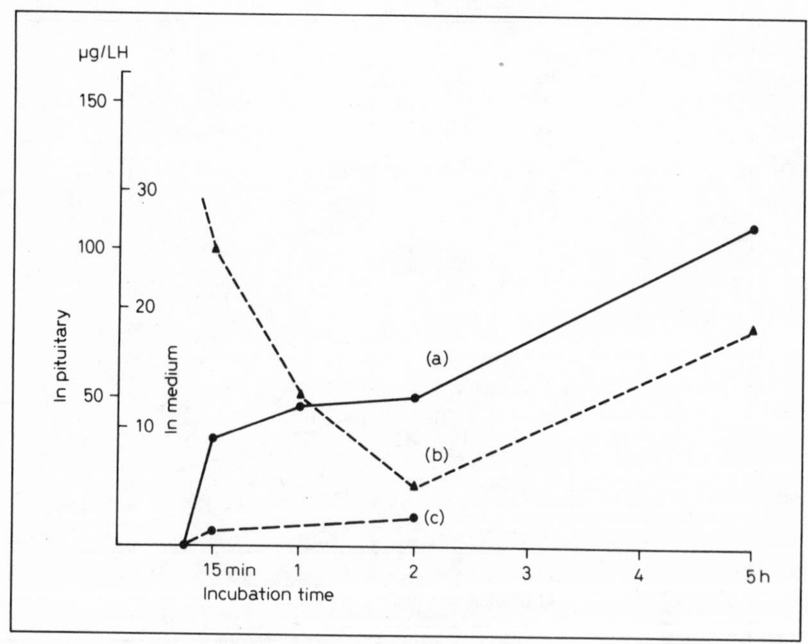

Fig. 2. Changes of LH content in pituitary and medium, incubated with synthetic LH-RF. (a) LH in medium – LH-RH group; (b) LH in pituitary – LH-RH group; (c) LH in medium – control.

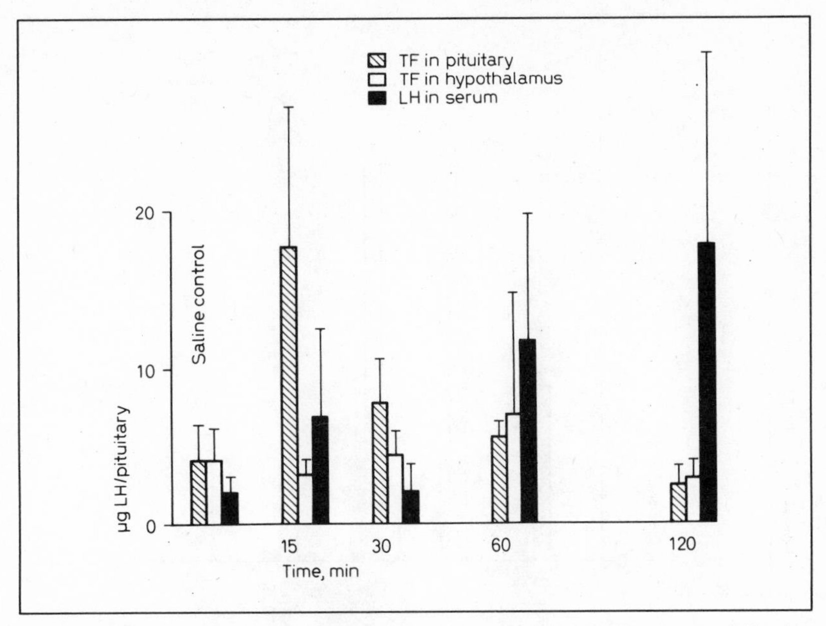

Fig. 3. Changes of serum LH, hypothalamic and pituitary LH-TF content, following intravenous injection of extracted LH-TF.

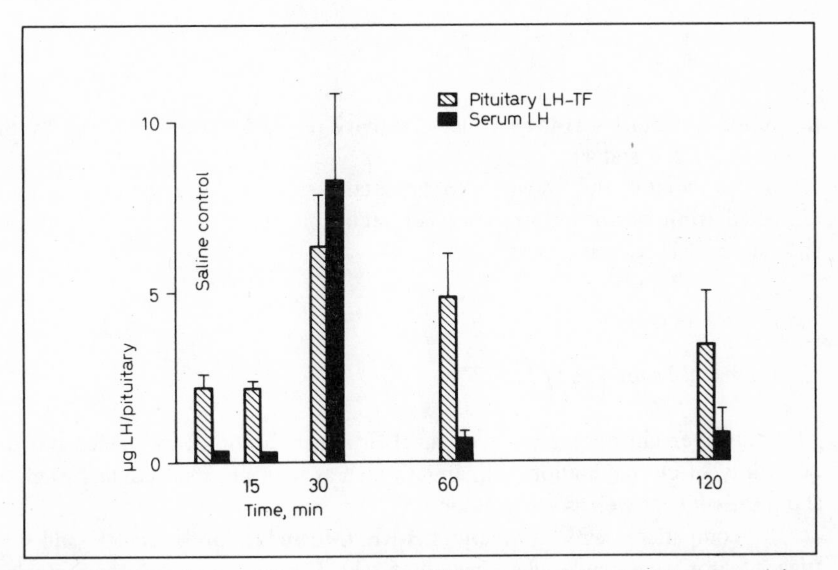

Fig. 4. Changes of serum LH and pituitary LH-TF content, following intravenous injection of synthetic LH-RF.

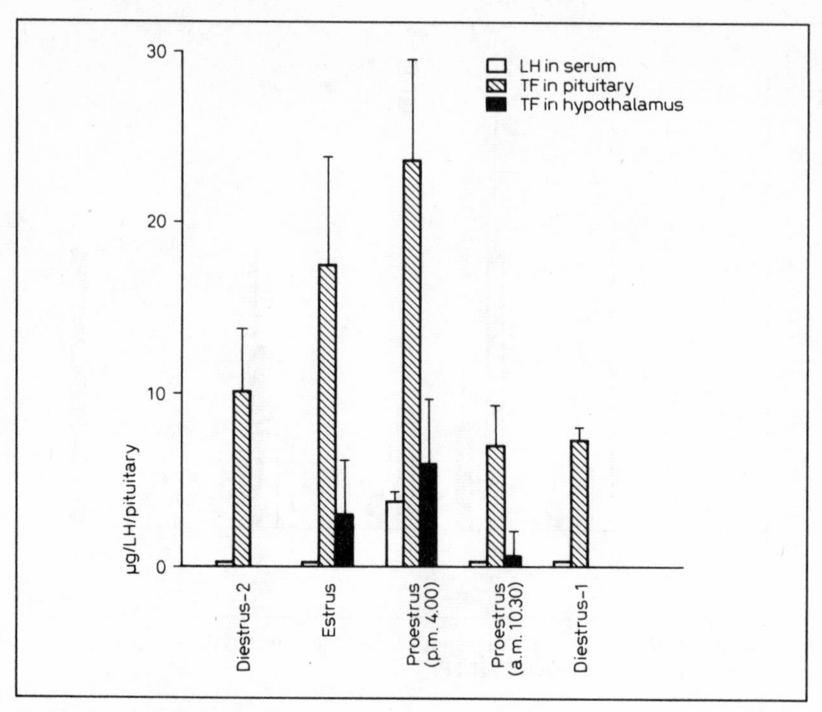

Fig. 5. LH-triggering activity of hypothalamus and pituitary at varying stages of the estrous cycle.

the blood. Pituitary LH-trigger factor activity reached the control level by the end of 2 h (fig. 3 and 4).

These *in vitro* and *in vivo* experimental results indicate that the material extracted from bovine pituitaries acts directly on the anterior pituitary of rats and releases LH from it.

Chemical Properties of LH-TF

Thin-layer chromatography of the LH-trigger factor extract ruled out the possibility of contamination with amines such as spermin, spermidine, histamine and putrescine, as well as vasopressor.

In comparison with synthetic LH-RF (Mochida), both LH-RF and LH-trigger factor were stable when heated at 100 °C for 10 min, but the biological activity of LH-trigger factor was gradually reduced at 100 °C for 30 min.

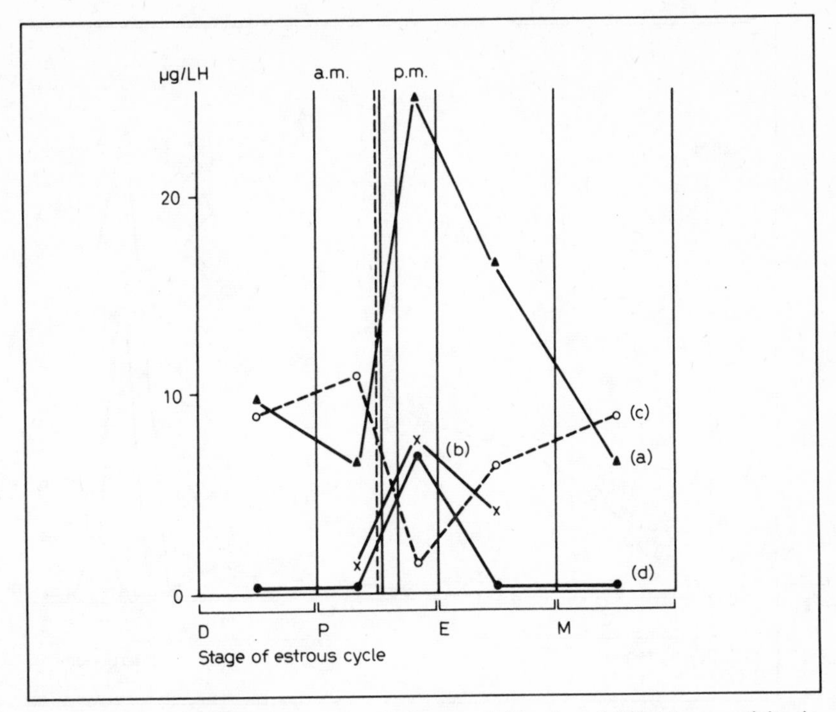

Fig. 6. Changes of LH contents in pituitary and blood, and LH trigger activity in hypothalamus and pituitary at each stage of cycle. (a) LH-TF in pituitary; (b) LH-TF in hypothalamus; (c) LH in pituitary, and (d) LH in serum (2 ml).

Synthetic LH-RF was inactivated by trypsin (synthetic LH-RF and trypsin were kindly supplied by Mochida Pharmaceutical Co., Tokyo) and α-chymotrypsin (kindly supplied as kymopsin by Esai Co., Tokyo) while LH-TF was not inactivated by trypsin, nor by α-chymotrypsin.

Physiological Significance of LH-TF

The levels of LH and LH-trigger factor in the pituitary and serum during the estrous cycle were measured using rats showing regular 4-day cycles.

LH-trigger factor activity in the pituitary was significantly higher on the afternoon of proestrus than at the other stages. In the same period, a rapid decrease in LH content in the pituitary and an increase in the serum were demonstrated, suggesting preovulatory LH surge (fig. 5 and 6).

LH and LH-TF in the pituitary and serum during pregnancy and puerperal period were also measured.

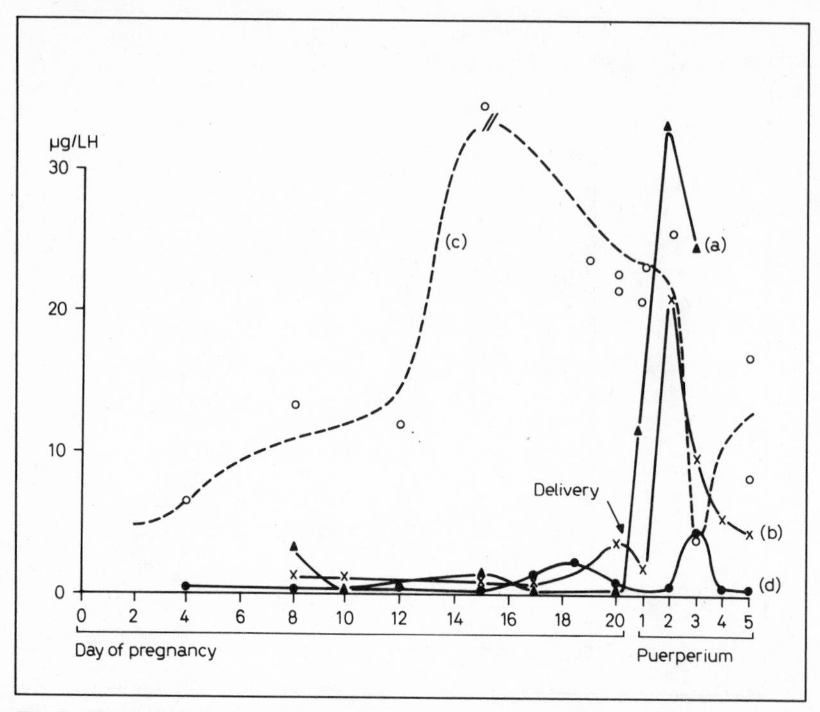

Fig. 7. Changes of LH contents in pituitary and blood, and LH trigger activity in hypothalamus and pituitary during pregnancy and puerperium. (a) LH-TF in pituitary; (b) LH-TF in hypothalamus; (c) LH in pituitary, and (d) LH in serum (2 ml).

LH in the pituitary increased rapidly from about the twelfth day of gestation, at which time placenta formation is completed, and remained nearly constant until the early stage of puerperium, then decreased abruptly thereafter. LH was undetectable in serum before the third day of puerperium, then was released. In the same period, remarkable increase of LH-trigger factor activity was found both in the pituitary and hypothalamus, although LH-TF in the pituitary was higher than that in the hypothalamus (fig. 7).

These data strongly suggest that LH-TF plays an important role in the releasing mechanism of LH synthesized in the pituitary by hypothalamic LH-RF.

Summary

The present results have demonstrated in rats that LH stored in the pituitary was released into the blood by another factor from LH-RF, and this new factor was named the LH-trigger factor (LH-TF).

LH-TF extracted from bovine pituitary was different chemically and biologically from synthetic LH-RF decapeptides (Mochida).

Changes in the content of LH-TF in various physiological conditions were demonstrated. Rapid increase of the LH-TF activity was always seen just before LH surge.

References

1 *Baba, Y.; Matsuo, H., and Schally, A.V.:* Structure of the porcine LH- and FSH-releasing hormone. II. Confirmation of the proposed structure by conventional sequential analyses. Biochem. biophys. Res. Commun. *44:* 459–463 (1971).

2 *Fukushima, M.; Kushima, K., and Noda, K.:* New ovulation agent. 5th Wld Congr. Gynec. Obstet., Sydney 1967, p. 346 (Butterworths/Australia, 1967).

3 *Guillemin, R.:* Hypothalamic factors releasing pituitary hormones. Recent Progr. Hormone Res. *20:* 89–130 (1964).

4 *McCann, S.M.; Dhariwal, A.P.S., and Porter, J.C.:* Regulation of the adenohypophysis. Ann. Rev. Physiol. *30:* 589–640 (1968).

5 *Samli, M.H. and Geshwind, I.I.:* Some effects of energy-transfer inhibitors and of Ca^{++}-free or K^{+}-enhanced media on the release of luteinizing hormone (LH) from the rat pituitary gland *in vitro.* Endocrinology, Springfield *82:* 225–231 (1968).

6 *Schally, A.V.; Arimura, A.; Kastin, A.J.; Matsuo, H.; Baba, Y.; Redding, T.W.; Nair, R.M.G.; Debeljuk, L., and White, W.F.:* Gonadotropin-releasing hormone: one polypeptide regulates secretion of luteinizing and follicle-stimulating hormones. Science *173:* 1036–1038 (1971).

7 *Schneider, H.P.G. and McCann, S.M.:* Estradiol and the neuroendocrine control of LH release *in vitro.* Endocrinology, Springfield *87:* 330–338 (1970).

Authors' address: Dr. *M. Fukushima,* Dr. *K. Kushima* and Dr. *K. Okuyama,* Department of Obstetrics and Gynecology, Akita University Medical School, *Akita* (Japan)

Psychoneuroendocrinology. Workshop Conf. Int. Soc. Psychoneuroendocrinology, Mieken 1973, pp. 178–186 (Karger, Basel 1974)

Existence of FSH-RF Distinct from LH-RH[1]

M. Igarashi, K. Taya and J. Ishikawa

Department of Obstetrics and Gynecology, School of Medicine, Gunma University, Maebashi

Introduction

Since the chemical structure of hypothalamic LH-RH/FSH-RH was determined in 1971, *Schally et al.* (8) claimed that this decapeptide is the only releasing factor (RF) controlling both LH and FSH secretion from the pituitary. *Guillemin* (4) also agreed to this 'one RF hypothesis'. On the other hand, FSH secretion from the pituitary is not always parallel to LH secretion in the normal menstrual cycle and other pathologic states such as amenorrheas and anovulatory cycles. The 'one RF hypothesis' finds difficulties in explaining the abovementioned dissociation in FSH and LH secretion. In our laboratory, the purification of beef hypothalamic FSH-RF and LH-RF has been investigated since the first demonstration (5) of FSH-RF existence in 1964 and evidence demonstrating that two distinct RFs, FSH-RF and LH-RF, control FSH and LH secretion, respectively, has been obtained since 1967. In the present paper, recent evidence suggesting the existence of FSH-RF distinct from LH-RH is reported.

Materials and Methods

The FSH- and LH-releasing activities *in vivo* were compared between 10 and 100 ng of synthetic LH-RH and the 5 crude rat stalk median eminence (SME) extract, with *in vivo* incubation assay method for FSH-RF and LH-RF, slightly modified from the method of *Bowers et al.* (1). The FSH- and LH-releasing activities *in vitro* were compared between the synthetic LH-RH and the partially purified beef hypothalamic FSH-RF, with *in vitro* assay method for FSH-RF and LH-RF, slightly modified the method of *Mittler and Meites* (7). The released FSH and LH were assayed with NIAMD rat FSH and LH radioimmunoassay

1 This research was partially supported by grant M73-35 from the Population Council, New York, N.Y.

kits. The partially purified FSH-RF was prepared from the beef hypothalamic extract in our laboratory using defatting, boiling, gel filtration on Sephadex G 25, phenol extraction and ion exchange chromatographies and was confirmed to contain no LH-RH decapeptide. Synthetic LH-RH was synthesized at the Protein Institute of Osaka University.

Results

(1) *In vivo* experiments: The intravenous injection of 10 ng of synthetic LH-RH induced a sudden increase and then a gradual decrease in both LH and FSH secretion in rats in duplicate experiments shown in figures 1 and 2. When the injected dose of LH-RH was increased to 100 ng, the changes in blood LH showed the same pattern as in 10 ng injection. However, blood FSH showed a small peak in 60 min in one (fig. 4) of two repeated experiments shown in figures 3 and 4. On the other hand, the injection of the 5 SME of hypothalamic extract induced the same pattern of LH secretion as injection of synthetic LH-RH in all of the three repeated experiments, while the patterns of FSH secretion showed the retarded peak in 40–70 min in all of the three repeated experiments shown in figures 5–7. Since the dose of LH-RH contaminated in these 5 SME of crude hypothalamic extract is calculated far below 10 ng from the values of the LH peak in table I, it can be concluded that the retarded FSH peak was induced by the other unknown substance (FSH-RF) which exists in crude hypothalamic extract and is distinct from the decapeptide LH-RH of *Schally et al.* (8).

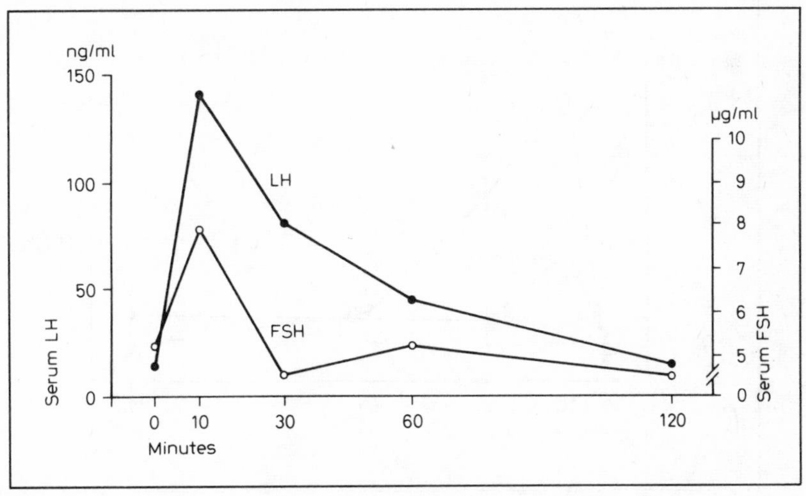

Fig. 1. Effect of intravenous injection of 10 ng LH-RH upon blood LH and FSH in rats (first experiment).

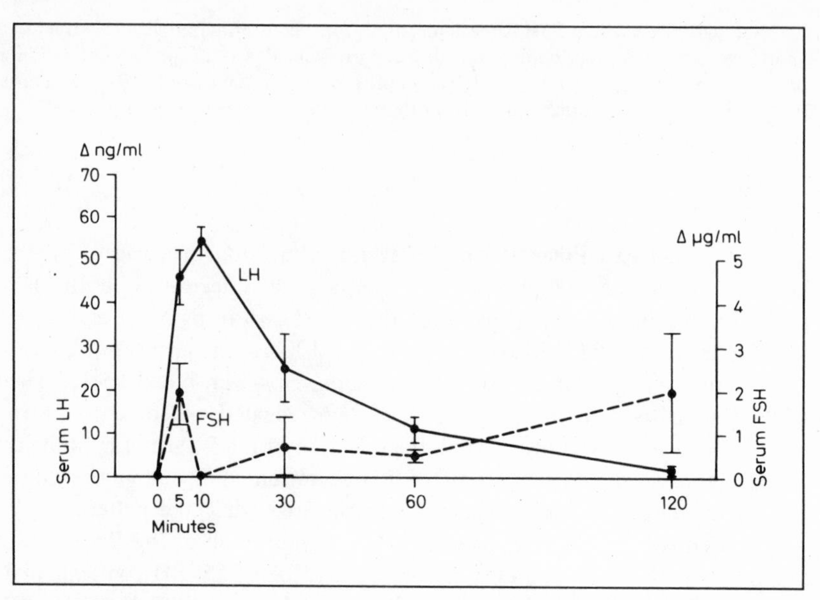

Fig. 2. Effect of intravenous injection of 10 ng LH-RH upon blood LH and FSH in rats (second experiment).

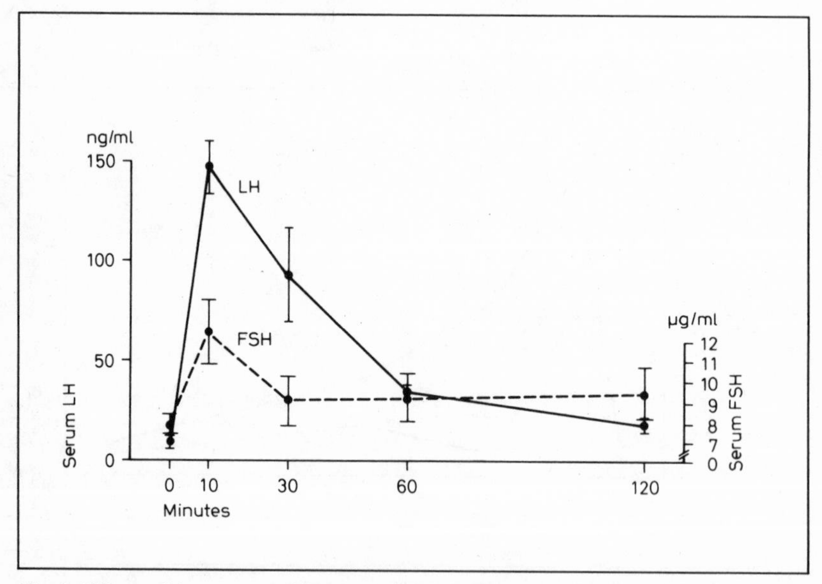

Fig. 3. Effect of intravenous injection of 100 ng LH-RH upon blood LH and FSH in rats (first experiment).

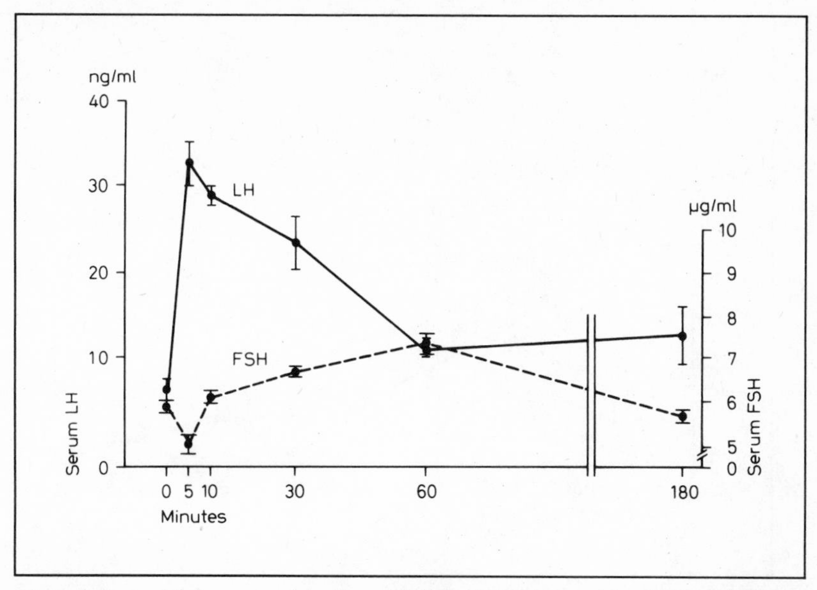

Fig. 4. Effect of intravenous injection of 100 ng LH-RH upon blood LH and FSH in rats (second experiment).

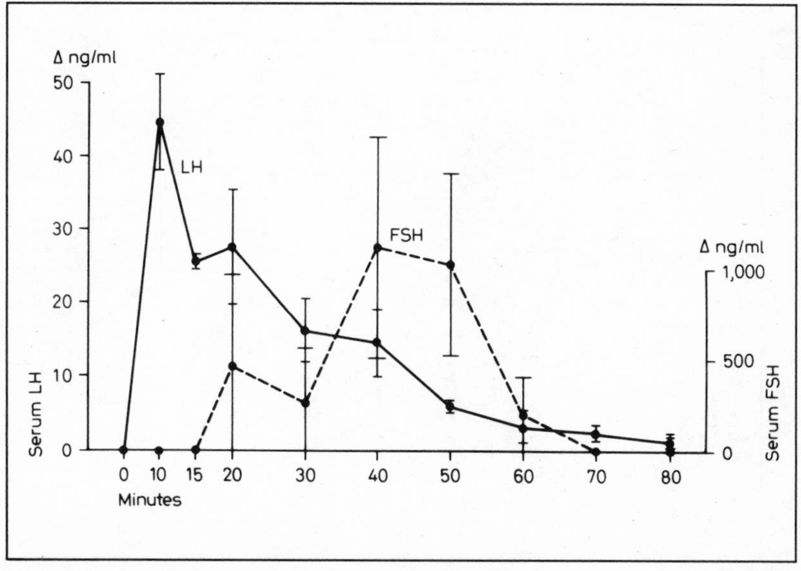

Fig. 5. Effect of intravenous injection of crude rat hypothalamic extract (5 SME) upon blood LH and FSH in rats (first experiment).

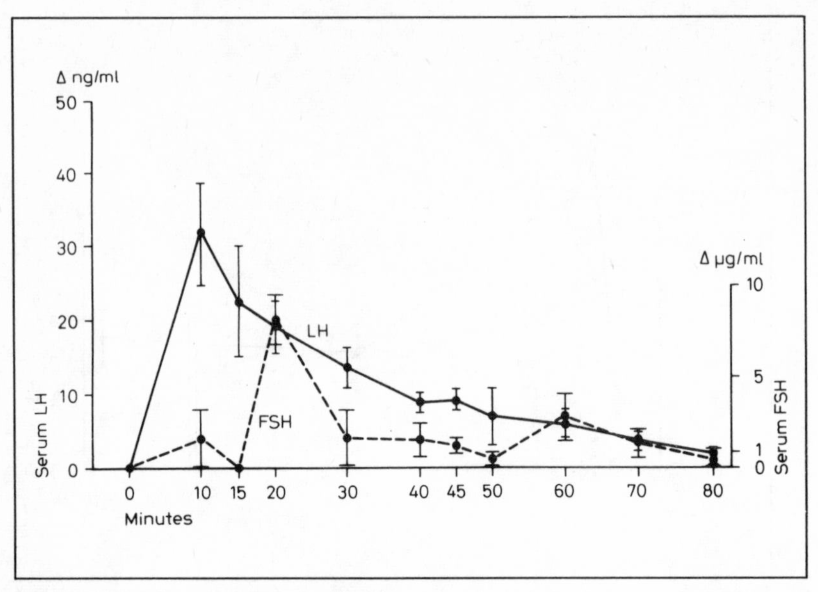

Fig. 6. Effect of intravenous injection of crude rat hypothalamic extract (5 SME) upon blood LH and FSH in rats (second experiment).

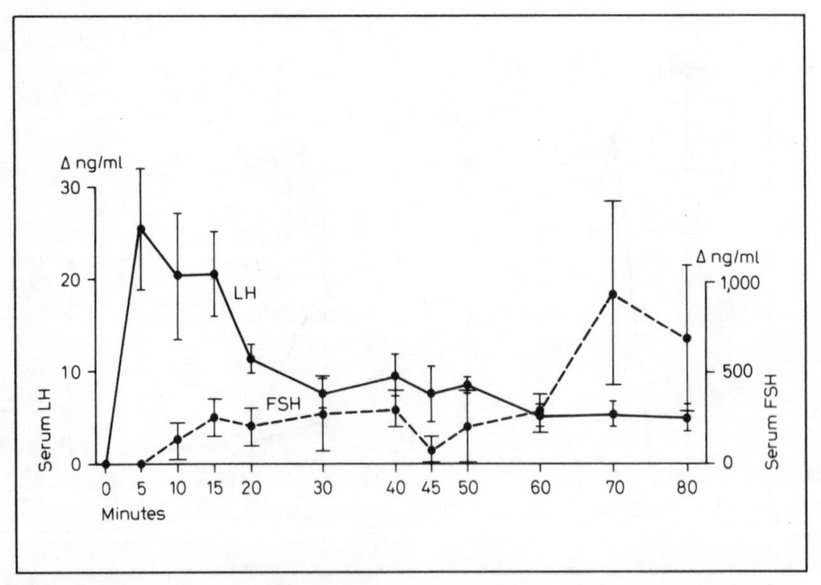

Fig. 7. Effect of intravenous injection of crude rat hypothalamic extract (5 SME) upon blood LH and FSH in rats (third experiment).

Table I. Relation between frequency of the retarded FSH peak (at 40–70 min) and the dose of LH-RH

Dose of RF	No. of experiment	Frequency of serum FSH peak, 40–70 min		LH peak value at 10 min, ng/ml
LH-RH 10 ng	2	0/2	0	62.0 ± 2.5
LH-RH 100 ng	2	1/2	50	146.0 ± 14.1
Rat crude SME, 5 SME	3	3/3	100	52.8 ± 5.2
				40.5 ± 9.4
				25.9 ± 4.1

(2) *In vitro* experiments: Figure 8 shows the released ratio of LH and FSH in a 2-hour incubation experiment using rat pituitary half together with synthetic LH-RH. The release ratio of LH was always superior to the FSH release ratio. Consequently, as shown in figure 9, the release ratio LH/FSH was always over 1.0, and moreover, this ratio increased, demonstrating a lineal dose response curve to the added dose of LH-RH. On the contrary, the incubation of the partially purified FSH-RF induced the LH and FSH release as shown in figure 10. The release ratio of FSH was always superior to the LH release ratio. Consequently, the release ratio LH/FSH was always below 1.0, as shown in figure 11.

Discussion

It is true that the decapeptide of *Schally et al.* (8) stimulates not only LH release, but FSH release in animal and human. However, the release ratio of LH is superior to the release ratio of FSH in almost all cases. In our laboratory, LH-RH was effective in the induction of human ovulation in amenorrheic or anovulatory women, only under the limited condition that the ovarian follicle was already mature, and ineffective when the ovarian follicle was immature. These results strongly suggest that the decapeptide of *Schally et al.* (8) is just LH-RF, but not FSH-RF.

Our present results show that crude hypothalamic extract contains a substance which is distinct from the decapeptide and stimulates FSH release and that some fraction, purified from the hypothalamus, distinct from LH-RH, showed strong FSH-releasing and weak LH-releasing activities. Quite recently, *Bowers et al.* (1); *Currie et al.* (3), and *Johansson et al.* (6) reported on the existence of FSH-RH distinct from the decapeptide of *Schally et al.* (8) in porcine hypothalamic extract. It is still unknown whether our partially purified

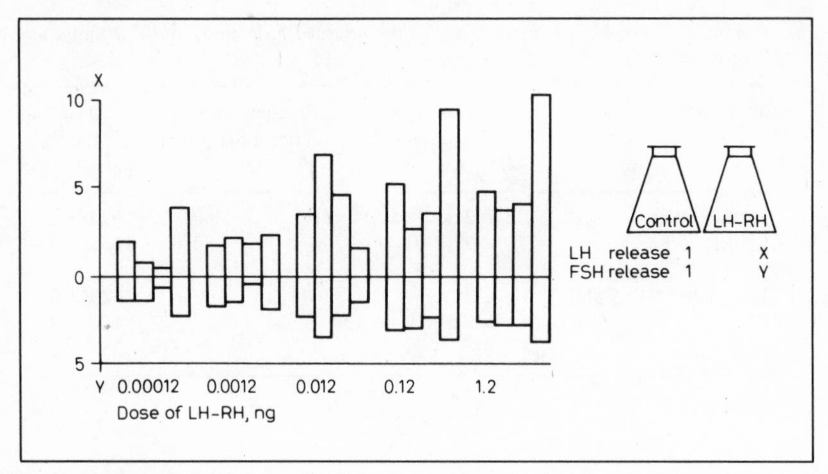

Fig. 8. Effect of LH-RH from 0.00012 ng to 1.2 ng *in vitro* upon LH release ratio (X) and FSH release ratio (Y).

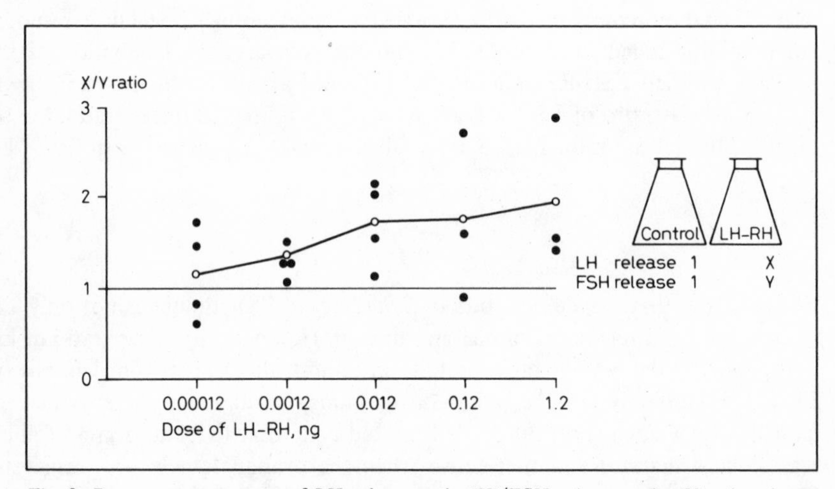

Fig. 9. Dose response curve of LH release ratio (X)/FSH release ratio (Y) when incubated with synthetic LH-RH.

FSH-RF is the same as the FSH-RH of the above-mentioned group (1, 6, 8). Although the detailed chemical structure of our FSH-RF and the above-mentioned group are still unpublished, our present data strongly support that FSH-RF is distinct from the decapeptide of *Schally et al.* (8) and has strong FSH-releasing and weak LH-releasing activities.

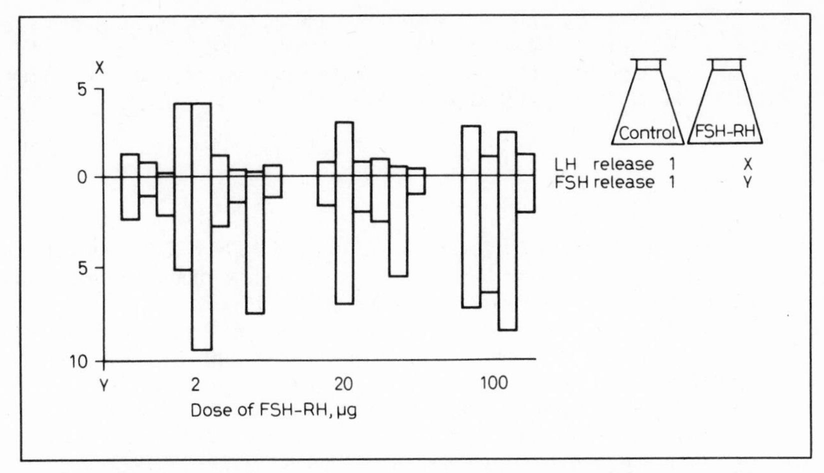

Fig. 10. Effect of partially purified beef hypothalamic FSH-RF from 2 to 100 μg *in vitro* upon LH release ratio (X) and FSH release ratio (Y).

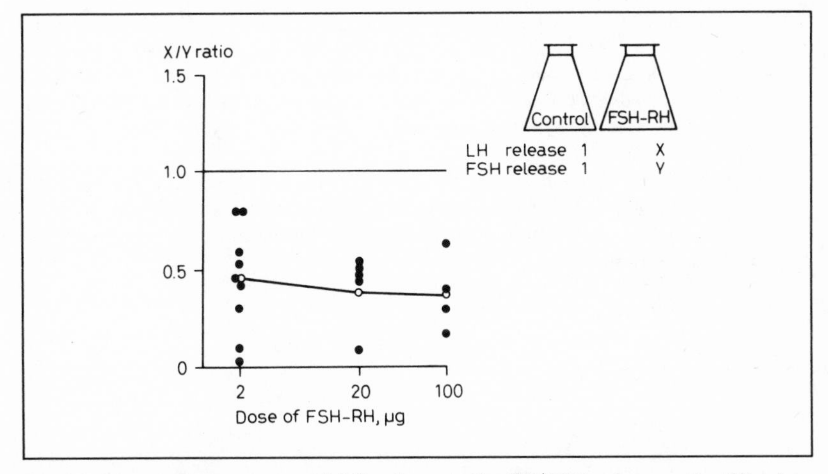

Fig. 11. Dose response curve of LH release ratio (X)/FSH release ratio (Y) when incubated with partially purified beef hypothalamic FSH-RF.

Summary

The intravenous injection of the crude 5 rat hypothalamic extract to the ovariectomized estrogen-progesterone blocked rats in the three repeated experiments always induced blood FSH peaks, which cannot be explained by the contaminated LH-RH. In *in vitro* rat pituitary incubation experiments, the LH-release to the FSH-release ratio was always over 1.0 when the synthetic LH-RH was incubated, while the same ratio was always less than 1.0 when the partially purified beef hypothalamic FSH-RF was incubated. These results strongly suggest that FSH secretion from the pituitary may be controlled mainly by the FSH-RF distinct from the LH-RH of *Schally et al.* (8).

References

1 *Bowers, C.Y.; Currie, B.L.; Johansson, K.N.G., and Folkers, K.:* Biological evidence that separate hypothalamic hormones release the follicle stimulating and luteinizing hormones. Biochem. biophys. Res. Commun. *50:* 20–26 (1973).

2 *Chang, J.K.; Sievertsson, H.; Bogentoft, C.; Currie, B.L.; Folkers, K., and Bowers, C.Y.:* Discovery of a new synthetic tetrapeptide having luteinizing-releasing hormone (LRH) activity. Biochem. biophys. Res. Commun. *40:* 409–413 (1971).

3 *Currie, B.L.; Johansson, K.N.G.; Folkers, K., and Bowers, C.Y.:* On the chemical existence and partial purification of the hypothalamic follicle stimulating hormone-releasing hormone. Biochem. biophys. Res. Commun. *50:* 14–19 (1973).

4 *Guillemin, R.:* Physiology and chemistry of the hypothalamic releasing factors for gonadotropins: a new approach to fertility control. Excerpta Medica International Congress, Series No. 234 b, p. 103 (1971).

5 *Igarashi, M. and McCann, S.M.:* A hypothalamic follicle stimulating hormone-releasing factor. *74:* 446–452 (1964).

6 *Johansson, K.N.G.; Currie, B.L.; Folkers, K., and Bowers, C.Y.:* Biosynthesis and evidence for the existence of the follicle stimulating hormone-releasing hormone. Biochem. biophys. Res. Commun. *50:* 8–13 (1973).

7 *Mittler, J.C. and Meites, J.: In vitro* stimulation of pituitary follicle-stimulating-hormone release by hypothalamic extract. Proc. Soc. exp. Biol. Med. *117:* 309–313 (1964).

8 *Schally, A.V.; Arimura, A.; Kastin, A.J.; Matsuo, H.; Baba, Y.; Redding, T.W.; Nair, R.M.G.; Debeljuk, L., and White, W.F.:* Gonadotropin-releasing hormone: one polypeptide regulates secretion of luteinizing and follicle-stimulating hormones. Science *173:* 1036–1938 (1971).

Authors' address: Prof. *M. Igarashi*, Dr. *K. Taya* and Dr. *J. Ishikawa*, Department of Obstetrics and Gynecology, School of Medicine, Gunma University, *Maebashi, Gunmaken, 371* (Japan)

Psychoneuroendocrinology. Workshop Conf. Int. Soc. Psychoneuroendocrinology, Mieken 1973, pp. 187–197 (Karger, Basel 1974)

Synthesis and Structure-Function Relationship of LH-RH

N. Yanaihara, C. Yanaihara, T. Hashimoto, N. Sakura, K. Miyazeki and T. Kaneko

Laboratory of Bioorganic Chemistry, Shizuoka College of Pharmacy, Shizuoka, and First Department of Internal Medicine, University of Tokyo, Faculty of Medicine, Tokyo

Introduction

One of the most elegant achievements of works by *Baba et al.* (1); *Matsuo et al.* (5); *Schally et al.* (10, 12) on hypothalamus-releasing hormones was the isolation and determination of the structure of porcine luteinizing hormone-releasing hormone (LH-RH). The structure of this releasing hormone was shown to be a decapeptide amide (fig. 1). Shortly after this structure had been reported, *Burgus et al.* (3) also succeeded in the isolation of LH-RH from ovine

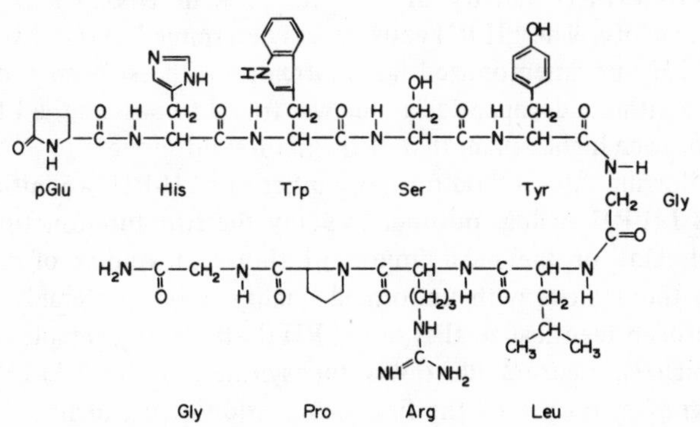

Fig. 1. Primary structure of LH-RH.

hypothalami and determined the structure as the same decapeptide amide. This decapeptide amide also possesses follicle stimulating hormone-releasing hormone (FSH-RH) activity (11).

After the structure of LH-RH was confirmed (1, 5, 10), this releasing hormone had attracted tremendous attention, not only of physiologists but also of peptide chemists, because of its biological importance. Since natural LH-RH can only be obtained in extremely small amounts, synthesis of this hormone was indispensable for further detailed investigation of its biological properties. We had, therefore, immediately started the synthesis of LH-RH by the conventional method for peptide synthesis in order to obtain enough amount of pure LH-RH (14). Using the established method for the synthesis of LH-RH, we prepared various LH-RH analogs for studies on structure-function relationship of LH-RH.

Synthesis and Structure-Function Relationship of LH-RH

The decapeptide amide was synthesized essentially by the stepwise chain elongation starting from the *C*-terminal prolylglycine dipeptide amide. Each of the constituent amino acids was introduced by the active ester method, except seryltyrosine (position 4 and 5) and histidine (position 2) which were coupled by the azide method. The carbobenzoxy group at the *N*-termini of the intermediates was removed first by HBr in acetic acid and, after introduction of seryltyrosine residues, by catalytic hydrogenation. Catalytic hydrogenation removed nitro group on arginine residue at position 8 at the same time. Carbobenzoxy nonapeptide amide and the final product, decapeptide amide, were purified by column chromatography on CM-Sephadex.

Our synthetic LH-RH preparation was compared with pure natural LH-RH in regard to LH-RH activity in cooperation with Drs. *Schally and Arimura,* Tulane University. The LH-RH activity was determined *in vivo* by stimulation of release of LH in ovariectomized rats pretreated with estrogen and progesterone (12). Our synthetic decapeptide amide was found to possess the LH-RH activity, the same or even higher than, that of the natural hormone.

By following this method for the synthesis of LH-RH, we initiated syntheses of various LH-RH analogs in order to study the structure-function relationship of LH-RH. One of the most important things in studies of this kind is to synthesize the analogs with uniformed quality, most preferably by using synthetic approach identical to that of LH-RH itself. As an example of syntheses of LH-RH analogs, figure 2 illustrates the synthesis of (Leu^3)-LH-RH and the chromatographic profile of the final purification of the analog is shown in figure 3. The LH-RH activities of all the synthetic analogs and fragments were compared with that of the natural pure LH-RH preparation in *Schally*'s laboratory.

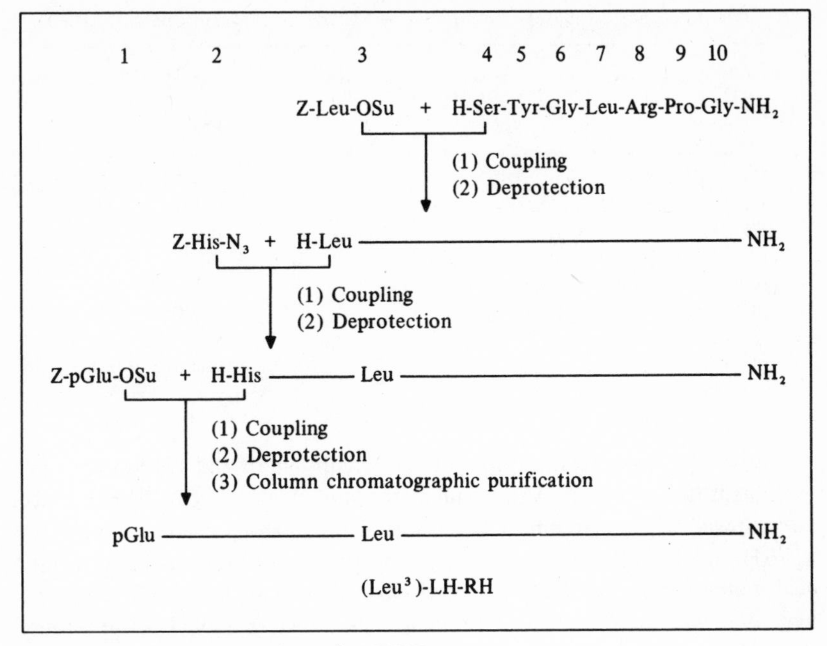

Fig. 2. Route for synthesis of (Leu^3)-LH-RH.

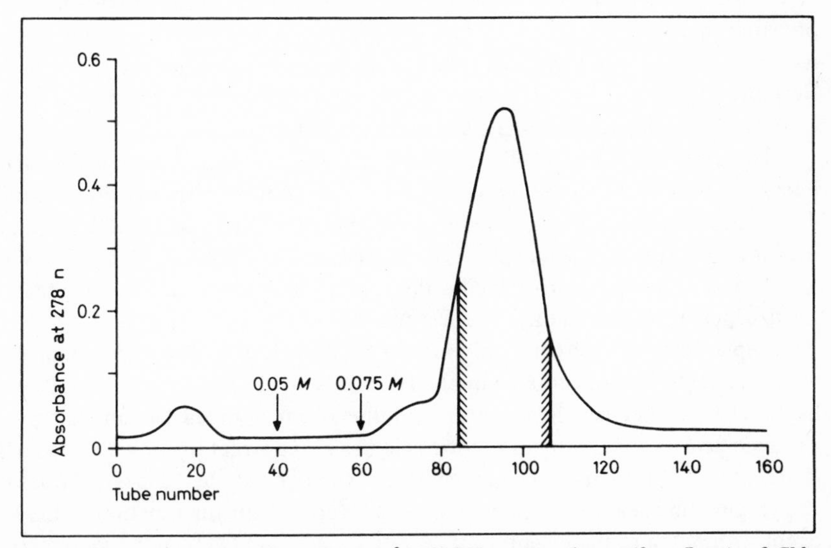

Fig. 3. Chromatography of crude (Leu^3)-LH-RH on a column (3 × 7 cm) of CM-Sephadex C-25 using NH_4OAc buffer (pH 6.5) as an eluent. Fractions of 17 g were collected.

Table I. Potency estimates of synthetic LH-RH fragments against natural LH-RH

Sample	LH-RH activity with 95-percent confidence limits, %
Natural LH-RH	Accepted as 100
Des-pGlu^{1}-LH-RH	< 0.002
Des-pGlu^{1}-des-His^{2}-LH-RH	Inactive
pGlu-His-Trp-Ser-Tyr-Gly-Leu-Arg-Pro-Gly-OH	0.078
pGlu-His-Trp-Ser-Tyr-Gly-Leu-OH	0.00097
pGlu-His-Trp-Ser-Tyr-Gly-OH	0.00054
pGlu-His-Trp-OH	Inactive, < 0.0004
pGlu-His-Trp-NH$_2$	Inactive, < 0.001

First of all, the relationship between chain length and LH-RH activity was examined. One amino acid shortening from the *N*-terminus of LH-RH molecule, i.e. removal of pyroglutamic acid residue, already caused nearly complete loss of LH-RH and FSH-RH activities, indicating the importance of the pyroglutamic acid moiety for the activities (14). For chain shortening from the *C*-terminus, we have prepared the *C*-terminal free analogs as shown in table I. Even removal of the amide group at the *C*-terminus resulted in considerable decrease in the activity (0.077 %). Three other shorter analogs were essentially inactive. On the other hand, *C*-terminus shortened analogs, with amide group at the carboxyl end, were prepared mainly by *Rivier et al.* (9). Removal of the *C*-terminal two amino acid residues, that is, prolylglycine, resulted in a drastic decrease of the activity. Recently, *Fujino et al.* (4) reported that des-Gly^{10}-LH-RH-ethylamide shows LH-RH activity higher than that of natural LH-RH.

The very low activity of the *C*-terminus free analogs is presumably attributed to that of the negative charge of its *C*-terminal carboxyl group which disturbs the molecule in constructing proper conformation as LH-RH or to that the binding ability to the receptor is decreased.

The next subject is to evaluate the role of each constituent amino acid of the decapeptide amide in physiological functions.

Replacement of pyroglutamic acid in position 1 with glutamic acid, proline or orotic acid residue, caused a marked decrease of LH-RH activity (16). Proline has a five-membered ring but it is not the five-membered lactam that pyroglutamic acid residue has. And orotic acid has a ring structure amide the same as lactam, but it is five- and not six-membered. The five-membered lactam structure of pyroglutamic acid in position 1 seems to be crucial for the function of LH-RH or for its binding to the receptor.

Substitution of histidine in position 2 also decreased the activity to a large extent. This implies that the histidine residue may be essential for the biological function of this hormone. Among our histidine-substituted analogs, only (Phe^2)-LH-RH showed fairly high activity, i.e. 1.4 %. This seems to be because phenylalanine residue possesses the same hydrophobic binding ability as histidine. Therefore, not only the acid-base property of the imidazole in histidine residue, but also aromatic properties of the structure in position 2 may be responsible for the LH-RH activity. Histidine-deleting analog, des-His^2-LH-RH, was reported inactive (6).

When tryptophan residue in position 3 was replaced by other amino acid residue, again the activity was markedly decreased (17). (Leu^3)-LH-RH was substantially inactive in doses as high as 5–25 μg, while (Phe^3)-LH-RH showed 0.43 % of the activity of natural LH-RH. The low but definite activity of this analog suggests that aromatic properties of the benzene structure in position 3 may, in part, substitute for the indole residue of tryptophan. Transposed analog in position 2 and 3, (Trp^2)(His^3)-LH-RH, and tryptophan-deleting analog, des-Trp^3-LH-RH, had extremely low activity, suggesting that conformational changes of LH-RH molecule have a marked effect on the LH-RH activity and that tryptophan-histidine transposition, or deletion of histidine or tryptophan

Table II. Potency estimates of synthetic analogs modified at position 1, 2, 3, 4 or 5 against natural LH-RH

Sample	LH-RH activity with 95-percent confidence limits, %
Natural LH-RH	Accepted as 100
(Pro^1)-LH-RH	0.0088 (0.0056–0.0154)
(Glu^1)-LH-RH	5.7 (2.66–23.8)
(Orotic $acid^1$)-LH-RH	0.005 (0.0019–0.013)
(Ser^2)-LH-RH	0.032 (0.02–0.06)
(Leu^2)-LH-RH	0.18 (0.086–0.35)
(Gln^2)-LH-RH	0.01 (0.0075–0.014)
(Phe^2)-LH-RH	1.4 (0.03–4.58)
(Gly^2)-LH-RH	0.000062
(Leu^3)-LH-RH	Inactive
(Phe^3)-LH-RH	0.43 (no limits)
(Trp^2)(His^3)-LH-RH	0.00935 (0.0012–46.2)
Des-His^2-des-Trp^3-LH-RH	0.014 (0.00437–0.0628)
Des-Trp^3-LH-RH	0.0026 (0.0058–36.1)
(Ala^4)-LH-RH	8.6 (5.4–13.5)
(Phe^5)-LH-RH	44.5 (19.1–97)
(Ala^4)(Phe^5)-LH-RH	1.01 (0.23–15.8)
(Leu^5)-LH-RH	5.02 (3.8–7.5)
(Tyr^3)(Trp^5)-LH-RH	0.01 (0.04–3.3)

residue, may result in the shift or elimination of vital active or binding site of the LH-RH molecule.

Table II summarizes the potencies of synthetic analogs modified in position 1, 2, 3, 4 or 5, as compared with natural LH-RH. The results with LH-RH analogs which were modified in position 1, 2 or 3 definitely indicate that pyroglutamic acid in position 1, histidine in position 2 and tryptophan in position 3, play most important roles for biological function of LH-RH. Particularly, almost complete loss of LH-RH activity in histidine or tryptophan-deleting analog indicates the importance of histidine and tryptophan for the LH-RH activity. Chemical modification of histidine or tryptophan in natural LH-RH was reported to cause inactivation of the hormone (2, 13).

Substitution in position 4 or 5 of LH-RH molecule seemed to be less serious for retention of the activity. (Ala^4)-LH-RH, (Phe^5)-LH-RH and (Ala^4)(Phe^5)-LH-RH still possess significant LH-RH activity. (Ala^4)(Phe^5)-LH-RH did not lose its biological activity in spite of the fact that acylation of natural LH-RH in pyridine leads to considerable inactivation. These findings virtually eliminate the hydroxyl groups of both serine and tyrosine residues as essential for the function of LH-RH.

Analogs in which glycine in position 6 of LH-RH molecule is replaced with *D*-alanine, β-alanine or sarcosine residue were assayed and the activities of these analogs were found to be 105, 43.5 and 14.7 %, respectively, of the activity of natural LH-RH. Since β-alanine has one more methylene group than glycine and sarcosine has steric hindrance, these results show that the structural changes at glycine in position 6 influence the LH-RH activity to a certain extent. The high activity of (*D*-Ala^6)-LH-RH may be due to resistance of the molecule against the action of endopeptidase, and steric effect of the side chain of *D*-alanine seems not to exist (table III).

Substitution of leucine in position 7 with other aliphatic amino acid gave little effect and the activities of such analogs remained significant. Activity of (Ile^7)-LH-RH was 102 % and the same as that of natural LH-RH, while (Val^7)-LH-RH was 9.7 % of natural LH-RH. Low activity of this compound may be due to the bulky side chain of valine.

In order to investigate the effect of basic guanido group of arginine residue in position 8, we synthesized (Gln^8)-LH-RH and (Leu^8)-LH-RH, which showed 4.88 and 0.76-percent LH-RH activity (15). The rather high activity of (Gln^8)-LH-RH, in which basic side chain of arginine is substituted with neutral amido group, is thought to be important. The result indicates strongly that the guanido group of arginine is not essential for the activity, but may contribute to the binding to the pituitary receptor.

On the other hand (Pro^8)(Arg^9)-LH-RH has extremely low activity. Transposition of the arginylproline sequence may cause the change of conformation of LH-RH molecule or shift of the binding site.

Table III. Potency estimates of synthetic analogs modified at position 6 against natural LH-RH

Sample	Structure at position 6	LH-RH activity with 95-percent confidence limits, %
Natural LH-RH	-NH-CH_2-CO-	Accepted as 100
(*D*-Ala[6])-LH-RH	-NH-CH-CO- (with CH_3 on CH)	105.0 (43.6–197.0)
(β-Ala[6])-LH-RH	-NH-CH_2-CH_2-CO-	43.5 (17.4–77.3)
(Sar[6])-LH-RH	-N(—CH_3)—CH_2-CO-	14.7 (1.2–38.5)

Table IV. Potency estimates of synthetic analogs modified at position 7, 8, 9 or 10 against natural LH-RH

Sample	LH-RH activity with 95-percent confidence limits, %
Natural LH-RH	Accepted as 100
(Ile[7])-LH-RH	102.00 (55.00–278.0)
(Val[7])-LH-RH	9.70 (2.40–21.3)
(Gln[8])-LH-RH	4.88 (2.50–13.0)
(Leu[8])-LH-RH	0.76 (0.19–8.8)
(Pro[8])(Arg[9])-LH-RH	0.02
(β-Ala[10])-LH-RH	3.40 (0.32–9.4)

Table V. FSH-RH activity of LH-RH analogs modified at position 8 as compared with natural LH-RH

Sample	FSH-RH activity *in vitro*	
	bioassay, %	RIA, %
Natural LH-RH	100.00[1]	100.00[1]
(Gln[8])-LH-RH	2.03	3.10
(Leu[8])-LH-RH	0.54	0.54

1 Accepted as 100 %.

LH-RH activity of (β-Ala[10])-LH-RH was 3.4 %. This analog has one more methylene group than LH-RH (table IV).

Table V shows FSH-RH activity of (Gln[8])-LH-RH and (Leu[8])-LH-RH. The FSH-RH activity of these analogs was decreased in a tendency similar to the case of the LH-RH activity.

Antigenic Determinants of LH-RH

In order to characterize the antigenic determinants of LH-RH to different antibodies, we prepared various LH-RH fragments and compared their cross-reactivities. The data obtained are summarized in table VI. Different antigenic determinants were demonstrated for three different antibodies which had been produced in rabbits by *Arimura and Schally.* In the case of No. 419, generated by PVP-adsorbed LH-RH, the major antigenic determinant located within tryptophyl-seryl-tyrosyl-glycyl-leucyl-arginyl-proline sequence, while leucyl-arginyl-prolyl-glycine amide is the determinant for the antiserum No. 742, generated by LH-RH conjugated with HSA at the *N*-terminus. The antigenic determinant of LH-RH for the antiserum No. 710, generated by LH-RH conjugated with BSA at the *C*-terminus, was rather similar to that in the case of No. 419, but histidine residue in position 2 seems more important for this cross-reaction. These immunological studies were performed in cooperation with *Arimura and Schally.*

On the basis of our studies and others so far performed, it can be suggested that the major functional part of LH-RH is the *N*-terminal tripeptide, pyro-glutamyl-histidyl-tryptophan, although the LH-RH activity of this tripeptide is extremely low, while the major binding site is the *C*-terminal heptapeptide amide. The antigenic determinant of LH-RH varies depending on the antisera used. Further studies on structure-function relationship of LH-RH are now under way.

MSH Release-Inhibiting Hormone

As a part of our studies on hypothalamic releasing and inhibiting hormones, we also performed syntheses of MSH release-inhibiting hormone (MIF). Prolyl-leucyl-glycine amide (MIF-I) and prolyl-histidyl-phenyl alanyl-arginyl-glycine amide (MIF-II) were isolated from bovine hypothalamic tissues and shown to possess MIF activity (7, 8). According to Dr. *Kastin,* VA Hospital of New Orleans, application to the pituitary of 100 μg MIF-II resulted in significant reduction in the melanocyte index of the frog, but no lightening of the toad or bullfrog.

Table VI. Cross-reactivity of LH-RH fragments and analogs with different anti-LH-RH antisera

	pGlu	His	Trp	Ser	Tyr	Gly	Leu	Arg	Pro	Gly	NH_2	Rabbit anti-LH-RH antisera		
												No. 419[1], %	No. 710[2], %	No. 742[3], %
1	—	—	—	—	(LH-RH)	—	—	—	—	—	NH_2	100.0	100	100
2	—	—	—	—	—	—	—	OH				0.0	+	0
3	—	—	—	—	—	—	OH					0.0	0	0
4	—	—	—	OH								0.0	0	0
5		—	—	—	—	—	—	OH				0.0	++	0
6			—	—	—	—	—	OH				0.0	0	0
7				—	—	—	—	OH				0.0	0	0
8							—	—	—	—	NH_2	0.0	0	20
9				—	—	—	—	—	—	—	NH_2	0.5	0	100
10	—	▭	▭	—	—	—	—	—	—	—	NH_2	0.5	0	100
11	—	▭	—	—	—	—	—	—	—	—	NH_2	10.4	0	100
12		—	—	—	—	—	—	—	—	—	NH_2	16.8	15	100
13			—	—	—	—	—	—	—	OH		0.8	0	0
14	—	—	—	—	—	—	—	—	—	—	OH	22.4 ~ 100.0	100	0
15			—	—	—	—	—	—	—	NH_2		13.0	0	0
16			—	—	—	—	—	—	—	—	NH_2	47.0 ~ 100.0	0	100
17	—	—	—	—	—	—	—	—	—	OH		47.0 ~ 100.0	100	0
18			—	—	—	—	—	—	NH_2			++	0	0

1 Generated by PVP-adsorbed LH-RH.
2 Generated by LH-RH conjugated with BSA at *C*-terminus.
3 Generated by LH-RH conjugated with HSA at *N*-terminus.

The histidyl-phenyl alanyl-arginyl-glycine sequence in MIF-II coincidently corresponds to the active core of MSH from which only tryptophyl residue is missing. With Dr. *Kastin,* we considered a possibility of overlooking a tryptophyl residue during the structural determination on a micro scale of MIF-II. For this reason, we synthesized prolyl-histidyl-phenyl alanyl-arginyl-tryptophyl-glycine amide as well as MIF-II. At a dose of 100 μg, MIF-II significantly lightened the lesioned *R. pipiens* as expected. The hexapeptide amide with tryptophan inserted was not active in the frog but did have a slight tendency to lighten the lesioned toad. Synthetic analog, pyroglutamyl-histidyl-phenyl alanyl-arginyl-glycine amide was inactive in all species.

References

1 *Baba, Y.; Matsuo, H., and Schally, A.V.:* Structure of the porcine LH- and FSH-releasing hormone. II. Confirmation of the proposed structure by conventional sequential analysis. Biochem. biophys. Res. Commun. *44:* 459–463 (1971).

2 *Baba, Y.; Arimura, A., and Schally, A.V.:* Studies on the properties of hypothalamic luteinizing hormone-releasing hormone. J. biol. Chem. *246:* 7581–7585 (1971).

3 *Burgus, R.; Butcher, M.; Amoss, M.; Ling, N.; Monahan, M.; Rivier, J.; Fellows, R.; Blackwell, R.; Vale, W., and Guillemin, R.:* Primary structure of the ovine hypothalamic luteinizing hormone-releasing factor (LRF). Proc. nat. Acad. Sci., Wash. *69:* 278–282 (1972).

4 *Fujino, M.; Kobayashi, S.; Obayashi, M.; Shinagawa, S.; Fukuda, T.; Kitada, C.; Nakayama, R., and Yamazaki, I.:* Structure-activity relationships in the *C*-terminal part of luteinizing hormone-releasing hormone (LH-RH). Biochem. biophys. Res. Commun. *49:* 863–869 (1972).

5 *Matsuo, H.; Baba, Y.; Nair, R.M.G.; Arimura, A., and Schally, A.V.:* Structure of the porcine LH- and FSH-releasing hormone. I. The proposed amino acid sequence. Biochem. biophys. Res. Commun. *43:* 1334–1339 (1971).

6 *Monahan, M.W.; Rivier, J.; Vale, W.; Guillemin, R., and Burgus, R.:* (Gly^2)LRF and des-His^2-LRF. The synthesis, purification, and characterization of two LRF analogs antagonistic to LRF. Biochem. biophys. Res. Commun. *47:* 551–556 (1972).

7 *Nair, R.M.G.; Kastin, A.J.,* and *Schally, A.V.:* Isolation and structure of hypothalamic MSH release-inhibiting hormone. Biochem. biophys. Res. Commun. *43:* 1376–1381 (1971).

8 *Nair, R.M.G.; Kastin, A.J., and Schally, A.V.:* Isolation and structure of another hypothalamic peptide possessing MSH-release inhibiting activity. Biochem. biophys. Res. Commun. *47:* 1420–1425 (1972).

9 *Rivier, J.; Vale, W.; Burgus, R.; Ling, N.; Amoss, M.; Blackwell, R., and Guillemin, R.:* Synthetic luteinizing hormone releasing factor analogs. Series of short-chain amide LRF homologs converging to the amino terminus. J. med. Chem. *16:* 545–549 (1973).

10 *Schally, A.V.; Arimura, A.; Baba, Y.; Nair, R.M.G.; Matsuo, H.; Redding, T.W., and Debeljuk, L.:* Isolation and properties of the FSH and LH-releasing hormone. Biochem. biophys. Res. Commun. *43:* 393–399 (1971).

11 *Schally, A.V.; Arimura, A.; Kastin, A.J.; Matsuo, H.; Baba, Y.; Redding, T.W.; Nair, R.M.G., and Debeljuk, L.:* Gonadotropin-releasing hormone. One polypeptide regulates secretion of luteinizing and follicle-stimulating hormones. Science *173:* 1036–1038 (1971).

12 *Schally, A.V.; Nair, R.M.G.; Redding, T.W., and Arimura, A.:* Isolation of the luteinizing hormone and follicle-stimulating hormone-releasing hormone from porcine hypothalami. J. biol. Chem. *246:* 7230–7236 (1971).
13 *Schally, A.V.; Baba, Y.; Redding, T.W.; Matsuo, H., and Arimura, A.:* Further studies on the enzymatic and chemical inactivation of hypothalamic FSH-RH. Neuroendocrinology *8:* 347–358 (1971).
14 *Yanaihara, N.; Yanaihara, C.; Sakagami, M.; Tsuji, K.; Hashimoto, T.; Kaneko, T.; Oka, H.; Schally, A.V.; Arimura, A., and Redding, T.W.:* Synthesis and biological evaluation of LH- and FSH-releasing hormone and its analogs. J. med. Chem. *16:* 373–377 (1973).
15 *Yanaihara, N.; Yanaihara, C.; Hashimoto, T.; Kenmochi, Y.; Kaneko, T.; Oka, H.; Saito, S.; Schally, A.V., and Arimura, A.:* Syntheses and LH- and FSH-RH activities of LH-RH analogs substituted at position 8. Biochem. biophys. Res. Commun. *49:* 1280–1291 (1972).
16 *Yanaihara, N.; Tsuji, K.; Yanaihara, C.; Hashimoto, T.; Kaneko, T.; Oka, H.; Arimura, A., and Schally, A.V.:* Syntheses and biological activities of analogs of luteinizing hormone-releasing hormone (LH-RH) substituted in position 1 or 2. Biochem. biophys. Res. Commun. *51:* 165–173 (1973).
17 *Yanaihara, N.; Hashimoto, T.; Yanaihara, C.; Tsuji, K.; Kenmochi, Y.; Ashizawa, F.; Kaneko, T.; Oka, H.; Saito, S.; Arimura, A., and Schally, A.V.:* Syntheses and biological evaluation of analogs of luteinizing hormone-releasing hormone (LH-RH) modified in position 2, 3, 4 or 5. Biochem. biophys. Res. Commun. *52:* 64–73 (1973).

Author's address: Dr. *N. Yanaihara,* Laboratory of Bioorganic Chemistry, Shizuoka College of Pharmacy, *Shizuoka* (Japan)

Psychoneuroendocrinology. Workshop Conf. Int. Soc. Psychoneuroendocrinology, Mieken 1973, pp. 198–205 (Karger, Basel 1974)

Studies on Long Feedback via Glucocorticoid and Short Feedback via ACTH

K. Takebe, M. Sakakura and A. Brodish

Second Department of Medicine, Hokkaido University School of Medicine, Sapporo, and Department of Physiology, College of Medicine, University of Cincinnati, Cincinnati, Ohio

Introduction

The suppression of plasma corticosterone by dexamethasone (DEX) under basal and stress conditions is well known. However, the site at which the blocking action of corticoids is exerted is still conflicting; some workers (1–4) have suggested that it is at the level of hypothalamus, while others (5–7) have reported that it is at the level of pituitary. Thus we tried to make sure of these points. Meanwhile, a few reports (8–11) suggest that the secretion of ACTH may be controlled by blood level of ACTH through a short feedback mechanism. However, there are very few reports which have made comparison of the ACTH effect to that of corticoid on hypothalamic-pituitary axis under basal and stress conditions. In order to also make sure of these points, we compared the effect of ACTH administration with that of corticoids on CRF activity in median eminence of hypophysectomized-adrenalectomized rats.

Materials and Methods

Male rats weighing 180–250 g: Sprague-Dawley (for experiments A and B) and Wistar strain (for experiment C) were used for these experiments. They were housed in animal quarters at a controlled temperature of 25 ± 1 °C with lighting regimen. Purina chow and water were provided *ad libitum*. The test materials containing hypothalamic CRF activity were obtained from rat hypothalamus. Immediately following decapitation, blocks of the median eminence were homogenized and extracted with ice-cold 0.01 *N* acetic acid saline (1 M.E./10 μl). After centrifugation at 3,000 r.p.m. for 15 min, an aliquot of 0.4 μl of the supernatant was used for assaying CRF activity in these experiments – *Hiroshige et al.* (12) – and the direct intrapituitary microinjection method with minor modifications developed by *Hiroshige et al.* (12) for CRF assay in all our experiments. The elevation of plasma corticosterone, following the direct injection of test materials into rat pituitaries, served as the index of response.

Experiment A. In order to determine whether or not the acting site of corticoid feedback was in hypothalamus, Sprague-Dawley male rats were injected various doses of DEX at 25 μg/100 g body weight, given at 22.00, and 0, 25, 50 and 75 μg/100 g body weight, 3 or 6 h prior to experiments, excepting the control rats. Control rats were administered saline instead of DEX. The crude extracts of median eminence were obtained from these rats after administration of DEX.

Experiment B. In order to know whether or not the acting site of corticoid feedback was in pituitary, Sprague-Dawley male rats bearing hypothalamic lesions were given different doses of DEX systematically i.p.; they were injected 25 μg/100 g body weight, given at 22.00, and 10, 25, 75 and 1,000 μg/100 g body weight, 4 h prior to experiments, excepting the control group. To avoid the complication of prior hypothalamic effects on pituitary sensitivity, rats bearing hypothalamic lesions were used as recipients for intrapituitary injection of hypothalamic extract. The same extract of median eminence was injected into the pituitary gland of lesioned rats at 3 p.m., and the increment of plasma corticosterone 20 min after the injection was estimated to determine the responsiveness of anterior pituitary.

Ventral hypothalamic lesions were made according to the methods developed by *Brodish* (13). The effectiveness of lesions was tested by subjecting these lesioned rats to ether stress, approximately 48 h after lesioning. Plasma corticosterone levels of less than 13 μg/100 ml, 20 min after the ether stress, indicated functional impairment of the hypothalamic-pituitary function.

Experiment C. In order to know whether or not there is a short feedback mechanism through ACTH, and to compare a long feedback via corticoids with short feedback via ACTH, hypophysectomized-adrenalectomized rats were injected synthetic ACTH-Zinc i.m. and DEX i.p. Two weeks after the operation of adrenalectomy-hypophysectomy, these animals received 0.1 mg/100 g body weight/day of ACTH-Z, because ACTH-Z of 0.05 mg or more, significantly suppressed CFR activity, as did 200 μg/100 g body weight/day of DEX. These experiments were started 6 h after the last injection. In order to compare the short feedback with long feedback mechanism under various stress conditions, we stressed hypophysectomized-adrenalectomized rats. Ether or ether-laparotomy with intestinal traction for 30 sec was used. For ether stress the animals were exposed to ether vapor for about 1 min and sacrificed 2 min after the start of stress. For ether and laparotomy stress the animals were subjected to stress of ether laparotomy for about 2 min with intestinal traction (30 sec) and sacrificed 2 min after the start of stress.

Results

The hypothalamic CRF activity of normal donor rats was significantly decreased by the administration of 50 μg/100 g body weight of DEX 6 h prior to experiments. When the dose of DEX was increased to 75 μg/100 g body weight, given at 22.00, and 3 or 6 h prior to decapitation, respectively, the CRF activity was markedly suppressed as compared with saline control group (fig. 1). These data suggest that the site of the inhibiting action of DEX may be above the hypothalamic level.

In pituitary responsiveness, pretreatment of recipient lesioned rats with 75 μg/100 g body weight of DEX caused a substantial and significant, although

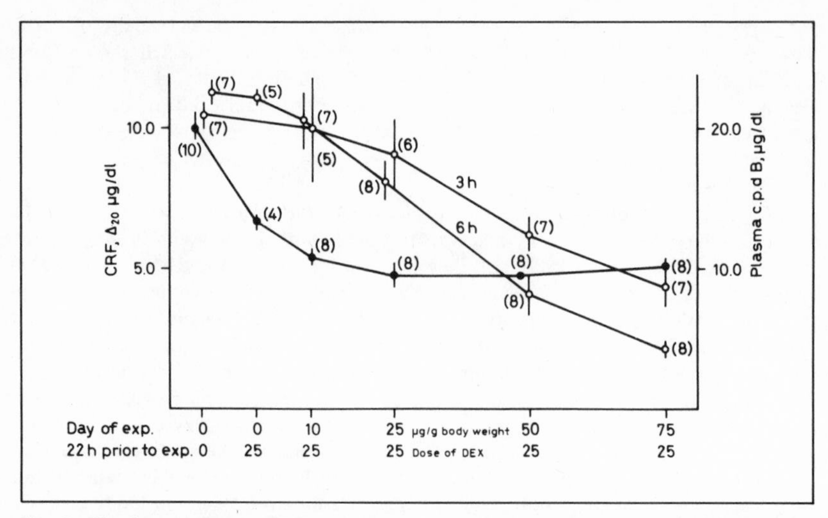

Fig. 1. The hypothalamic CRF activity and plasma corticosterone (c.p.d. B) levels in normal rats administered DEX. CRF activity = the level of c.p.d. B 20 min after administration of test materials – initial c.p.d. B level; • = c.p.d. B; ○ = CRF activity; figures in parentheses = numbers of rats; vertical bar indicates SE of the mean.

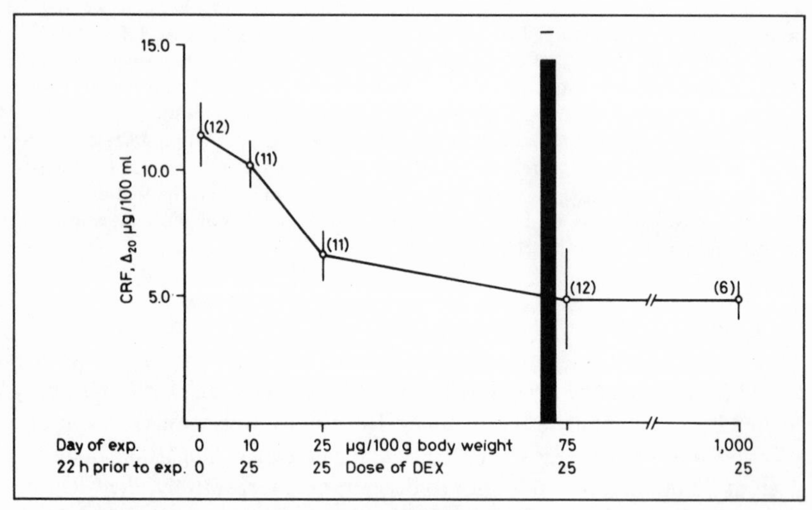

Fig. 2. Microinjection of the hypothalamic CRF into the anterior pituitary of hypothalamic lesioned rats. Black column shows the responsiveness in the normal rats administered DEX (25 μg, given at 22.00, and 75 μg/100 g body weight 4 h prior to experiment) to the same extract of median eminence. There is no significant difference in responsiveness between black column group and lesioned group (without DEX); figures in parentheses = numbers of rats; vertical bar indicates SE of the mean.

lesser, decline in responsiveness, whereas pretreatment with 10 μg/100 g body weight of DEX showed virtually no effect on pituitary sensitivity when compared to saline-pretreated recipient lesioned rats (fig. 2). The data of these experiments demonstrate that (a) both pituitary and hypothalamus are sites at which DEX exerts its effect; (b) relatively small doses of DEX are effective at the pituitary level. The hypothalamic CRF activity obtained from rats bearing hypophysectomy and adrenalectomy following ACTH-Z administration for 5 days, was more significantly suppressed than that of the saline group. Hypothalamic CRF activity of adrenalectomized-hypophysectomized rats injected ACTH-Z for 1 or 2 days was not inhibited, but CRF activity in rats administered DEX (200 μg/100 g body weight) for 1 day was significantly suppressed (fig. 3).

These results suggest that the secretion of ACTH may be controlled by blood level of ACTH via a short feedback mechanism, and the appearance time of ACTH short feedback is slower than that of corticoid long feedback.

Meanwhile, under basal conditions, CRF activity following a single injection of DEX was significantly ($p < 0.05$) more decreased than that of ACTH-Z group (for 5 days) (fig. 3).

However, CRF activity of hypophysectomized-adrenalectomized rats administered both ACTH-Z (5 days) and DEX (1 day) was not different from that of

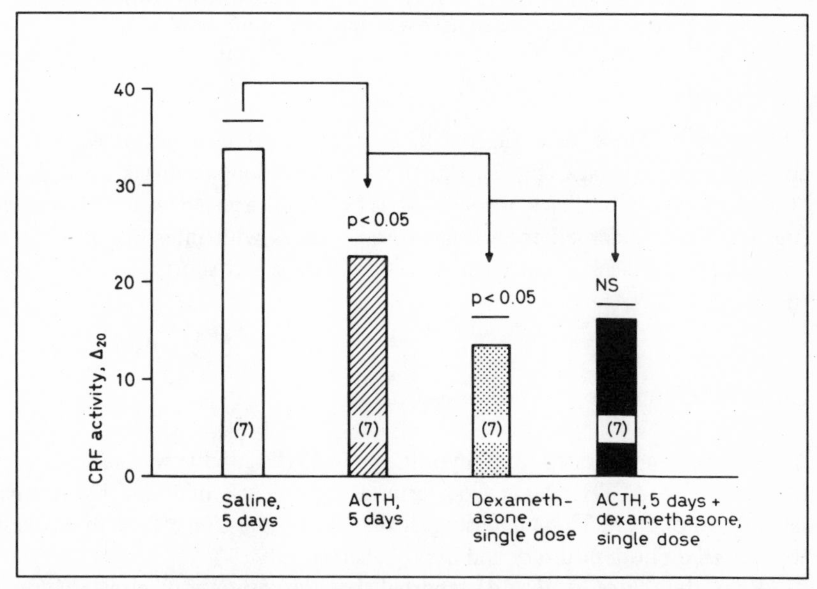

Fig. 3. Effect of ACTH-Z (5 days), DEX (1 day) and combined treatment of ACTH-Z (5 days) and DEX (1 day) on CRF activity in the median eminence under control conditions; figures in parentheses = numbers of rats; horizontal bar indicates SE of the mean.

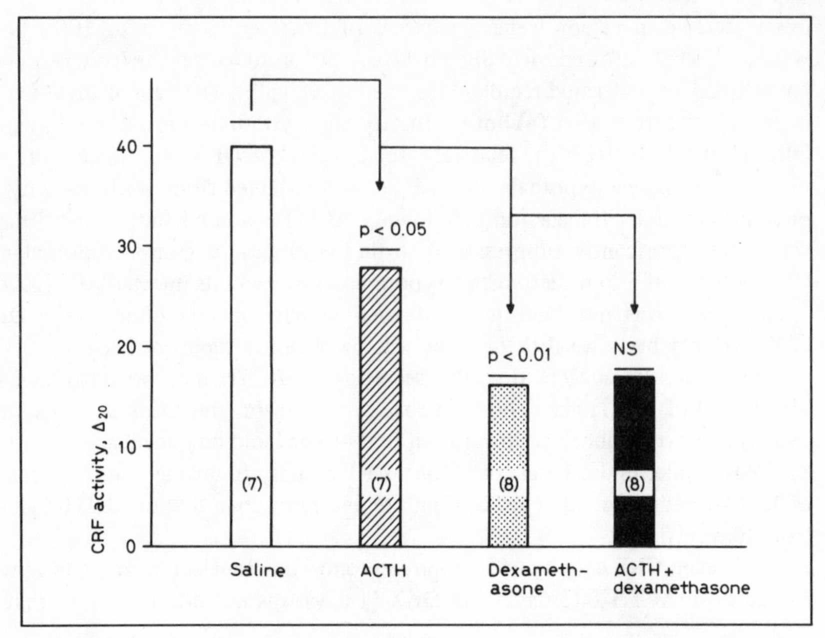

Fig. 4. Effect of ACTH-Z (5 days), DEX (1 day) and combined treatment of ACTH-Z (5 days) and DEX (1 day) on CRF activity in median eminence under ether stress; figures in parentheses = numbers of rats; horizontal bar indicates SE of the mean.

DEX (1 day). These data suggest that ACTH short feedback is weaker than corticoid long feedback (fig. 3). Under ether stress, long feedback through DEX (1 day) and short feedback through ACTH (5 days) exerted clearly their effects (fig. 4). While under ether and laparotomy stress with intestinal traction for 30 sec, long feedback mechanism was still effective but short feedback was not (fig. 5).

Discussion

It was demonstrated that hypothalamic CRF activity was suppressed by DEX and also ACTH release from pituitary gland was inhibited by relatively small doses of DEX. These data suggest that the suppression effects of corticoids are exerted on both pituitary and hypothalamus.

Recently, *Yates et al.* (14) reported that the response of pituitary-adrenocortical system to CRF was inhibited until the dose of DEX, given s.c. as pretreatment, reached a value of approximately 5 μg/100 g body weight.

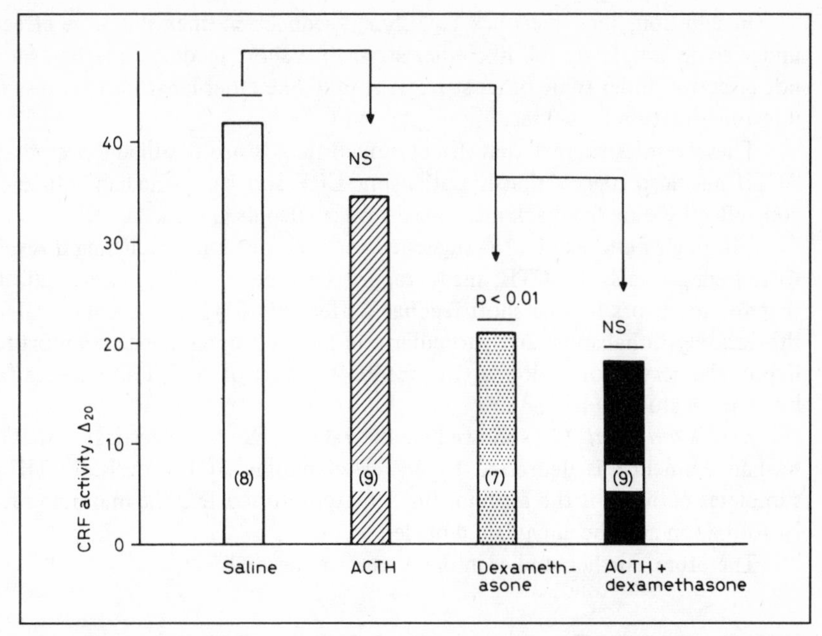

Fig. 5. Effect of ACTH-Z (5 days), DEX (1 day) and combined treatment of ACTH-Z (5 days) and DEX (1 day) on CRF activity in the median eminence under ether and laparotomy with intestinal traction-stress; figures in parentheses = numbers of rats; horizontal bar indicates SE of the mean.

In this study (experiment B), the response of pituitary to a crude extract of median eminence was gradually decreased by increased doses of DEX administered to lesioned rats until the dose reached 50 μg/100 g body weight and was maximally effective for the prevention of acute ACTH release from pituitary. The reason for this discrepancy in the dose of DEX between our data and that of *Yates et al.* (14) is not clear, but may be due to the difference in methods.

Motta et al. (8) found that permanent ACTH implants into the median eminence were effective in significantly suppressing blood corticosterone levels, but ACTH implants into other brain regions or its pituitary were ineffective. Recently *Motta et al.* (9), *Seiden and Brodish* (10) and *Takebe et al.* (11) reported that the increase of hypothalamic CRF activity induced by hypophysectomy and adrenalectomy was prevented following administration of ACTH.

In our data, the CRF activity following ACTH-Z administration for 5 days was significantly suppressed ($p < 0.05$) and the treatment with a single dose of DEX markedly inhibited the hypothalamic CRF activity more than that with ACTH-Z for 5 days.

In addition, long feedback and short feedback mechanisms were effective under some simple stimuli like ether stress, but short feedback mechanism was not effective under some other stronger stimuli like ether-laparotomy stress with intestinal traction for 30 sec.

These results suggest that effects of corticoids are dominant over that of ACTH and also suggest that hypothalamic CRF activity in median eminence is controlled by dual feedback mechanisms via corticoids and via ACTH.

Although *Fand et al.* (15) suggested that the pituitary itself might respond to changing levels of ACTH, many reports have been interpreted as indicating that the receptors for the short feedback effect of ACTH are mainly located in the basal hypothalamus, and particularly in the median eminence. Support for a hypothalamic site of action of the feedback effect of ACTH also comes from histological studies (16–18).

Kawakami et al. (19) showed in the rabbit that the electrical activity of median eminence is depressed by increased plasma ACTH levels. ACTH also stimulates activity of the septum, the somatomotor cortex, the midbrain reticular formation and the amygdaloid nuclei.

Therefore, further studies in these areas are needed.

Conclusion

By direct microinjection into the anterior pituitary of lesioned rats, we demonstrated that the site of corticoid feedback action is both in the hypothalamus and pituitary. However, it is not known which site plays the main role in suppression by corticoids. Meanwhile, the short feedback effect of ACTH was observed in the median eminence of hypophysectomized-adrenalectomized rats under specified conditions, but it is difficult to know whether or not the ACTH short feedback is observed under physiological conditions. Therefore, the CRF in the median eminence is under a dual feedback control via corticoids and ACTH.

It can be said, however, that the effect of corticoids is dominant over that of ACTH.

References

1 *Egdahl, R.H.:* The acute effects of steroid administration on pituitary adrenal secretion in the dog. J. clin. Invest. *43:* 2178–2183 (1964).

2 *Stark, E.; Grevai, A.; Acs, Zs.; Szalay, K.Sz., and Varga, B.:* The site of the blocking action of dexamethasone on ACTH secretion: *in vivo* and *in vitro* studies. Neuroendocrinology *3:* 275–284 (1968).

3 *Hedge, G.A. and Smelik, P.G.:* The action of dexamethasone and vasopressin on hypothalamic CRF production and release. Neuroendocrinology *4:* 242–253 (1969).

4 *Bohus, B. and Strashimirou, D.:* Localization and specificity of corticosteroid 'feedback receptors' at the hypothalamo-hypophyseal level comparative effects of various steroids implanted in the median eminence or the anterior pituitary of the rat. Neuroendocrinology *6:* 197–209 (1970).
5 *Kendall, J.W. and Allen, C.:* Studies on the glucocorticoid feedback control of ACTH secretion. Endocrinology, Springfield *82:* 397–405 (1968).
6 *Arimura, A.; Saito, T., and Schally, A.V.:* Assays for corticotropin-releasing factor (CRF) using rats treated with morphine, chlorpromazine, dexamethasone and nembutal. Endocrinology, Springfield *81:* 235–245 (1967).
7 *Russell, S.M.; Dhariwal, A.P.S.; McCann, S.M., and Yates, F.E.:* Inhibition by dexamethasone of the *in vivo* pituitary response to corticotropin-releasing factor (CRF). Endocrinology, Springfield *35:* 512–521 (1969).
8 *Motta, M.; Mangili, G., and Martini, L.:* A short feedback loop in the control of ACTH secretion. Endocrinology, Springfield *77:* 392–395 (1965).
9 *Motta, M.; Fraschini, F.; Piva, F., and Martini, L.:* Hypothalamic and extrahypothalamic mechanisms controlling adrenocorticotrophin secretion; in The investigation of hypothalamic-pituitary-adrenal function, pp. 3–18 (Cambridge Univ. Press, 1968).
10 *Seiden, G. and Brodish, A.:* Physiological evidence for short-loop feedback effects of ACTH on hypothalamic CRF. Neuroendocrinology *8:* 154–164 (1971).
11 *Takebe, K.; Sakakura, M., and Mashimo, K.:* Continuance of diurnal rhythmicity of CRF activity in hypophysectomized rats. Endocrinology, Springfield *9:* 1515–1520 (1972).
12 *Hiroshige, T.; Kunita, H.; Yoshimura, Y., and Itoh, S.:* An assay method for corticotropin-releasing activity microinjection in the rat. Jap. J. Physiol. *18:* 179–189 (1968).
13 *Brodish, A.:* Diffuse hypothalamic system for the regulation of ACTH secretion. Endocrinology, Springfield *73:* 727–731 (1963).
14 *Yates, F.E.; Russell, S.M.; Dallman, M.F.; Hedge, G.A.; McCann, S.M., and Dhariwal, A.P.S.:* Potentiation by vasopressin of corticotropin-release induced by corticotropin-releasing factor. Endocrinology, Springfield *88:* 3–15 (1971).
15 *Fand, S.B.; Ehmann, C.W., and Anderson, J.R.:* Aspects of ACTH release indicated by histochemical studies of human anterior pituitary in organ culture. Program of the Forty-Eighth Meeting of the Endocrine Society, p. 60 (1966).
16 *Castor, C.W.; Baker, B.L.; Ingle, D.J., and Li, C.H.:* Effect of treatment with ACTH or cortisone on anatomy of the brain. Proc. Soc. exp. Biol. Med. *76:* 353–357 (1951).
17 *Mühlen, K. und Ockenfels, H.:* Morphologische Veränderung in Diencephalon und Telencephalon nach Störungen des Regelkreises Adenohypophyse-Nebennierende. I. Ergebnisse beim Meerschweinchen nach Verabreichung vor naturlichen und synthetischen ACTH. Z. Zellforsch., Abt. Histochem. *85:* 124–144 (1968).
18 *Aleksanyan, Z.A.:* Role of the pituitary in generation of the slow electrical activity ot the hypothalamus in the frog. Bull. exp. Biol. Med., Moscow *62:* 854–856 (1967).
19 *Kawakami, M.; Koshino, T., and Hattori, Y.:* Changes in EEG of the hypothalamus and limbic system after administration of ACTH, Su-4885 and ACH in rabbits with special reference to neurohumoral feedback regulation of pituitary-adrenal system. Jap. J. Physiol. *16:* 551–569 (1966).

Authors' addresses: Dr. *K. Takebe* and Dr. *M. Sakakura,* Second Department of Medicine, Hokkaido University School of Medicine, *Sapporo* (Japan); Dr. *A. Brodish,* Department of Physiology, College of Medicine, University of Cincinnati, *Cincinnati, OH 45219* (USA)

Psychoneuroendocrinology. Workshop Conf. Int. Soc. Psychoneuroendocrinology, Mieken 1973, pp. 206–213 (Karger, Basel 1974)

Hormonal Factors Regulating Brain Development[1]

Paola S. Timiras and Arlene S. Luckock

Department of Physiology-Anatomy, University of California, Berkeley, Calif.

The regulatory role of several hormones in the development of the central nervous system (CNS) has been well established clinically and experimentally. For example, studies in man and in animals have demonstrated that thyroid hormones are necessary at an early age for normal maturation of the brain (10, 34); that, in a number of animal species, sex steroids are responsible for the differentiation into female or male type limbic and hypothalamic centers controlling gonadotropin secretion (2, 15, 27); and that adrenal steroids *in vivo* and *in vitro* influence the proliferation of glial cells and thereby accelerate or delay brain maturation depending on the dose of the hormone and the age of the animal or tissue at the time of its administration (31, 33). On the basis of these findings, it has been suggested that hormones have 'organizing' actions on CNS development, and that such actions become manifest at specific ages when the organism as a whole, or the CNS as a selected system, grow most rapidly and are particularly vulnerable to internal and external environmental factors (7, 9, 27, 28, 32). Thus, the concept of 'critical periods', originally formulated by embryologists to designate the early circumscribed periods of organ or tissue differentiation, can be extended to include a number of developmental stages occurring after organogenesis and culminating in the morphologic, functional and chemical maturation of the organ or system considered. On this basis, hormones can be viewed as exerting complex organizing actions at critical periods, the effects of which depend on the hormone and the brain structure studied, the timetable of development characteristic for a chosen species or system, and the functional or behavioral pattern tested.

With specific respect to the role of thyroid hormones, the critical age at which these hormones exert their organizing action on the brain and, in particular, on the cerebral cortex, generally encompasses the late fetal and the neonatal periods in humans and, in the rat, can be circumscribed precisely from the 10th

1 This investigation was supported by NIH-HD-101 and NSF GB-28202.

to the 15th postnatal days (10, 14). In rats, we know that when thyroid hormones are lacking, or insufficient during this critical period, the brain fails to develop normally in terms of structure (11, 21, 24), function (3, 16, 20, 23), and chemistry (1, 4, 6, 12), and, conversely, that thyroid hormones administered to the hypothyroid animal within this critical age period, restore normal brain maturation and assure functional competence at later ages. Once the critical age has passed, administration of thyroid hormones, even in large doses, fails to prevent the severe impairment in physical and mental development that inevitably results when hypothyroidism in early infancy remains untreated. Critical periods during pre- and postnatal development also have been identified for other hormones besides those of the thyroid gland, as well as for a variety of environmental influences; thus, the future potential of any given organ, tissue or function, in addition to genetic factors, will depend on environmental inputs, and the interplay between the two becomes maximally effective at a specific critical time during ontogeny.

Even though the age dependency of hormonal actions on brain development has been substantiated by accumulating evidence, the mechanisms that underlie the increased sensitivity of the brain to hormones at critical periods remain, for the most part, unknown. It is the purpose of the present study to explore changes that occur with age in the thyroid-brain relationship which may explain, at least in part, the critical role of these hormones during specific periods of brain development. The experimental animal chosen is the rat, in which the critical age for the organizing effects of thyroid hormones on the brain has been ascertained to occur postnatally within a period circumscribed to a few days. A first hypothesis to be tested is whether the uptake of thyroxine (the major thyroid hormone) by the brain is greater during the critical age period when the hormone is most effective, and whether it is greater in the brain than in other organs not depending on thyroxine for maturation at this time. A second hypothesis to be tested is whether the age-dependent effects of thyroxine are due to subtle changes occurring in selected cell components inasmuch as several developmental changes take place in the brain and particularly in the cerebral hemispheres at this period; indeed, days 10–15 postnatally represent a period of considerable structural (i.e., dendritic proliferation and synaptogenesis), functional (i.e., eye-opening) and biochemical (i.e., rapid myelinogenesis) activity in the rat.

Comparison of the Developmental Pattern of Thyroxine Uptake in the Brain and Other Tissues

The developmental pattern of thyroxine uptake in the cerebral hemisphere (the brain area known to be profoundly affected by thyroid hormones during the first 3 weeks of postnatal life in the rat) was compared with that in liver, which responds to thyroxine at all ages, and spleen, which is unresponsive to thyroxine at all ages. Days 6, 12 and 22 were selected for investigation in all experiments: day 12 postnatally is considered to be representative of the critical period of thyroxine's effect on brain maturation, and days 6 and 22 were chosen to represent stages before and after the critical period, respectively.

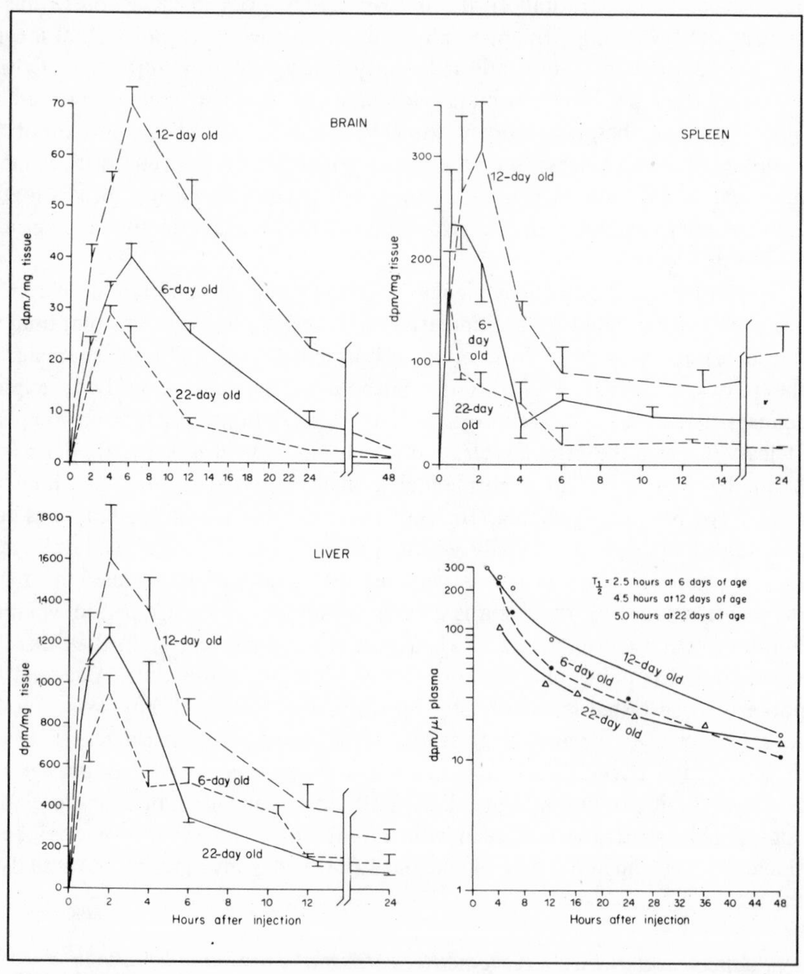

Fig. 1. Uptake of ^{14}C-thyroxine by brain, liver and spleen at three ages. Each animal was injected intraperitoneally with the labeled hormone (0.1 μc/g body weight). Each point represents the mean of three or more determinations, and standard errors are represented by bars. Lower right: disappearance of ^{14}C-thyroxine from plasma at three ages. Experimental conditions are as stated above. Each point represents the mean of three or more determinations. $T_{1/2}$ is measured from the early phase of the disappearance curve.

Female Long-Evans rats were injected intraperitoneally with 0.1 μc of ^{14}C-thyroxine/g body weight. At various time intervals following injection of the label, animals were killed by decapitation. The cerebral hemispheres, liver, and spleen were rapidly removed, and 10-percent homogenates made in 0.32 molar sucrose. Blood samples also were collected, and the plasma separated by centrifugation. Aliquots of the tissue homogenates and the plasma were counted in a liquid scintillation system in a Packard Tri-Carb Spectrometer and corrected for quenching. The identity of the labeled substance found in the brain was determined by butanol extraction and descending paper chromatography. When compared with known markers, the labeled substance was identified as thyroxine. Comparison of thyroxine uptake at the three ages studied indicates that the brain takes up a significantly greater amount of the label at day 12 than at day 6 or 22, whether the results are expressed in terms of tissue weight, as in figure 1, or in terms of protein. These results would suggest that the brain does indeed show an increased removal of thyroxine from the circulation during the critical period in which thyroxine is required for normal maturation. However, the accumulation of label by the liver and spleen is also significantly greater at day 12 than at the other two ages, as shown in figure 1. The only difference to be noted is that the 22-day-old liver takes up more of the hormone than does the 6-day-old liver, a situation which is reversed in brain and spleen. The results, therefore, point to a nonspecific increase in the uptake of thyroxine by all three organs studied at day 12. The half-life of the labeled hormone in the plasma increases steadily from day 6 to day 22, as shown in figure 1.

Developmental Changes in the Subcellular Distribution of Thyroxine in the Brain

The results presented above indicate that the immature brain does increase its uptake of thyroxine during this critical period, but that it is not unique among tissues in its pattern of thyroxine uptake. To test the possibility that developmental changes in the brain's response to thyroxine are the result of changes in the subcellular distribution of thyroxine during the course of brain maturation, subcellular fractions were prepared by the procedure of differential centrifugation, in which the 900 *g* pellet is defined as the nuclear fraction, the 11,000 *g* pellet as the crude mitochondrial fraction, the 105,000 *g* pellet as the microsomal fraction, and the 105,000 *g* supernatant as the soluble fraction.

In the 6-day-old brain (fig. 2), most of the label was found in the soluble fraction of the cell, followed by the mitochondrial, nuclear and microsomal fractions. At 12 days of age (fig. 2), the height of the critical period, a marked change occurs in the subcellular distribution of thyroxine, with the greater proportion of the label found in the mitochondrial fraction, followed by the nuclear, soluble and microsomal fractions. In the 22-day-old brain (fig. 2), the subcellular distribution of thyroxine again returns to the pattern previously observed in the 6-day-old brain, with most of the label being found in the soluble fraction, followed by the mitochondrial, nuclear and microsomal fractions.

The data in figure 2 have been expressed as d.p.m./g of tissue, and it is therefore conceivable that the differences observed could be related to changes occurring in the relative sizes of the various subcellular compartments with

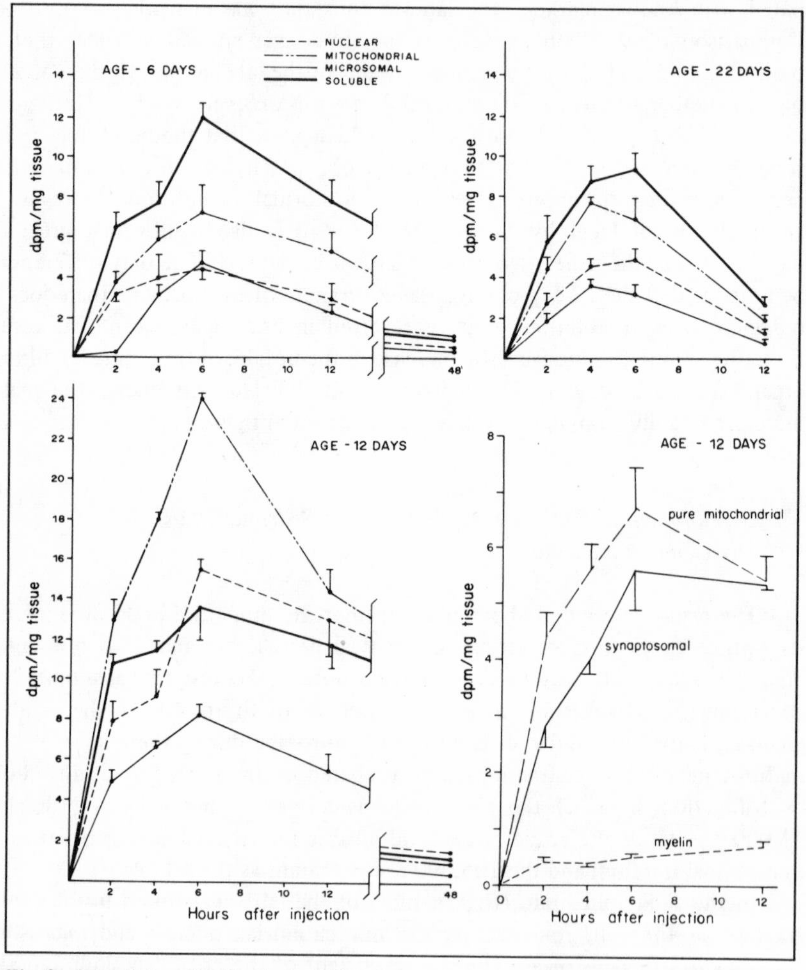

Fig. 2. Subcellular distribution of ^{14}C-thyroxine in the brain at three ages. Experimental conditions and representations are as in figure 1. Definitions of the subcellular fractions are given in the text.

development. In order to account for this possible difference, the results also were expressed as d.p.m./mg protein. When expressed in this manner, the soluble fraction has the highest levels of radioactivity at all three ages studied; however, there are significant increases in the labeling of all particulate fractions at day 12, especially the crude mitochondrial fraction. This fraction includes myelin, synaptosomes and mitochondria. In order to assess the relative contribution of each of these subfractions to the increased uptake of thyroxine by the crude mitochondrial fraction at day 12, the crude fraction was separated into the three subfractions by means of density gradient centrifugation. The subfractions were examined electron microscopically and found to be reasonably pure. When the radioactivity in the three subfractions was counted separately, it was found that most of the label in the crude mitochondrial fraction at day 12 is localized in the synaptosomal and mitochondrial compartments (fig. 2).

Discussion

Based on the results presented, it seems very probable that the mechanism by which the developing brain alters its response to thyroxine during the critical period may involve changes in thyroxine-binding sites at the level of subcellular membranes. Other investigators have reported the uptake of significant amounts of thyroxine and triiodothyronine by various tissues and subcellular components (8, 17, 18, 25). *Hillier* (19) has found that phospholipid membranes (liposomes) *in vitro* bind thyroxine in a manner similar to that of biological membranes. Thyroxine-phospholipid interactions also have been theoretically implicated in the mechanism of action of thyroxine (13). *Manuelidis* (22) has found a marked association of labeled thyroxine with cell membranes in autoradiographic studies. The experiments of *Valcana and Timiras* (29, 30) demonstrate that neonatal hypothyroidism results in a later-appearing deficiency of certain functions normally associated with cellular membranes. *Cragg* (5) has found abnormal membranous bodies in neuronal and glial cytoplasm and in synapses of hypothyroid rats, and has suggested that they may be the result of normal lipid synthesis in the absence of normal protein synthesis. Thus, since neonatal thyroid deficiency is associated with abnormal structure and function of membranes, an increased association of thyroxine with the particulate fractions of the cell during the critical period for thyroid hormone effects on cerebral development, as demonstrated in this investigation, is not unexpected.

In addition, since the mitochondria seem to undergo a marked increase in their affinity for thyroxine during the period when thyroxine most profoundly affects the developing brain, it seems possible, as other investigators have suggested (26), that the mitochondria may be intimately involved in the mechanism of action of thyroxine.

It may be concluded from these experiments that the developing brain increases its sensitivity to thyroxine during the critical period, both by the removal of more of the hormone from the circulation and by an alteration in thyroxine-binding sites on subcellular membranes.

References

1 *Balázs, R.; Cocks, W.A.; Eayrs, J.T., and Kovacs, S.:* Biochemical effects of thyroid hormones on the developing brain; in *Hamburgh and Barrington* Hormones in development, pp. 357–379 (Appleton-Century-Crofts, New York 1971).

2 *Barraclough, C.A.:* Hormones and the ontogenesis of pituitary regulating mechanisms; in *Sawyer and Gorski* Steroid hormones and brain function, pp. 149–159 (University of California Press, Berkeley 1971).

3 *Bradley, P.B.; Eayrs, J.T., and Schmalbach, K.:* The electroencephalogram of normal and hypothyroid rats. Electroenceph. clin. Neurophysiol. *12:* 467–477 (1960).

4 *Cocks, J.A.; Balázs, R., and Eayrs, J.T.:* The effect of thyroid hormones on the biochemical maturation of the rat brain. Biochem. J. *111:* 18P (1969).

5 *Cragg, B.G.:* Synapses and membranous bodies in experimental hypothyroidism. Brain Res. *18:* 297–307 (1970).

6 *Dalal, K.B.; Valcana, T.; Timiras, P.S., and Einstein, E.R.:* Regulatory role of thyroxine on myelinogenesis in the developing rat. Neurobiology *1:* 211–224 (1971).

7 *Denenberg, V.H.:* The development of behavior (Sinauer, Stamford, Conn. 1971).

8 *Dimino, M.J. and Hoch, F.L.:* Localization of endogenous and exogenous thyroid hormone in rat liver mitochondria. Fed. Proc. *31:* 213 (1972).

9 *Dobbing, J.:* Vulnerable periods in developing brain; in *Davison and Dobbing* Applied neurochemistry, pp. 287–316 (Blackwell, Oxford 1968).

10 *Eayrs, J.T.:* Thyroid and central nervous system development. The scientific basis of medicine annual reviews, pp. 317–339 (Athlone, London 1966).

11 *Eayrs, J.T.:* Thyroid and developing brain: anatomical and behavioral considerations; in *Hamburgh and Barrington* Hormones in development, pp. 345–355 (Appleton-Century-Crofts, New York 1971).

12 *Geel, S.E. and Timiras, P.S.:* The role of hormones in cerebral protein metabolism; in *Lajtha* Protein metabolism of the nervous system, pp. 335–354 (Plenum, New York 1970).

13 *Gruenstein, E. and Wynn, J.:* A molecular mechanism of action of thyroxin: modification of membrane phospholipid by iodine. J. theor. Biol. *26:* 343–363 (1970).

14 *Hamburgh, M:* The role of thyroid and growth hormones in neurogenesis; in *Moscona and Monroy* Current topics in developmental biology, pp. 109–148 (Academic Press, New York 1969).

15 *Harris, G.W.:* Sex hormones, brain development and brain function. Endocrinology, Springfield *75:* 627–648 (1964).

16 *Hatotani, N. and Timiras, P.S.:* Influence of thyroid function on the postnatal development of the transcallosal response in the rat. Neuroendocrinology *2:* 147–156 (1967).

17 *Hillier, A.P.:* The uptake and release of thyroxine and triiodothyronine by the perfused rat heart. J. Physiol., Lond. *199:* 151–160 (1968).

18 *Hillier, A.P.:* The uptake of thyroxine and triiodothyronine by perfused hearts. J. Physiol., Lond. *203:* 665–674 (1969).

19 *Hillier, A.P.:* The binding of thyroid hormones to phospholipid membranes. J. Physiol., Lond. *211:* 585–597 (1970).
20 *Lansing, R.W. and Trunnell, J.B.:* Electroencephalographic changes accompanying thyroid deficiency in man. J. clin. Endocrin. *23:* 470–480 (1963).
21 *Legrand, J.:* Influence de l'hypothyroidisme sur la maturation du cortex cérébelleux. C.R. Acad. Sci. *261:* 544–547 (1965).
22 *Manuelidis, L.:* Studies with electron microscopic autoradiography of thyroxine ^{125}I in organotypic cultures of the CNS. II. Sites of cellular localization of thyroxine ^{125}I. Yale J. Biol. Med. *45:* 501–518 (1972).
23 *Meisami, E.; Valcana, T., and Timiras, P.S.:* Effects of neonatal hypothyroidism on the development of brain excitability in the rat. Neuroendocrinology *6:* 160–167 (1970).
24 *Mitskevich, M.S. and Moskovkin, G.N.:* Some effects of thyroid hormone on the development of the central nervous system in early ontogenesis; in *Hamburgh and Barrington* Hormones in development, pp. 437–452 (Appleton-Century-Crofts, New York 1971).
25 *Schwartz, H.L.; Bernstein, G., and Oppenheimer, J.H.:* Effect of phenobarbital administration on the subcellular distribution of ^{125}I-thyroxine in rat liver: importance of microsomal binding. Endocrinology, Springfield *84:* 270–276 (1969).
26 *Sokoloff, L.:* Role of mitochondria in the stimulation of protein synthesis by thyroid hormones; in *San Pietro, Lamborg and Kenney* Regulatory mechanisms for protein synthesis in mammalian cells, pp. 345–367 (Academic Press, New York 1968).
27 *Timiras, P.S.:* Estrogens as 'organizers' of CNS function; in *Ford* Influence of hormones on the nervous system, pp. 242–254 (Karger, Basel 1971).
28 *Timiras, P.S.:* Developmental physiology and aging (Macmillan, New York 1972).
29 *Valcana, T. and Timiras, P.S.:* Effect of hypothyroidism on ionic metabolism and Na-K activated ATP phosphohydrolase activity in the developing rat brain. J. Neurochem. *16:* 935–943 (1969).
30 *Valcana, T. and Timiras, P.S.:* Effect of thyroid hormones on ionic metabolism of the developing rat brain; in *Hamburgh and Barrington* Hormones in development, pp. 453–463 (Appleton-Century-Crofts, New York 1971).
31 *Vernadakis, A.:* Hormonal factors in the proliferation of glial cells in culture; in *Ford* Influence of hormones on the nervous system, pp. 42–55 (Karger, Basel 1971).
32 *Vernadakis, A. and Timiras, P.S.:* Disorders of the nervous system; in *Assali* Pathophysiology of gestation, vol. 3, pp. 233–304 (Academic Press, New York 1972).
33 *Vernadakis, A. and Woodbury, D.M.:* Effects of cortisol on maturation of the central nervous system; in *Ford* Influence of hormones on the nervous system, pp. 85–97 (Karger, Basel 1971).
34 *Wilkins, L.:* The diagnosis and treatment of endocrine disorders in childhood and adolescence; 3rd ed. (Thomas, Springfield 1965).

Authors' address: Dr. *Paola S. Timiras* and Dr. *Arlene S. Luckock,* Department of Physiology-Anatomy, University of California, *Berkeley, CA 94720* (USA)

Psychoneuroendocrinology. Workshop Conf. Int. Soc. Psychoneuroendocrinology, Mieken 1973, pp. 214–220 (Karger, Basel 1974)

Studies on the Mechanism of Action of Cold in Producing an Increase of Thyroid Hormone Secretion

T. Yamada and T. Onaya

Department of Medicine, Institute of Adaptation Medicine, Shinshu University School of Medicine, Matsumoto

It is well known that exposure of animals to cold results in thyroid hypertrophy, enhanced thyroidal radioiodine uptake and increased thyroidal radioiodine release (1, 2). Therefore, it is generally felt that accelerated thyroid hormone production and release are the major factors responsible for the sustained chronic phase of cold adaptation, characterized by increased metabolism. The theories of the mechanism of action, which have been proposed for acute thyroidal response to cold, are very conflicting. For instance, *Brown-Grant* (3) has suggested that, during exposure to cold, an increased rate of peripheral utilization of thyroid hormone resulting in a lowered blood level of the hormone may act via the feedback mechanism to produce increased thyroid activity. *Bottari* (4) hypothesized that increased TSH secretion takes place through the central nervous mechanism and not through the pituitary-thyroid feedback mechanism. Furthermore, *Andersson et al.* (5, 6) have proposed that hypothalamic cells are sensitive to changes of temperature and that a low blood temperature stimulates such cells. Because of these uncertainties, we have attempted to study an acute thyroidal response to cold by estimating plasma labeled thyroid hormone and intrathyroidal colloid droplets.

Parameters of Thyroid Hormone Secretion

The thyroid hormone secretion was evaluated by two parameters, calculation of intrathyroidal colloid droplets (7) and measurement of plasma labeled thyroid hormone secreted from the prelabeled thyroid gland (8). Figure 1 indicates the difference in the time-response patterns of the two parameters occur-

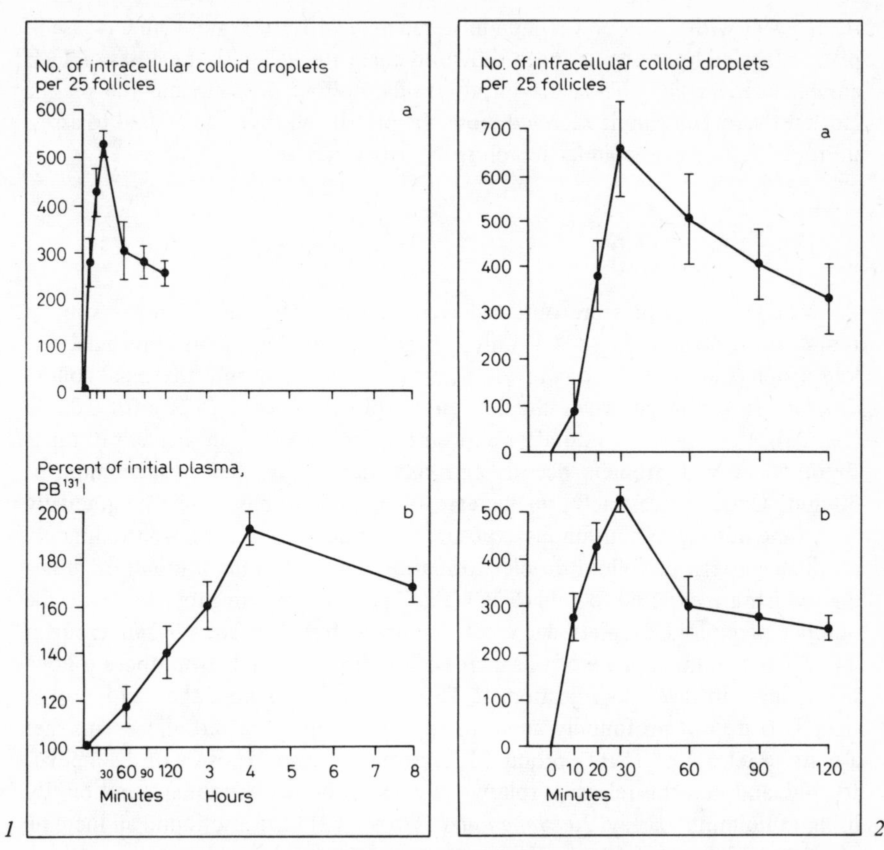

Fig. 1a and b. Comparison of two thyroid parameters after TSH (20 mU) administration: *a* indicates intrathyroidal colloid droplets after TSH injection: *b* indicates an increase of plasma labeled thyroid hormone secreted from the prelabeled thyroid gland.

Fig. 2a and b. An increase of intrathyroidal colloid droplets in response to cold is indicated: *a* indicates the thyroid from rats; *b* indicates the thyroid from the guinea pigs.

ring in response to administration of 20 mU of TSH to rats. As shown in figure 1a, an increase in intrathyroidal colloid droplets occurs within 10 min and reaches the maximum 30 min after the administration of TSH. This increase of intrathyroidal colloid droplets subsides rapidly thereafter. In contrast, a significant increase in plasma labeled hormone occurs later and reaches its maximum 4 h after injection of TSH (fig. 1b). After certifying this difference in their time response curves, a dose response curve is further compared after administration of graded doses of TSH or thyrotropin releasing hormone (TRH). It has been shown that the magnitude of an increase in intrathyroidal colloid droplets correlates well with doses of TSH administered *in vivo.* Furthermore, an increase of plasma labeled hormone has been shown to correlate well with the doses of TRH administered *in vivo.* Thus the intrathyroidal colloid droplets and the plasma labeled thyroid hormone secreted from the prelabeled thyroid are used to assess thyroidal response to cold in the following experiments.

Effect of Cold on Intrathyroidal Colloid Droplets

When the thyroids are obtained from the control rats which are kept at room temperature (27–28 °C), only a negligible number of intrathyroidal colloid droplets are found. In contrast, a marked increase in intrathyroidal colloid droplets is produced when the rats are exposed to cold (5 °C) for 30 min (fig. 2a). As early as 10 min after exposure to cold (5 °C), an increase of intrathyroidal colloid droplets occurs, and this increase reaches its maximum at 30 min. Quite interestingly, an increase of colloid droplets subsides gradually with time in spite of continuous exposure of the animals to cold. Approximately a similar pattern of colloid droplet formation is found in the thyroids of guinea pigs which are exposed to cold (5 °C) (fig. 2b). After reaching its maximum, the number of colloid droplets decreases gradually, but they are still numerous at 2 h. Since the initial phase of this increase is indistinguishable from those following a single intravenous injection of TSH or TRH, and since the factors other than TSH do not profoundly augment thyroid hormone secretion, it seems that a massive release of TSH is produced immediately after exposure of the animals to cold, and that this release is followed by a slow but supranormal secretion. By using radioimmunoassay, *Hershman and Pittman* (9) have also found an increase of plasma TSH within 10 min after exposure of the rats to cold.

Possible Increase of TRH Secretion in Response to Cold (fig. 3)

It is well established that hypothalamic lesions not only slow thyroidal radioiodine release but also block goiter formation produced by antithyroid drugs (10). *Knigge and Bierman* (11) and *Van Beugen and Van der Werff ten Bosch* (12) have suggested that hypothalamic lesions prevent an increase in thyroidal radioiodine release induced by cold. Thus, a possible role of the hypothalamus in eliciting thyroidal response to cold is studied by measuring plasma labeled thyroid hormone secretion from the prelabeled thyroid gland. Plasma labeled thyroid hormone increases progressively with time and a 1.8-fold increase is found at 4 h when the animals are exposed to cold. In cold-exposed animals with hypothalamic lesions, however, no significant increase in plasma labeled thyroid hormone is found throughout the experimental period.

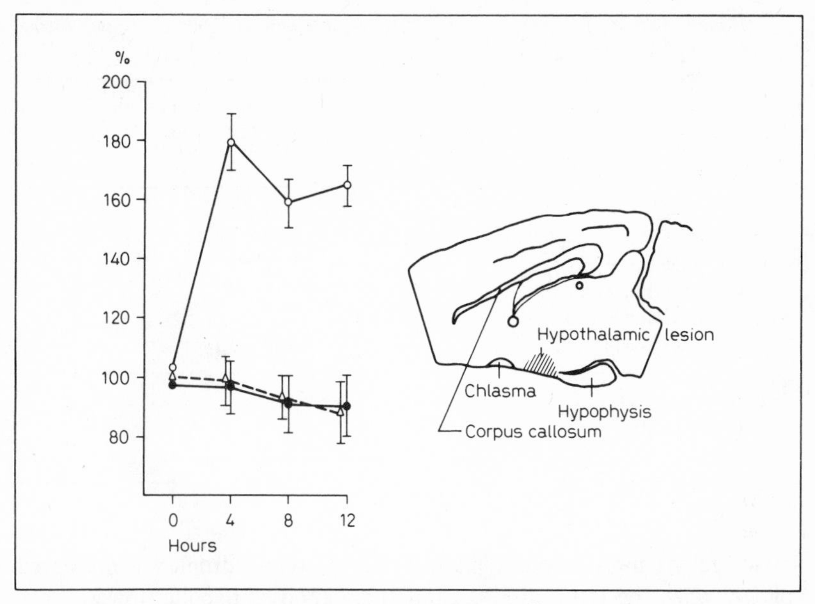

Fig. 3. Effect of hypothalamic lesion on plasma PB ^{131}I in cold environment. An increase of plasma labeled hormone is produced when intact guinea pig with labeled thyroid is exposed to cold. This response is blocked by hypothalamic lesion. ○ = Intact, cold (5 °C); ● = hypothalamic lesion, cold; △ = intact, room (25–27 °C).

Effect of Cerebellopontin Transection on Thyroidal Response to Cold

It is not known how the cold stimulus is transferred from the skin receptor to the hypothalamus where it manifests its stimulatory effect. As suggested by *Brown-Grant* (3), it may be considered that plasma thyroid hormone concentration is decreased because of increased peripheral utilization of thyroid hormone under cold environment, and this decrease of plasma thyroid hormone in turn activates hypothalamic TRH center. It is also considered that blood temperature may be decreased during cold exposure and this decrease of temperature may stimulate TRH release, since cooling of the hypothalamus stimulates the secretion of TSH (5, 6). However, these are not plausible explanations, since no significant decreases in plasma thyroid hormone or in blood temperature occur within 10 min after exposure of the animals to cold (13, 14). Finally, it may be that cold stimulation originating from the skin receptor is transferred to the hypothalamus via the spinal cord and this cold stimulus augments TRH secretion. To study this hypothesis, cerebellopontin transection is made without disrupting the blood supply to the central nervous system. As shown in table I, a significant increase of intrathyroidal colloid droplets is found after exposure of

Table I. Effect of midbrain transection on intracellular colloid droplet formation in guinea pigs exposed to cold

Group	No. of animals	No. of colloid droplets per 25 follicles	p values from group 1
1 Room	6	2 ± 1[1]	–
2 Cold	6	212 ± 17	$p < 0.001$
3 Transection + room	4	2 ± 1	–
4 Transection + cold	6	8 ± 5	–
5 Transection + TRF (1 μg) + room	3	203 ± 15	$p < 0.001$

1 Mean ± SE.
The thyroids are obtained 30 min after cold exposure.

intact guinea pigs to cold. This increase of colloid droplets is prevented completely when the transectioned animals are exposed to cold. However, a marked increase of intrathyroidal colloid droplets is found after intravenous injection of TRH in transectioned animals. This indicates that blood supply to the central nervous system is left intact, and that the pituitary can respond normally to TRH in transectioned animals. It is therefore concluded that the cold stimulus originating from the skin is transferred to the hypothalamus via the spinal cord and stimulates TRH secretion.

A Negative Dopaminergic Mechanism in the Hypothalamic Control of TRH Secretion in Response to Cold

Considerable evidence has recently accumulated that brain biogenic amines are involved in the release of certain hypothalamic hypophysiotropic hormones. *Grimm and Reichlin* (15) have recently demonstrated that norepinephrine, rather than dopamine, increases the release of ^{3}H-labeled TRH from mouse hypothalamus fragments *in vitro.* However, early studies *in vivo* on neurotransmitter control of thyroid function are contradictory in the rabbit (16) and in the rat (17). Therefore, effects of L-dopa and a dopamine-β-hydroxylase inhibitor on thyroidal response to cold are studied. Following the administration of L-dopa which causes an early increase in brain dopamine level (18), it is observed that acute thyroidal endocytotic response to cold is significantly inhibited (table II). Diethyldithiocarbamate (DDC), a dopamine-β-hydroxylase inhibitor which blocks the conversion of dopamine to norepinephrine (19), also significantly

Table II. Effects of L-dopa or/and diethyldithiocarbamate (DDC) on endocytotic response to cold

Groups	No. of animals	No. of colloid droplets per 25 follicles	p values
A Room temperature	4	2 ± 1[1]	–
B DDC	5	3 ± 1	–
C Cold	5	438 ± 28	–
D L-dopa + cold	5	221 ± 28 C–D,	$p < 0.01$
E DDC + cold	5	114 ± 42 C–E,	$p < 0.01$
F DDC + L-dopa + cold	5	16 ± 13 D–F,	$p < 0.01$

1 Mean ± SE.
Rats were acclimatized to a high room temperature (30–31 °C) for 7 days and then exposed to cold for 30 min. L-dopa (10 mg) and DDC (50 mg) were injected i.p., 30 min and 1 h, respectively, prior to cold exposure.

depressed TRH release in response to cold. The combined treatment with L-dopa and DDC, which causes a selective increase in brain dopamine level, further produced a marked inhibition of acute thyroidal response to cold. These findings have clearly demonstrated that a negative dopaminergic mechanism is involved in the hypothalamic secretion of TRH in response to cold in the rat. These drugs do not inhibit thyroidal response to exogenous TRH.

Summary

Acute thyroid hormone secretion in response to cold is assessed by measuring intrathyroidal colloid droplets and plasma labeled thyroid hormone. An increase of thyroid hormone secretion is produced within 10 min after exposure of the animals to cold but this response is prevented by hypothalamic lesions or cerebellopontin transection. Since the pituitary can respond normally in such operated animals, it is concluded that cold stimulus, originating from the skin transferred to the hypothalamus via the spinal cord, stimulates TRH secretion. Further, it has been demonstrated that a negative dopaminergic mechanism is involved in the hypothalamic control of TRH secretion in response to cold.

References

1 *Smith, R.E. and Hoijer, D.J.:* Metabolism and cellular function in cold acclimation. Physiol. Rev. *42:* 60–151 (1962).
2 *Knigge, K.M.:* Thyroid function and plasma binding during cold exposure of the hamster. Fed. Proc. *22:* 755–760 (1963).
3 *Brown-Grant, K.:* The 'feedback' hypothesis of the control of thyroid function. Ciba Found. Colloq. Endocrin. *10:* 97–116 (1957).

4 *Bottari, P.M.:* The concentration of thyrotrophic hormone in the blood of rabbit under different experimental conditions. Ciba Found. Colloq. Endocrin. *11:* 52–69 (1957).
5 *Andersson, B.; Ekman, L.; Gale, C.C., and Sundsten, J.W.:* Activation of the thyroid gland by cooling of preoptic area in the goat. Acta physiol. scand. *54:* 191–192 (1962).
6 *Andersson, B.; Ekman, L.; Gale, C.C., and Sundsten, J.W.:* Thyroidal response to local cooling of the preoptic 'heat loss center'. Life Sci. *1:* 1–11 (1962).
7 *Onaya, T. and Solomon, D.H.:* Effects of chlorpromazine and prostaglandin on *in vitro* thyroid activation by thyrotropin, long-acting thyroid stimulator and dibutyryl cyclic AMP. Endocrinology, Springfield *85:* 1010–1017 (1969).
8 *Yamada, T.; Kajihara, A.; Onaya, T.; Kobayashi, I.; Takemura, Y., and Shichijo, K.:* Studies on acute stimulatory effect of cold on thyroid activity and its mechanism in the guinea pig. Endocrinology, Springfield *77:* 968–976 (1965).
9 *Hershman, J.M. and Pittman, J.A., jr.:* Utility of the radioimmunoassay of serum thyrotropin in man. Ann. intern. Med. *74:* 481–490 (1971).
10 *Greer, M.A.:* Studies on the influence of the central nervous system on anterior pituitary function. Recent Progr. Hormone Res. *8:* 67–98 (1957).
11 *Knigge, K.M. and Bierman, S.M.:* Evidence of central nervous system influence upon cold-induced acceleration of thyroidal ^{131}I-release. Amer. J. Physiol. *192:* 625–630 (1958).
12 *Van Beugen, L. and Van der Werff ten Bosch, J.J.:* Effects of hypothalamus and of cold on thyroid activity in the rat. Acta endocrin., Kbh. *38:* 585–597 (1961).
13 *Forster, R.E. and Ferguson, T.B.:* Relationship between hypothalamic temperature and thermoregulatory effects in unanesthetized cat. Amer. J. Physiol. *207:* 736–739 (1964).
14 *Lomax, P.; Malveaux, E., and Smith, R.E.:* Brain temperatures in the cat during exposure to low environmental temperature. Amer. J. Physiol. *207:* 736–739 (1964).
15 *Grimm, Y. and Reichlin, S.:* Thyrotropin-releasing hormone (TRH). Neurotransmitter regulation of secretion by mouse hypothalamic tissue *in vitro.* Endocrinology, Springfield *93:* 626–631 (1973).
16 *Harrison, T.S.:* Some factors influencing thyrotropin release in the rabbit. Endocrinology, Springfield *68:* 466–478 (1961).
17 *Greer, M.A.; Yamada, T., and Iino, S.:* The participation of the nervous system in the control of thyroid function. Ann. N.Y. Acad. Sci. *86:* 667–675 (1960).
18 *Everett, G.M. and Borcherding, J.R.:* L-dopa: effect on concentration of dopamine, norepinephrine, and serotonin in brains of mice. Science *168:* 849–850 (1970).
19 *Donoso, A.O.; Bishop, W.; Fawcett, C.P.; Krulich, L., and McCann, S.M.:* Effects of drugs that modify brain monoamine concentrations on plasma gonadotropin and prolactin levels in the rat. Endocrinology, Springfield *89:* 774–784 (1971).

Authors' address: Prof. *T. Yamada* and Assoc. Prof. *T. Onaya,* Department of Medicine, Institute of Adaptation Medicine, Shinshu University School of Medicine, *Matsumoto 390* (Japan)

Psychoneuroendocrinology. Workshop Conf. Int. Soc. Psychoneuroendocrinology, Mieken 1973, pp. 221–231 (Karger, Basel 1974)

Role of Biogenic Amines and Peptides on Pituitary TSH Secretion with Special Reference to Action of Thyrotropin-Releasing Hormone (TRH)

M. Sakoda and T. Kusaka

Second Department of Internal Medicine, Kobe University School of Medicine, Kobe

It has been well documented that biogenic amines and peptides play an essential role in neuroendocrine regulation of organic homeostasis. An action of hypothalamic-releasing hormone, postulated as a biogenic peptide, is now being tested in man as a mediator between the central nervous system and anterior pituitary gland (2, 5, 6).

Also biogenic amines such as L-dopa and 5-hydroxytryptophan (5-HTP) are being utilized in the studies of the controlling mechanism of pituitary tropic hormone secretion (1, 4, 7).

In this paper, the diagnostic and therapeutic application of thyrotropin-releasing hormone (TRH) to cases of hypothalamic pituitary disorders and the preliminary observations on the influence of L-dopa and 5-HTP on pituitary functions, is presented.

Clinical Subjects

In table I the diagnoses of 120 cases of hypothalamic pituitary disorders tested by TRH, are shown. These subjects consist of hypothalamic disorders and pituitary failures, including craniopharyngioma, ectopic pinealoma, meningioma, idiopathic diabetes insipidus, anorexia nervosa, and pituitary adenoma.

From these 120 cases, 30 well documented and pathologically determined cases of hypothalamic pituitary disorders were chosen and TRH tests were performed on them.

TRH preparation used in this observation was synthesized after *Gillessen* (3). TRH was administered intravenously in doses of 50–100 μg and plasma TSH responses were estimated by radioimmunoassay using double antibody technique (7).

Table I. Pathologic condition of clinical subjects

	No. of cases
Pathologic condition involving hypothalamus (56 cases)	
A. Tumors	31
Craniopharyngioma	12
Pinealoma	11
Meningioma	5
Metastatic tumor	1
Glioblastoma multifome	1
Hemangioblastoma	1
B. Idiopathic diabetes insipidus	10
C. Anorexia nervosa	6
D. Encephalitis and meningitis	3
E. Head injuries	2
F. Froehlich's syndrome	1
G. Others	3
Pathologic condition involving pituitary gland (64 cases)	
A. Tumors	39
Eosinophilic adenoma	20
Chromophobe adenoma	17
Mixed adenoma	2
B. Dwarfism	11
C. Suspicion of tumor	9
D Sheehan's syndrome	2
E. Others	3
Total	120

TRH Test in Hypothalamic Pituitary Disorders

These 30 pathologically well-determined cases included seven cases of craniopharyngioma, two of ectopic pinealoma, one of meningeoma, 16 of pituitary chromophobe adenoma, and four of acromegaly.

These 30 cases were tentatively divided into three groups, i.e. (a) ten cases of hypothalamic dysfunction; (b) 16 of pituitary chromophobe adenoma, and (c) four of acromegaly.

The response patterns of TRH-induced TSH increase obtained by preliminary observations in healthy volunteers were classified into five types as follows.

(a) Normal response: peak value of TSH increase being between 4.8 μU and 23.1 μU/ml after the i.v. administration of 50–100 μg of TRH.

(b) Subnormal response: peak value being below 4.8 μU, the minimum TSH increment observed in healthy volunteers.

(c) Absent response: no TSH increase after TRH administration.

(d) Supernormal response: peak value being above 23.1 μU, the maximum TSH increment observed in healthy volunteers.

(e) Delayed response: peak time of TSH increase coming 30 min or more after the administration of TRH.

The response patterns of TRH-induced TSH increase, in the three groups of hypothalamic pituitary disorders, are presented in table II.

Normal and supernormal response frequently appeared in the group of hypothalamic dysfunction, while subnormal or absent response appeared seldom.

Conversely, in the group of pituitary chromophobe adenoma, subnormal or absent response was frequently observed, while about half of this group showed normal TSH reserve.

The group of acromegaly showed normal or subnormal TSH increase.

From the viewpoint of the mode of TSH increase, subnormal or absent TSH response appeared with extremely high incidence in pituitary chromophobe adenoma, and supernormal response was frequently observed in cases of hypothalamic dysfunction.

On the other hand, time-response relationship was characteristically observed in hypothalamic dysfunction cases. Namely, eight of ten cases of hypothalamic disorders showed delayed TSH increase, while in only five of 16 cases of pituitary adenoma the response was delayed. Three of four cases acromegaly showed a slight delay of response. To sum-up these findings:

(1) Almost all the cases of hypothalamic disturbances indicated a delayed response with or without supernormal TSH increase. (These cases include the so-called hypothalamic hypothyroidism.)

Table II. TRH induced TSH response in 30 cases with hypothalamic pituitary disorders who were determined histologically

Patient groups	Normal	Sub-normal	Absent	Super-normal	Total
Lesions of the hypothalamus [10] Craniopharyngioma [7] Ectopic pinealoma [2] Meningeoma [1]	5 (5)	1 (0)	1 (0)	3 (3)	10 (8)
Pituitary chromophobe adenoma [16]	8 (3)	4 (1)	3 (0)	1 (1)	16 (5)
Acromegaly [4]	2 (1)	2 (2)	0	0	4 (3)

Figures in parentheses ( ) = delayed response. Figures in square brackets [ ] = number of cases.

(2) Pituitary chromophobe adenoma showed a varied response; however, there is a relatively high incidence of subnormal or absent TSH response.

(3) The group of acromegalics showed normal or subnormal response.

TRH Treatment

TRH is thought to be useful in therapeutic as well as diagnostic applications, especially to the patients with hypothyrotropic hypothyroidism.

An attempt was made to evaluate the possibility of therapeutic use of TRH in cases of the above-mentioned hypothyrotropic hypothyroidism, with special reference to its effects on endocrine function and psychosomatic condition.

Firstly, the thyrotropic effect of single oral administration of TRH in doses ranging from 1 to 10 mg on healthy volunteers was observed.

Plasma TSH increase was obtained from 2 to 4 h after the oral administration of 4, 5 and 10 mg of TRH.

Serum T_4 and PBI levels also rose 6–8 h after oral TRH administration in doses of 3 mg or more.

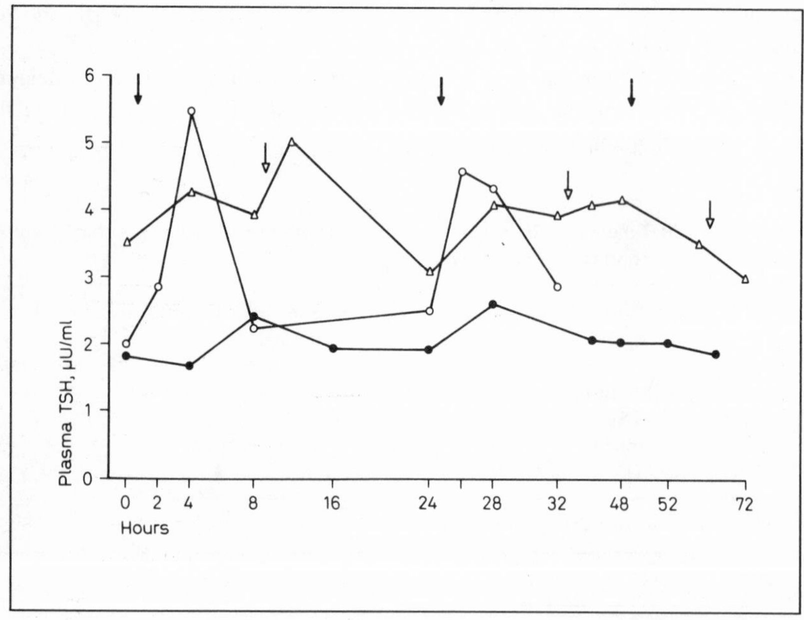

Fig. 1. Change in plasma TSH after repeated oral administration of TRH in healthy subjects. Solid arrow shows administration of 2, 4 and 6 mg of TRH, and open arrow 2 mg, respectively; △ = 2 mg (n =2); ● = 4 mg (n =5); ○ = 6 mg (n =2).

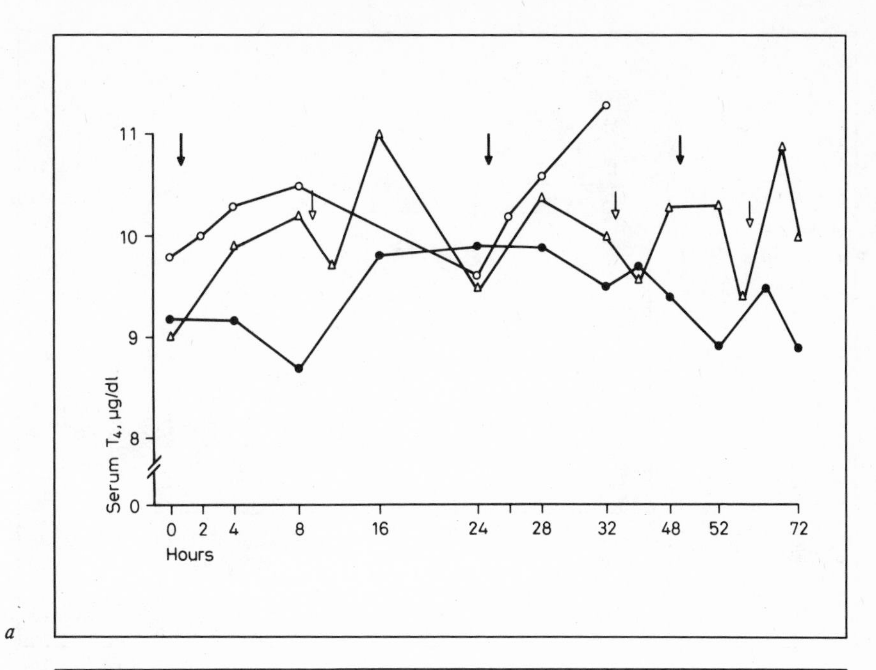

a

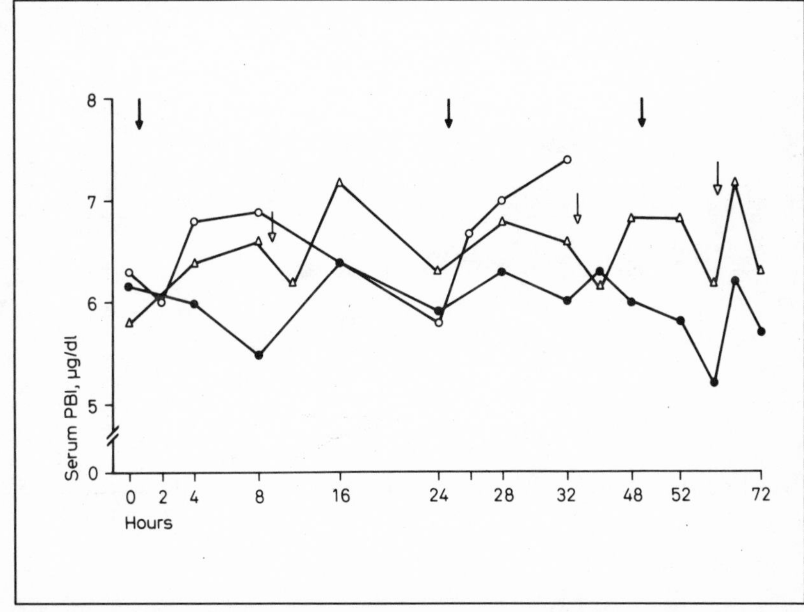

b

Fig. 2. Change in serum T_4 and PBI after repeated oral administration of TRH in healthy subjects. Solid arrow shows administration of 2, 4 and 6 mg of TRH, and open arrow 2 mg, respectively; △ = 2 mg (n =2); ● = 4 mg (n = 5); ○ = 6 mg (n = 2).

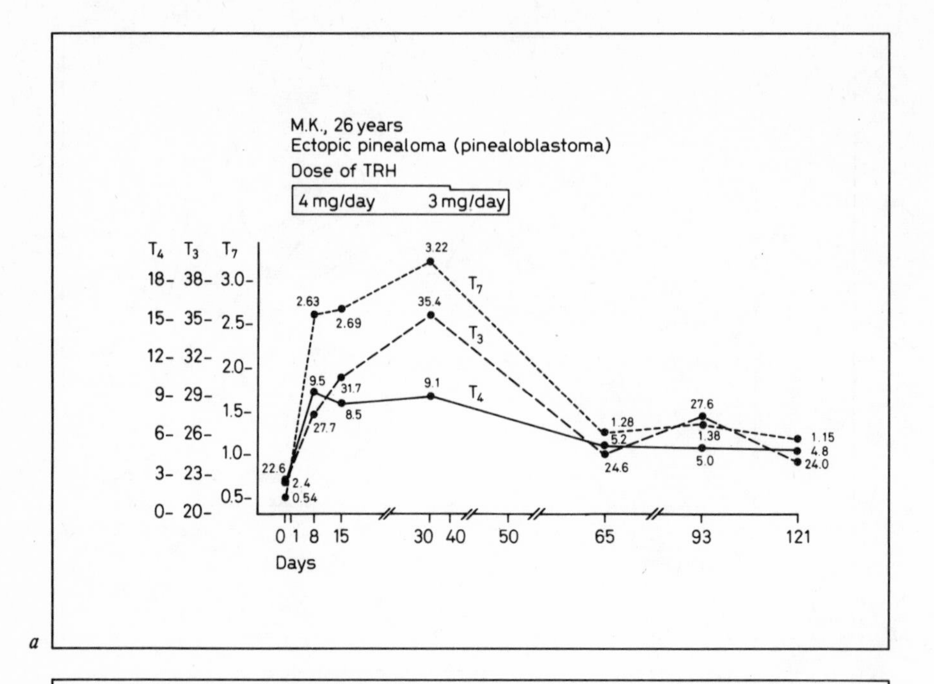

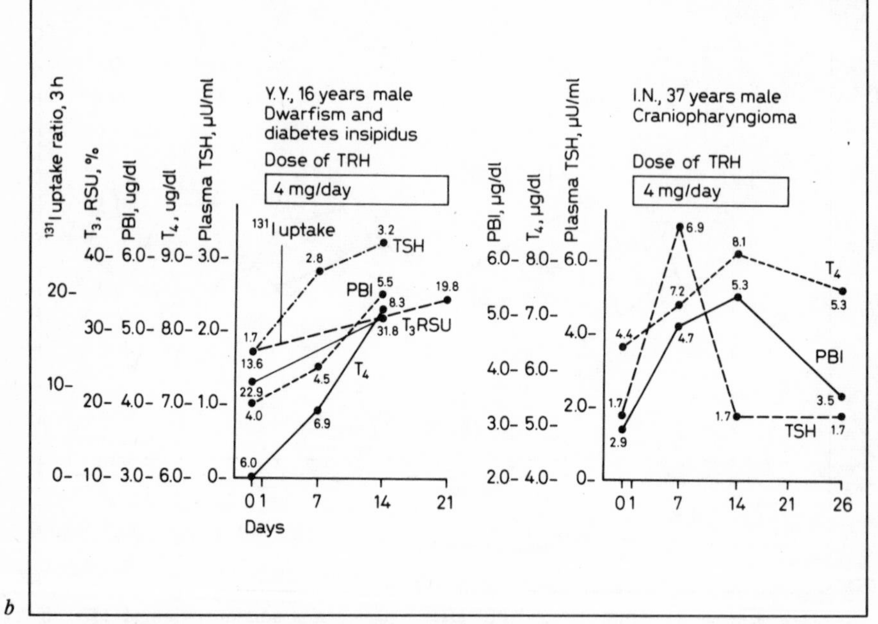

Fig. 3. Effect of repeated oral administration on pituitary-thyroid function in cases with hypothalamic hypothyroidism and pituitary failure.

These findings were already reported by us (7).

Repeated oral administration of TRH was given to healthy volunteers. TRH in doses of 2, 4 and 6 mg was administered orally from two to six times, and observation continued for 2–3 days.

Plasma TSH increases 2–4 h after each administration of TRH as indicated in figure 1.

Figure 2 shows the effect of the repeated oral TRH administration on serum T_4 and PBI levels. Serum T_4 and PBI levels rose 6–8 h after each administration, a rising pattern similar to that of plasma TSH, the maximum ($\triangle$) T_4 being 2.4–5.6 μg/dl.

In figure 3, the results of repeated oral TRH administration to hypothyrotropic patients are presented.

The TRH administration concerned three patients with hypothyroidism due to craniopharyngioma, ectopic pinealoma, and idiopathic hypopituitarism.

The I.N. and M.K. cases showed sufficient pituitary TSH reserve to TRH test, in which plasma TSH, rising from a low basal level, reached the normal range. Thus, both may be postulated as belonging to the category of the so-called hypothalamic hypothyroidism. The Y.Y. case, which has subclinical hypothyroidism, showed subnormal TSH increase to TRH test.

Four milligram of TRH was administered daily to these three patients for 3 weeks. Gradual increases of serum T_4, T_3, and PBI were obtained, and these increases reached their peaks 3 weeks after the onset of the oral administration.

Although the I.N. case showed a decrease of blood thyroid hormone concentration after the withdrawal of TRH administration, the M.K. case showed a continued high level of serum T_4, the level being within the normal range, even 2 and 6 weeks after the withdrawal. Before the TRH treatment, the M.K. and I.N. cases showed psychomotoric depression due to hypothalamic hypothyroidism, giving speechless, gloomy, asthenic, and tame impressions.

Both cases, however, became talkative from 3 to 4 weeks after the onset of TRH treatment, namely, the M.K. case became argumentative, while in the I.N. case his family members witnessed a tendency to rapid action and bulimia. These cases are still being followed up with finer observations.

From these findings TRH treatment is thought to be effective in psychosomatic improvement.

Effect of L-Dopa and 5-HTP on Pituitary Function

In this part the preliminary observations on the effects of L-dopa and 5-HTP on pituitary tropic hormone secretion in cases of parkinsonism and depressive psychoses is presented.

Plasma HGH concentration of healthy subjects was elevated by the oral administration of L-dopa in doses of 0.5 or 1 g, the elevation starting 60 min after the administration, while plasma TSH level remained unchanged.

In cases of parkinsonism treated by therapeutic doses of L-dopa for at least 1 month, an additional single administration of L-dopa induced varied plasma HGH response.

Differences of HGH response among these patients were documented with special reference to the relation between their clinical severity and the total amount of therapeutic dose of L-dopa administered (fig. 4).

Judged after Yahr, almost all the cases that showed early and sufficient HGH increase were not clinically severe cases, and had short duration of clinical course and L-dopa treatment.

Conversely, the cases that showed insufficient or delayed HGH increase were clinically quite severe in Yahr's classification and their clinical courses were long. Also they had received large doses of L-dopa. Figure 5 shows TRH-induced TSH responses of those who were treated with chronic L-dopa treatment. A slight TSH increase was observed in those patients.

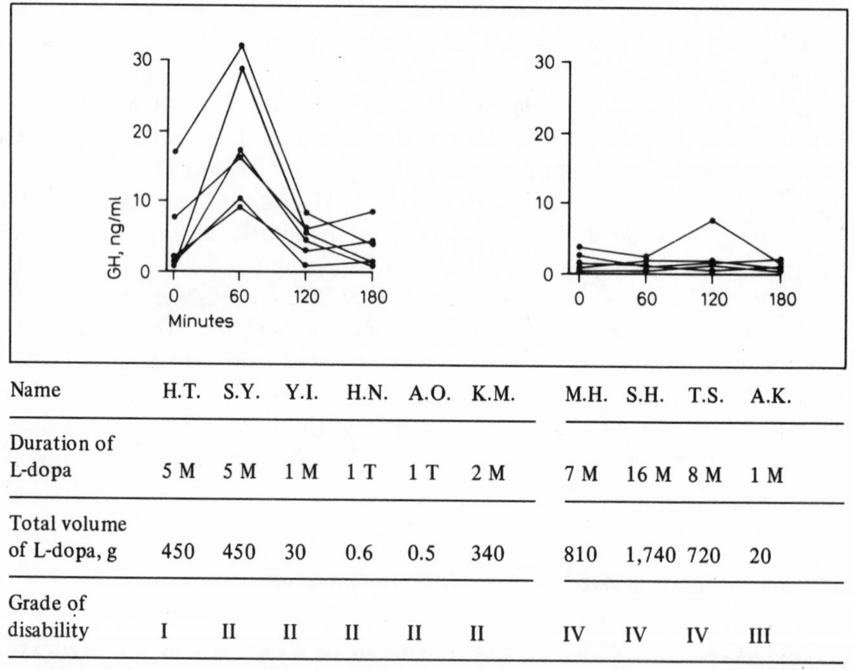

Name	H.T.	S.Y.	Y.I.	H.N.	A.O.	K.M.	M.H.	S.H.	T.S.	A.K.
Duration of L-dopa	5 M	5 M	1 M	1 T	1 T	2 M	7 M	16 M	8 M	1 M
Total volume of L-dopa, g	450	450	30	0.6	0.5	340	810	1,740	720	20
Grade of disability	I	II	II	II	II	II	IV	IV	IV	III

Fig. 4. Effect of L-dopa on plasma HGH in cases with parkinsonism with special reference to duration, total volume of L-dopa and clinical severity.

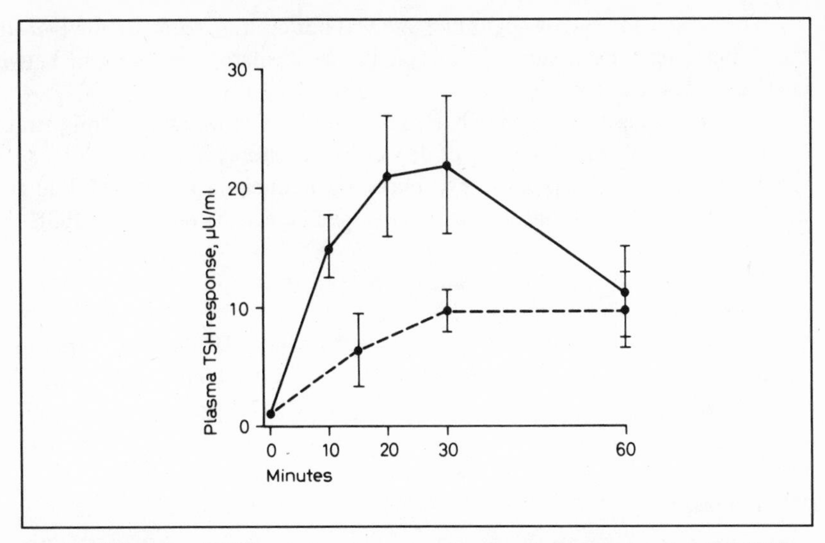

Fig. 5. TRH induced TSH response in cases with chronic L-dopa treatment. Solid line shows the TSH response of healthy subjects (five cases) and the dotted line indicates the TSH response of cases treated with L-dopa. Maximum △ TSH; normal = 22.2 ± 5.8; dopa + TRF = 8.2 ± 4.2.

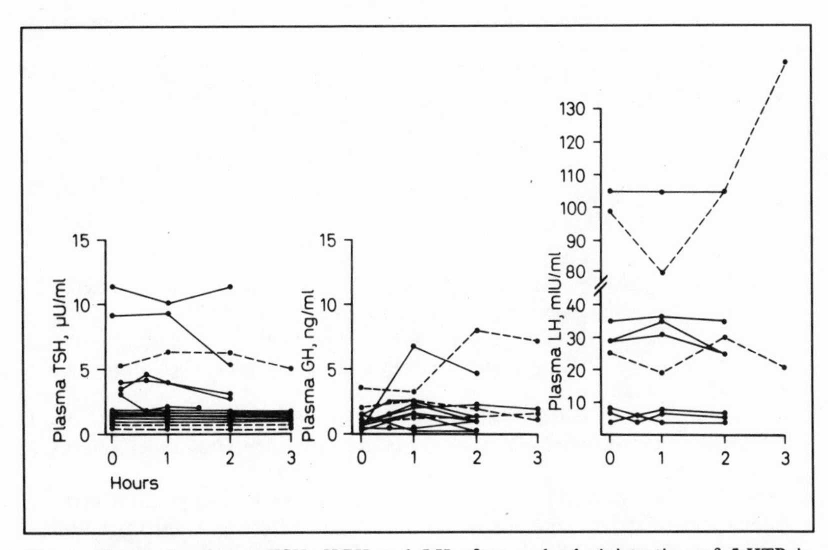

Fig. 6. Change in plasma TSH, HGH and LH after oral administration of 5-HTP in patients with depressive psychoses and parkinsonism. Solid lines indicate the response of cases tested by 100 mg of 5-HTP, and dotted lines show the response of cases tested by 200 mg of 5-HTP.

The reason for such a slight increase was not clear. It seemed, however, that there might have been some confusion in the regulatory mechanism between HGH and TSH secretion.

Finally, changes in plasma HGH, TSH, and LH following the administration of 5–HTP were observed in cases of depressive psychoses (fig. 6).

In eight cases of depressive psychoses, the administration of 5-HTP in doses of 50, 100 and 200 mg did not show any specific effects on plasma HGH, LH, and TSH.

In three cases of parkinsonism, the administration of 5-HTP did not influence pituitary tropic hormones either.

Conclusively, in cases of depressive psychoses, 5–HTP administered under the condition mentioned above seems to be ineffective on pituitary HGH, TSH, and LH secretion.

Summary

(1) Thyrotropin-releasing hormone (TRH), a peptide of hypothalamic origin, induced varied plasma TSH response. Delayed TSH increase was observed in hypothalamic dysfunction with or without supernormal TSH increase, while subnormal or absent TSH response frequently appeared in pituitary adenoma.

(2) Repeated oral administration of TRH is thought to be effective in psychosomatic and endocrine improvement in cases of hypothyrotropic hypothyroidism.

(3) Oral administration of L-dopa induced plasma HGH increase and inhibited TRH effect on plasma TSH.

(4) 5-HTP in doses of 100 and 200 mg did not show any specific effect on plasma HGH, LH, and TSH in cases of parkinsonism and depressive psychoses.

References

1 *Boyd, A.E.; Levovitz, H.E., and Pfeiffer, J.B.:* Stimulation of human-growth-hormone secretion by L-dopa. New Engl. J. Med. *283:* 1425–1429 (1970).

2 *Gual, C.; Kastin, A.J., and Schally, A.V.:* Clinical experience with hypothalamic-releasing hormone. I. Thyrotropin-releasing hormone. Recent Progr. Hormone Res. *28:* 173–200 (1972).

3 *Gillessen, D. von; Felix, A.M.; Lergier, W. und Studer, R.O.:* Synthese des Thyrotropin-Releasing-Hormons (TRH) und verwandter Peptide. Helv. chim. Acta *53:* 63–72 (1970).

4 *Kamberi, I.A.; Mical, R.S., and Porter, J.C.:* Effect of anterior pituitary perfusion and intraventricular injection of catecholamines and indoleamines on LH release. Endocrinology, Springfield *87:* 1–12 (1970).

5 *Kastin, A.J.; Gual, C., and Schally, A.V.:* Clinical experience with hypothalamic-releasing hormone. II. Luteinizing hormone-releasing hormone and other hypophysiotropic-releasing hormone. Recent Progr. Hormone Res. *28:* 201–227 (1972).

6 *Ormston, B.J.; Kilborn, J.R.; Garry, R.; Amos, J., and Hall, R.:* Further observations of the effect of synthetic thyrotropin-releasing hormone in man. Brit. med. J. *ii:* 199–202 (1971).
7 *Sakoda, M.; Otsuki, M.; Fukatsu, H.; Baba, S., and Hiroshige, N.:* Effect of thyrotropin-releasing factor (TRF) on pituitary TSH secretion in man. Folia Endocrin, jap. 1046–1060 (1972).
8 *Sakoda, M.; Otsuki, M.; Hiroshige, N.; Kanao, K.; Yagi, A., and Honda, M.:* Effect of synthetic thyrotropin-releasing factor (TRF) on pituitary TSH secretion in man. Endocrin. jap. *17:* 541–549 (1970).
9 *Spaulding, S.W.; Burrow, G.N.; Donabedian, R., and Melvin Van Woert:* L-dopa suppression of thyrotropin-releasing hormone response in man. J. clin. Endocrin. *35:* 182–185 (1972).

Authors' address: Dr. *M. Sakoda* and Dr. *T. Kusaka,* Second Department of Internal Medicine, Kobe University School of Medicine, 12 Kusunoki-cho, 7-Chome, *Ikuta-Ku, Kobe* (Japan)

Psychoneuroendocrinology. Workshop Conf. Int. Soc. Psychoneuroendocrinology, Mieken 1973, pp. 232–242 (Karger, Basel 1974)

Studies on Growth Hormone-Releasing Activity in Bovine Hypothalami

S. Sawano, T. Matsuno, M. Yamazaki, Y. Baba and K. Takahashi

Okinaka Memorial Institute for Medical Research; Department of Endocrinology, Toranomon Hospital; Institute for Adult Diseases, Asahi Life Foundation; College of General Education, University of Tokyo; Sankyo Central Research Laboratories, and Department of Neurochemistry and Psychology, Tokyo Metropolitan Institute for Neurosciences, Tokyo

Recent studies have shown that the partially purified extracts from rat and porcine hypothalami possess an ability to stimulate the release of immunoreactive growth hormone (IRGH) from isolated rat pituitaries *in vitro* (*Wilber et al.*, 1971). More recently, *Malacara et al.* (1972) reported that the partially purified ovine GH-releasing factor (GRF) raised the levels of plasma GH in rats pretreated with estrogen and progesterone. We have been attempting to extract and purify GRF from bovine hypothalami. The results indicated that the partially purified GRF had an activity to stimulate the secretion of IRGH both *in vitro* and *in vivo.* The synthesis of GH was also stimulated by GRF.

Materials and Methods

Preparation of bovine hypothalamic extract. One batch of 3,000 bovine hypothalami (SME) was homogenized in 0.1 *M* acetic acid. After the supernatant was lyophilized, the lyophilized extract was defatted with acetone and petroleum ether. The defatted powder was again homogenized in 2 *M* acetic acid at 45 °C. The supernatant was applied on a Sephadex G-25 fine-beaded column (7.4 × 220 cm), equilibrated with 1.0 *M* acetic acid at 4 °C. The fraction size was 120 ml. The transmittance at 280 n, pressor activity (*Dekanski,* 1952) and specific conductivity were measured in each fraction. The concentration of potassium in fractions was determined by flame photometer. The active fractions following Sephadex G-25 gel filtration were further applied on CM Sephadex C-25 column (3 × 12 cm) equilibrated with 0.05 *N* ammonium acetate buffer at pH 6.0. The fraction size was 17 ml.

In vitro *assay for GH-releasing activity.* Male Wistar rats, weighing 150–180 g, were used. Two hemipituitaries were placed in 2 ml of Krebs-Ringer-bicarbonate (KRB) medium containing 2 mg/ml glucose at pH 7.4 and preincubated for 30 min at 37 °C in an atmosphere of 95 % O_2–5 % CO_2. The medium was then removed and 2 ml of fresh medium containing the test material was added. The incubation was continued for an additional

30 min unless mentioned otherwise. Two corresponding hemipituitaries were incubated in medium alone in a similar way and served as the control. The concentration of rat GH (RGH) in the incubation medium was measured by a double antibody radioimmunoassay developed by NIAMD (*Birge et al.*, 1967). GH-releasing activity (GRF activity) was expressed as 'percent of control' (medium RGH concentration released from 1 mg pituitary with the test material ÷ RGH concentration released from 1 mg control pituitary) × 100.

Test for GH-releasing activity of GRF in the dog. Three well-trained adult mongrel dogs, weighing 8–10 kg, were used under unanesthetized condition. After an overnight fast, the test materials dissolved in 1 ml physiological saline, were injected intravenously. Blood samples were obtained from the brachial vein at the time indicated in the results. The plasma levels of dog GH (CGH) were measured by a specific double antibody radioimmunoassay (*Tsushima et al.*, 1971).

Test for GH synthesis-stimulating activity of GRF. Two hemipituitaries or one whole pituitary from male Wistar rats, weighing 180–220 g, were incubated in 0.3 ml of KRB medium containing 0.15 μc of ^{14}C-leucine (344 mc/m*M*) for 6 h. In a few experiments, 1.5 μc of ^{3}H-uridine (24 c/m*M*) was added. After incubation, aliquots of the pituitary homogenates and of incubation media were subjected to polyacrylamide gel electrophoresis (*Ornstein*, 1964; *Davis*, 1964). Each gel was stained with 0.25-percent amido black for 10 min. After removing excess dye, the GH band was cut out. The gel slice was dried and was placed at the bottom of a glass counting vial. Thirty percent hydrogen peroxide (0.3 ml) was added to the vial, and then the vial was capped firmly, placed in a constant-temperature bath at an angle to ensure continual bathing of the slice by the peroxide, and heated at 70 °C until solubilization of the gel was complete. BBS-3 (Bio-Solv solubilizer No. 184838, Beckman) and toluene-based scintillation fluid were added to the vial in 1.5 and 10 ml quantitities, respectively. The results were expressed in terms of d.p.m. of incorporated ^{14}C-leucine in GH per two hemipituitaries or per mg pituitary. Significance of the differences was calculated by paired t-test or Student's t-test. Five to seven flasks per group were used throughout the studies.

Results

GRF activity in bovine hypothalami. The elution pattern following Sephadex G-25 is represented in figure 1. GRF activity of the fractions is shown in figure 2. Fractions 41–44 and 45–48 caused a significant increase in medium RGH (490 ± 50 %; $p < 0.01$ versus control and 382 ± 35 %; $p < 0.01$ versus control, respectively). Fraction 69–76 also had a smaller GRF activity (212 ± 26 %; $p < 0.02$). However, potassium concentration in fraction 69–76 at 1 SME equivalent/ml of distilled water was 6.9 mEq/l, while that in fraction 41–48 at 1 SME equivalent/ml was negligible. This result may indicate that the activity seen in fraction 69–76 is attributable to a high concentration of potassium ion, so further evaluation was not performed in these fractions.

A relationship between the dose of fraction 45–48 and its GRF activity was investigated during 30-min or 4-hour incubation periods. As shown in table I, the

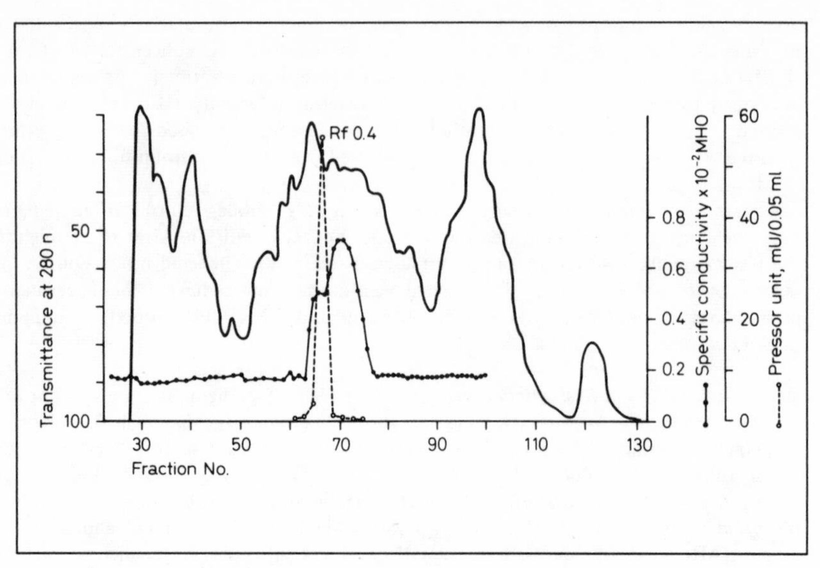

Fig. 1. The elution pattern following Sephadex G-25 gel filtration of bovine hypothalami.

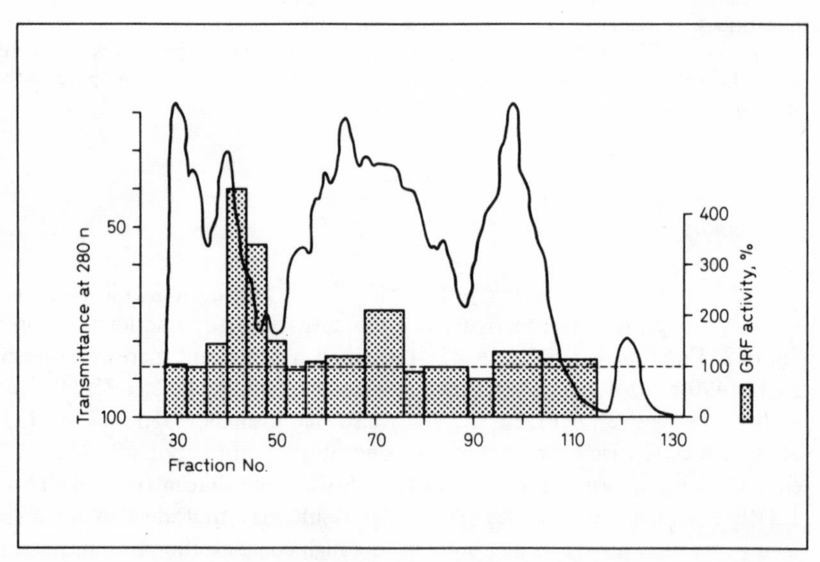

Fig. 2. Growth hormone-releasing activity in Sephadex G-25 fractions shown in figure 1 is represented on the right ordinate as percent of control. The horizontal dotted line indicates 100 % of control. Stippled vertical bars indicate fractions used for *in vitro* assay at 1 SME equivalent/ml medium. The bars indicate the mean.

Table I. Effect of graded doses of GRF on release of RGH during 30-min or 4-hour incubation intervals

Dose of GRF[1]		Medium RGH, percent of control			
μg/ml	SME equivalent/ml	30 min	p[2]	4 h	p[2]
10	0.2	118 ± 20[3]	NS	–	–
30	0.6	228 ± 31	0.01	116 ± 36	NS
50	1.0	346 ± 58	0.01	–	–
90	1.8	454 ± 75	0.01	167 ± 21	0.05
150	3.0	435 ± 32	0.01	145 ± 12	0.05

1 GRF used was fraction 45–48 following Sephadex G-25 gel filtration.
2 p versus paired control.
3 Mean ± SE.
50 μl of the incubation medium was collected after the initiation of incubation and the incubation was continued for 4 h.

minimum effective dose of GRF was 30 μg/ml, equivalent to 0.6 SME/ml. The maximum GRF activity was observed at dose 90 μg/ml, equivalent to 1.8 SME/ml, while GRF activity was markedly reduced at 4-hour incubation intervals.

Fractions 41–44 and 45–48 following Sephadex G-25 were combined and subsequently purified by CM Sephadex C-25 column (fig. 3). The greatest GRF activity was found in fraction 3–17–4 (310 ± 45 %; $p < 0.01$). The biological activity of fraction 3–17–4 was also examined in the dog (fig. 4). The fraction at the dose of 100 SME equivalent, approximately 200 μg, was given at 0 time. In dog No. 1, the basal level of plasma CGH was 2 ng/ml. An initial sharp peak of plasma CGH (9 ng/ml) appeared 5 min after the injection and returned to the preinjection level at 30 min. A second CGH rise was also noticed at 180 min. In dog No. 2, the levels of plasma CGH increased from the undetectable level (less than 1 ng/ml) to 5.5 ng/ml 15 min after the injection and similarly returned to the basal level at 30 min. In dog No. 3, plasma CGH levels did not change until 60 min, but increased sharply to 20 ng/ml at 90 min.

GH-synthesis-stimulating activity of GRF. Figure 5 shows the uptake of ^{14}C-leucine into rat pituitaries. Gel slice No. 5 indicates the GH band. 10 μg of GRF (fraction 45–48) stimulated the incorporation of ^{14}C-leucine into the GH band compared to control. No clear change was observed in other gel slices. These results clearly indicate that GRF has GH synthesis-stimulating activity and this activity is specific to GH.

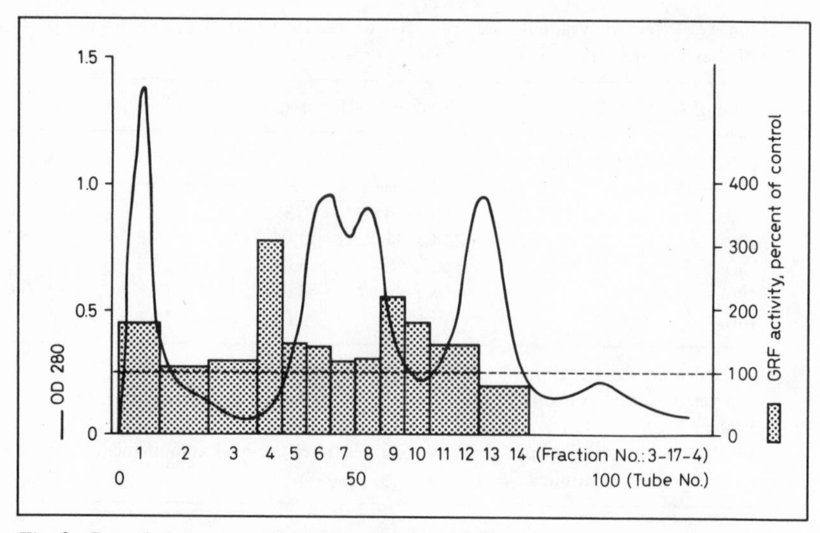

Fig. 3. Growth hormone-releasing activity in CM Sephadex C-25 fractions is represented on the right ordinate as percent of control. The horizontal dotted line indicates 100 % of control. Stippled vertical bars indicate fractions used for *in vitro* assay at 4 SME equivalent/ml medium. The bars indicate the mean.

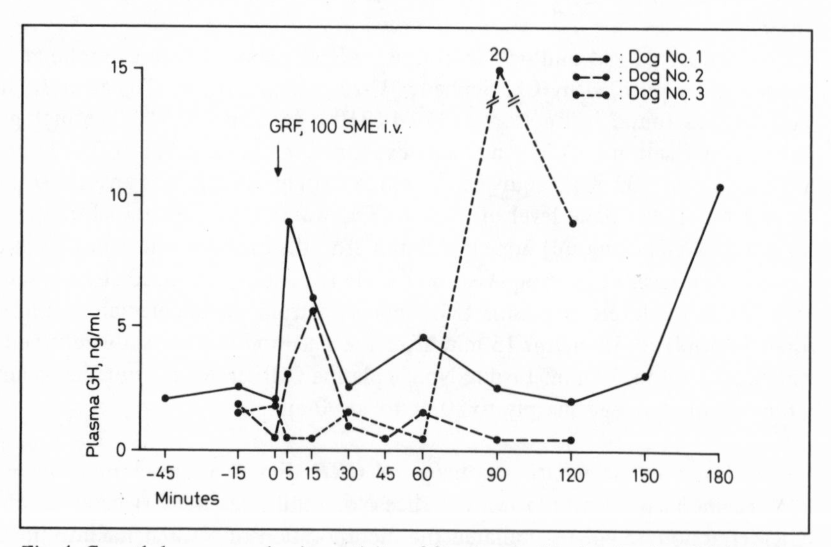

Fig. 4. Growth hormone releasing activity of fraction 3–17–4 following CM Sephadex C-25 in three conscious dogs. Blood samples for the measurement of plasma GH were collected at the time indicated on the abscissa. Fraction 3–17–4 at 100 SME equivalent was injected intravenously at 0 time.

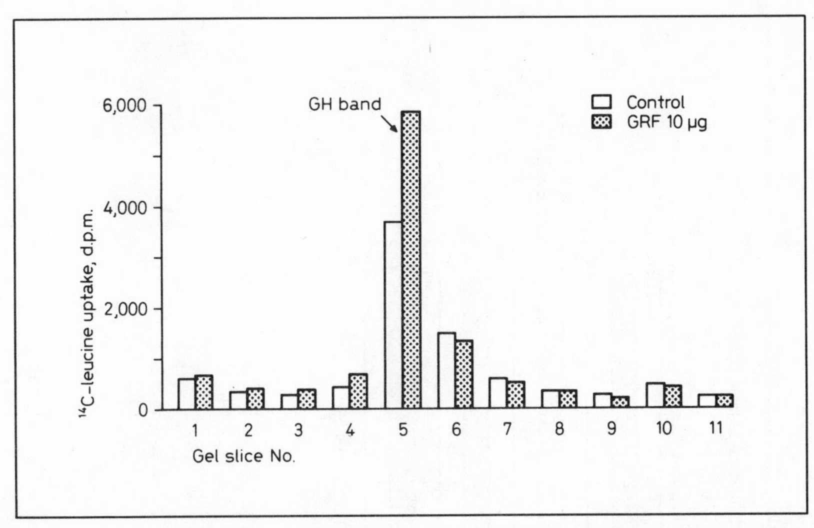

Fig. 5. In vitro incorporation of ^{14}C-leucine into pituitary gland proteins during 6-hour incubation. Two hemipituitaries were incubated with 10 µg of GRF (fraction 45–48) and two corresponding hemipituitaries served as the control. After polyacrylamide gel electrophoresis of the labeled gland homogenates, each gel was cut into ten slices. Slice No. 5 is the GH band. Each paired bar indicates the mean of triplicate flasks. Open bars represent control, and stippled bars represent addition of GRF into the incubation medium.

Table II. Effect of GRF on the synthesis of RGH during various incubation intervals

Group	GRF[1], µg	Incubation interval, h	^{14}C-leucine uptake (d.p.m./mg pituitary)			
			medium	p	pituitary	p
Control	–	0.5	45 ± 3[2]	–	227 ± 17	–
Expl.	10	0.5	44 ± 3	NS	385 ± 30	0.01
Control	–	1.0	106 ± 9	–	377 ± 29	–
Expl.	10	1.0	112 ± 9	NS	823 ± 87	0.01
Control	–	3.0	157 ± 26	–	824 ± 78	–
Expl.	10	3.0	189 ± 22	NS	1,614 ± 217	0.01
Control	–	6.0	157 ± 8	–	691 ± 32	–
Expl.	10	6.0	277 ± 12	0.01	1,605 ± 242	0.01

1 GRF preparation used was fraction 45–48 following Sephadex G-25 gel filtration.
2 Mean ± SE.

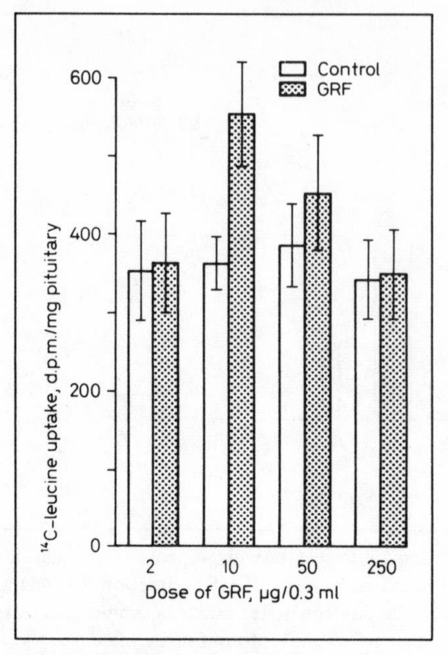

Fig. 6. Effects of various doses of GRF (fraction 45–48) on *in vitro* incorporation of ^{14}C-leucine into pituitary GH during 6-hour incubation. Two hemipituitaries were incubated with 2–250 μg of GRF and two corresponding hemipituitaries served as the control. In each pair the open bar indicates control, and the stippled bar indicates addition of GRF into the incubation medium. The bars represent the mean and the brackets represent SE.

The effects of various doses of GRF on the synthesis of GH are displayed in figure 6. The incorporation of ^{14}C-leucine into control pituitaries was 356 ± 34 d.p.m./mg pituitary. Addition of 10 μg of GRF stimulated the value significantly to 550 ± 64 d.p.m./mg pituitary ($p < 0.05$). As the dose increased, however, the stimulatory effect reduced.

A relationship between the GH synthesizing activity of 10 μg of GRF and the incubation intervals was studied next. The pituitaries were incubated for 30 min, 1, 3 and 6 h (table II). When GRF was added, a significant increment of newly synthesized GH in the pituitary was observed as early as 30 min after the initiation of incubation as compared with control. GH synthesis was stimulated by GRF until 3 h and reached plateau thereafter. On the other hand, the release of the synthesized GH into medium was not stimulated by GRF until 3 h and increased significantly 6 h later.

Effects of actinomycin D (Act D) and cycloheximide (CH) on the release and synthesis of GH by GRF. GRF activity of 100 μg/ml of GRF was not

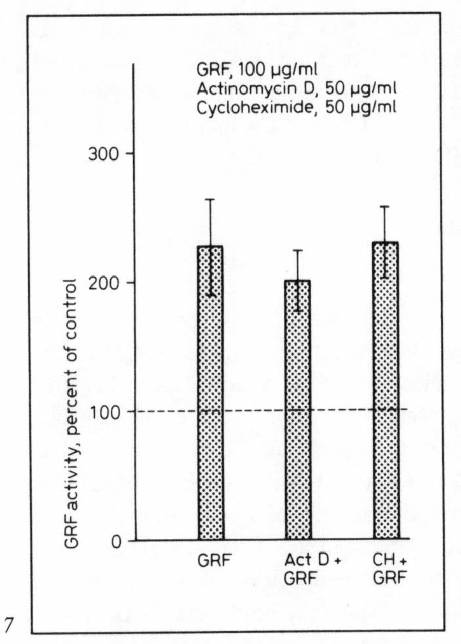

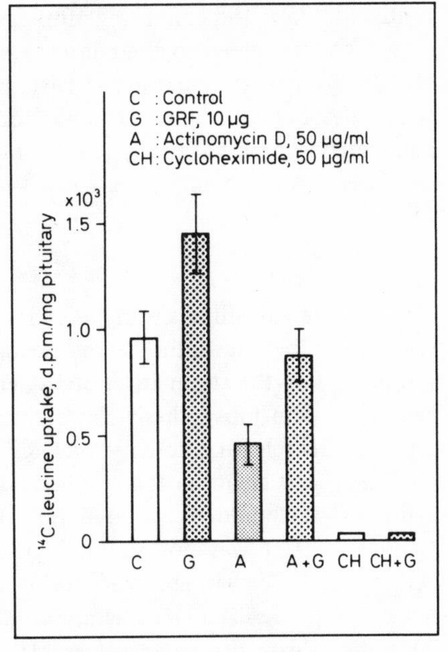

Fig. 7. Growth hormone-releasing activity of GRF (fraction 45–48) at the dose of 2 SME equivalent/ml in the presence of either 50 µg/ml of actinomycin D (Act D) or 50 µg/ml of cycloheximide (CH). The horizontal dotted line indicates 100 % of control. Incremental change in medium RGH is represented as percent of control on the ordinate. The bars indicate the mean and the brackets indicate SE.

Fig. 8. Effect of 10 µg of GRF (fraction 45–48) on *in vitro* incorporation of ^{14}C-leucine in the presence of either 50 µg/ml of actinomycin D (Act D) or 50 µg/ml of cycloheximide (CH). One whole pituitary was incubated for 6 h. The bars represent the mean and the brackets represent SE.

affected in the presence of either 50 µg/ml of Act D or 50 µg/ml of CH ($p < 0.01$ versus each corresponding control) (fig. 7). When the pituitaries were incubated with ^{3}H-uridine in the presence of 50 µg/ml of Act D for 6 h, the incorporation of ^{3}H-uridine into the pituitary GH decreased from the control value of 63 ± 5 d.p.m./mg pituitary to 14 ± 2 d.p.m./mg pituitary ($p < 0.001$). This indicated that the dose of Act D used was sufficient to inhibit RNA synthesis. The pituitaries were then incubated with 50 µg/ml of Act D or 50 µg/ml of CH (fig. 8). In this series of experiments, 10 µg of GRF (G) increased the incorporation of ^{14}C-leucine into pituitary GH from the control value (C) of 963 ± 165 d.p.m./mg pituitary to 1,450 ± 187 d.p.m./mg pituitary ($p < 0.05$). Addition of 50 µg/ml of Act D (A) reduced the ^{14}C-leucine uptake to 463 ± 88 d.p.m./mg pituitary (C versus A; $p < 0.05$). While the value increased significantly to 875 ± 165 c.p.m./mg pituitary in the presence of both GRF and Act D (A + G) as compared to the control (A) (A + G versus A; $p < 0.05$). The degree of GH synthesis-stimulating activity of GRF was almost the same with and without Act D. On the other hand, addition of CH into the incubation medium completely suppressed the synthesis of GH (CH).

Discussion

The present study clearly demonstrates that the partially purified bovine GRF can stimulate both the release of IRGH and the synthesis of GH in the pituitary. For the stimulation of IRGH release, GRF does not require the process of protein biosynthesis, since the presence of either Act D or CH did not affect IRGH-releasing activity of GRF. On the other hand, it is indicated that GRF may participate in the synthesis of GH at the translational level of protein synthesis. It should be pointed out that the release of IRGH and of newly synthesized GH was not simultaneously stimulated by GRF. IRGH-releasing activity of GRF was observed during a shorter incubation interval, when the release of synthesized GH was not stimulated. This may imply that GRF stimulates the release of readily releasable GH in the pituitary. By prolonging the incubation interval, synthesized GH may be accumulated in a different pool where GRF could not affect. A significant increase in the release of synthesized GH in the presence of GRF was first noticed 6 h later and the increase might not be caused by the direct effect of GRF, but secondarily by the increased accumulation of synthesized GH. Our results are inconsistent with the observations reported by *Stachura et al.* (1972). They found that their ovine GRF was able to stimulate the release of newly synthesized GH, even for 1-hour incubation intervals, but not to affect the release of IRGH. The discrepancy remains unexplained. The failure of demonstrating a dose response curve for GH synthesis may result from the enhanced toxicity of the preparation in high concentrations. Although there is no direct evidence, the possible existence of some toxic substance (s) in GRF preparation is considered. A similar result was reported by *Stachura et al.* (1972).

It is quite important to demonstrate whether or not the GRF preparation having the activity *in vitro* can raise the levels of plasma IRGH. *Machlin et al.* (1968) evaluated their preparations on the basis of an increase in plasma IRGH levels of sheep. A similar attempt was made by *Malacara et al.* (1972), who observed that their ovine GRF caused a rise of plasma IRGH levels in rats pretreated with estrogen-progesterone. On the other hand, the decapeptide which was isolated from porcine hypothalami (*Schally et al.*, 1971) was active in the tibial epiphyseal bioassay, but did not raise plasma IRGH. The results herein, indicated that our partially purified GRF had a biological activity in the dog *in vivo,* although the limited availability of the preparation prevented us from performing further experiments. The dog was selected for the experimental animal, because the effect of anesthetics is excluded by using unanesthetized dogs and the changes in plasma CGH levels induced by various stimuli were very similar to those observed in humans (*Tsushima et al.*, 1971).

Very recently GH-release inhibiting factor (GIF) has been isolated from ovine hypothalami (*Brazeau et al.*, 1972). However, GIF activity was not detected in the fractions following Sephadex G-25 gel filtration by the *in vitro* method used in the present study.

Summary

Extraction and partial purification of GRF from bovine hypothalami were performed. The partially purified GRF stimulated the release of IRGH from isolated rat pituitaries *in vitro* and also raised the levels of plasma IRGH in dogs. In addition, GRF stimulated the synthesis of GH in the pituitary, but did not evoke the release of newly synthesized GH during a short incubation interval. GH-releasing activity of GRF was not affected in the presence of either actinomycin D or cycloheximide. Actinomycin D caused a decrease in GH synthesis, but GH-synthesizing activity of GRF was not abolished in the presence of actinomycin D. Possible mechanism of the action of GRF was discussed.

Acknowledgement

We are indebted to the National Institute of Arthritis, Metabolism and Digestive Diseases Rat Pituitary Hormone Distribution Program for the supply of rat growth hormone immunoassay kits.

References

Birge, C.A.; Peake, G.T.; Mariz, I.K., and Daughaday, W.H.: Radioimmunoassayable growth hormone in the rat pituitary gland: effect of age, sex and hormonal state. Endocrinology, Springfield *81:* 195–204 (1967).

Brazeau, P.; Vale, W.; Burgus, R.; Ling, N.; Butcher, M.; Rivier, J., and Guillemin, R.: A hypothalamic polypeptide that inhibits the secretion of pituitary growth hormone. Science *179:* 71–79 (1972).

Davis, B.J.: Disc electrophoresis. II. Method and application to human serum proteins. Ann. N.Y. Acad. Sci. *121:* 404–425 (1964).

Dekanski, J.: Quantitative assay of vasopressin. Brit. J. Pharmacol. *7:* 567–572 (1952).

Machlin, L.J.; Takahashi, Y.; Horino, M.; Hertelendy, F.; Gordon, R.S., and Kipnis, D.: Regulation of growth hormone secretion in non-primate species; in *Pecile and Müller* Growth hormone, pp. 292–305 (Excerpta Medica, Amsterdam 1968).

Malacara, J.M.; Valverde-R., C.; Reichlin, S., and Bollinger, J.: Elevation of plasma radioimmunoassayable growth hormone in the rat induced by porcine hypothalamic extract. Endocrinology, Springfield *91:* 1189–1198 (1972).

Ornstein, L.: Disc electrophoresis. I. Background and theory. Ann. N.Y. Acad. Sci. *121:* 321–349 (1964).

Schally, A.V.; Baba, Y.; Nair, R.M.G., and Bennett, C.D.: The amino acid sequence of a peptide with growth hormone-releasing activity isolated from porcine hypothalamus. J. biol. Chem. *246:* 6647–6650 (1971).

Stachura, M.E.; Dhariwal, A.P.S., and Frohman, L.A.: Growth hormone synthesis and release *in vitro.* Effects of partially purified ovine hypothalamic extract. Endocrinology, Springfield *91:* 1071–1078 (1972).

Tsushima, T.; Irie, M., and Sakuma, M.: Radioimmunoassay for canine growth hormone. Endocrinology, Springfield *89:* 685–693 (1971).

Wilber, J.F.; Nagel, T., and White, W.F.: Hypothalamic growth hormone-releasing activity (GRA). Characterization by the *in vitro* rat pituitary and radioimmunoassay. Endocrinology, Springfield *89:* 1419–1424 (1971).

Author's address: Dr. *S. Sawano,* Department of Endocrinology, Toranomon Hospital, 2 Akasaka Aoi-cho, *Minato-ku, Tokyo* (Japan)

Psychoneuroendocrinology. Workshop Conf. Int. Soc. Psychoneuroendocrinology, Mieken 1973, pp. 243–250 (Karger, Basel 1974)

Role of Monoamines in the Control of Growth Hormone and Prolactin Secretion in Man and Rats

H. Imura and Y. Kato

Third Division, Department of Medicine, Kobe University School of Medicine, Kobe

Introduction

Evidence has accumulated in recent years suggesting the important role of brain amines in regulating secretion of the anterior pituitary hormones. In 1968 we first observed that plasma human growth hormone (HGH) response to oral glucose loading was abnormal in patients with pheochromocytoma and that adrenergic-blocking agents affected plasma HGH levels in these patients (7, 19). Subsequently we studied the effect of various adrenergic, dopaminergic and serotonergic agents on plasma HGH levels in man (8–11). The present report consists of a review of our previous studies, and a description of our new experiments designed to clarify the role of monoaminergic mechanisms, in regulating secretions of human prolactin (HPr), rat growth hormone (RGH) and rat prolactin (RPr).

The Role of Monoamines in Regulating Secretions of HGH and HPr in Normal Subjects

Intravenous infusion of 10 mg of propranolol, a β-adrenergic blocking agent, caused a rise in plasma HGH in most of the subjects tested. (fig. 1). This rise in plasma HGH elicited by propranolol was completely suppressed by a combined administration of 0.36 mg of isoproterenol, a β-adrenergic stimulating agent, as shown in figure 1. The HGH rise elicited by propranolol was also suppressed by 50 g of glucose given orally immediately before the start of propranolol infusion (8). These results did not agree with those reported by *Blackard and Heidingsfelder* (1) who observed no significant change in plasma HGH during the infusion of propranolol. The discrepancy between their experiments and ours may be explained either by racial difference or by difference in dosage of propranolol per kg body weight. *Massara and Camanni* (16) observed a rise in plasma HGH after propranolol infusion in Orientals but not in Caucasians.

Intravenous infusion of 25 mg of methoxamine, an α-stimulant, caused a rise in plasma HGH in most of the subjects tested (fig. 1). This effect of methoxamine was significantly blunted by a concomitant administration of 20 mg of phentolamine, an α-blocking agent, as shown in figure 1.

Plasma HGH response to insulin-induced hypoglycemia was significantly blunted by the infusion of phentolamine and enhanced by the infusion of propranolol (1, 8, 22). The stimulatory effect of propranolol and the inhibitory effect of phentolamine were also observed in plasma HGH response to various stimuli, as shown in table I. All these results suggest that the adrenergic mechanism is involved in HGH secretion and that α-receptors stimulate HGH secretion, whereas β-receptors inhibit it. The only exception for this hypothesis is an observation by *Lucke and Glick* (15) that a rise in plasma HGH occurring during sleep was influenced neither by propranolol nor by phentolamine.

Boyd et al. (2) reported that oral administration of L-dopa raised plasma HGH in patients with parkinsonism. We observed that the intravenous infusion of 100 mg of L-dopa caused a significant plasma HGH rise in most of the normal subjects tested (9). This rise in plasma HGH was not influenced by oral adminis-

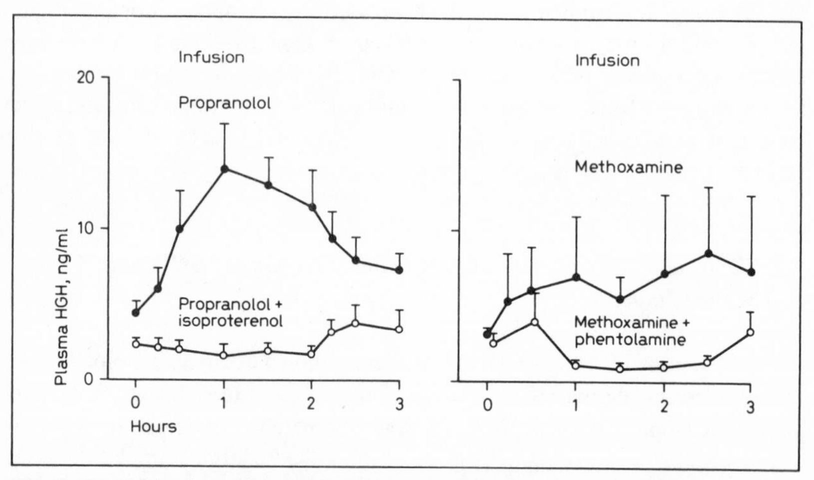

Fig. 1. Effect of adrenergic-stimulating or -blocking agents on plasma HGH levels in normal subjects. The left panel shows that the rise in plasma HGH induced by the intravenous infusion of 10 mg of propranol was completely inhibited by the combined infusion of 0.36 mg of isoproterenol. The right panel shows that the plasma HGH rise elicited by the infusion of 25 mg of methoxamine was significantly suppressed by the combined infusion of 20 mg of phentolamine. Means ± SEM are shown.

Table I. Effect of α-adrenergic blockade by phentolamine or β-adrenergic blockade by propranolol on plasma HGH response to various stimuli

Stimuli	α-Blockade	β-Blockade
Insulin-induced hypoglycemia	inhibit	enhance
Arginine	inhibit	enhance
Glucagon		enhance
Vasopressin	inhibit	
Exercise	inhibit	enhance
L-dopa	inhibit	enhance
5-HTP	inhibit	enhance
Sleep	no effect	no effect

tration of 50 g of glucose (9). We also observed that the intravenous infusion of propranolol enhanced plasma HGH response to L-dopa (9). *Kansal et al.* (13) reported the inhibitory effect of phentolamine on plasma HGH response to L-dopa. L-dopa is known to be converted to dopamine and, in part, to norepinephrine in the central nervous system. The fact that adrenergic-blocking agents affect plasma HGH response to L-dopa may suggest the role of norepinephrine converted from dopamine in the control of HGH secretion. However, *Takahashi et al.* (21) observed that fusaric acid, a dopamine β-hydroxylase inhibitor, did not influence plasma GH response to L-dopa in dogs. It is possible, therefore, that dopamine has a stimulatory effect on HGH secretion. Further studies are required to solve the problem.

In order to elucidate the role of serotonin, we observed the effect of 5-hydroxytryptophan (5-HTP) on plasma HGH levels. Oral administration of 150 mg of 5-HTP caused a rise in plasma HGH in most of the subjects tested (10). This rise in plasma HGH caused by 5-HTP was significantly inhibited by oral glucose loading (10). It was also suppressed by phentolamine or by cyproheptadine (17, 18) and enhanced by propranolol (17).

All these results indicate that adrenergic, dopaminergic and serotonergic mechanisms are involved in HGH secretion, although the exact relationship among these monoaminergic mechanisms needs further clarification.

The role of monoamines in regulating HPr secretion was also studied (our unpublished observation). Intravenous infusion of 0.36 mg of isoproterenol caused a significant rise in plasma HPr in most of the subjects tested. Intravenous infusion of 30 g of arginine caused a slight but significant rise in plasma HPr. This rise in plasma HPr was significantly enhanced by the infusion of phentolamine, an α-blocking agent, although phentolamine alone had no significant effect. These results suggest that the secretions of HPr and HGH are regulated in opposite ways by the adrenergic mechanism (table II).

Table II. Effect of adrenergic, dopaminergic and serotonergic agents on plasma levels of growth hormone and prolactin in man and rats

	Man		Rat	
	HGH	HPr	RGH	RPr
Adrenergic agents				
α-Stimulating agents	enhance	–	inhibit	–
β-Blocking agents	enhance	–	inhibit	–
β-Stimulating agents	inhibit	enhance	enhance	enhance
α-Blocking agents	inhibit	enhance	enhance	enhance
Dopaminergic agents				
L-dopa	enhance	inhibit	inhibit	inhibit
Chlorpromazine[1]	inhibit	enhance	enhance	enhance
Serotonergic agent				
5-HTP	enhance	enhance	no effect	enhance

1 Chlorpromazine is listed in this table as an antidopaminergic agent, although it was reported to have antiadrenergic action also.

It is well known that oral administration of L-dopa lowers plasma HPr levels in normal subjects (6). On the contrary, a significant rise in plasma HPr was observed following the oral administration of 150 mg of 5-HTP, as shown in figure 2 (14b). Cyproheptadine, a serotonergic antagonist, significantly blunted plasma HPr response to 5-HTP (14b). These results suggest a possible stimulatory role of serotonin in regulating HPr secretion.

The results of our experiments in man are summarized in table II. Secretions of HGH and HPr seem to be regulated in an opposite way by the adrenergic and dopaminergic mechanisms, although serotonergic mechanism appears to enhance secretion of both hormones.

The Role of Monoamines in Regulating Secretions of RGH and RPr in Rats

The effect of various monoaminergic agents on plasma RGH and RPr levels was studied in urethane-anesthetized male rats, by drawing blood serially from the jugular vein. Intravenous injection of isoproterenol elicited a significant rise in RGH. This rise was suppressed completely by the combined administration of propranolol (3, 14a). Plasma RGH was also increased following the injection of phentolamine. This effect of phentolamine was abolished by the concomitant administration of phenylephrine, an α-stimulating agent (14a). These results suggest that the adrenergic mechanism is involved in RGH secretion and that α-receptors suppress RGH secretion, whereas β-receptors enhance it.

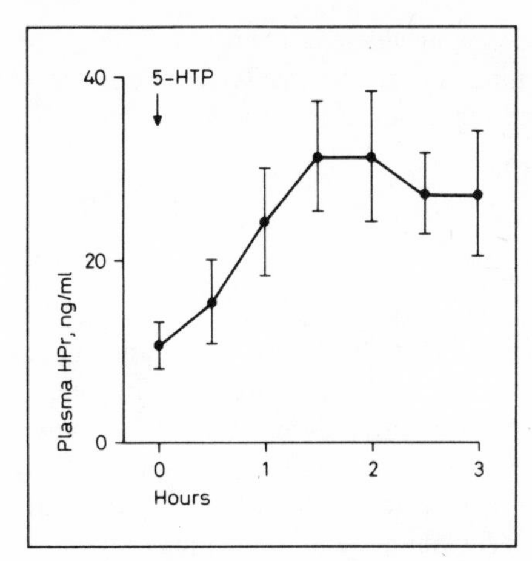

Fig. 2. Effect of the oral administration of 150 mg of 5-HTP on plasma prolactin levels in 13 normal subjects. Means ± SEM are shown.

Intravenous injection of chlorpromazine, a possible dopaminergic antagonist, significantly raised plasma RGH. This rise was blocked by the combined administration of L-dopa (3, 14a). It appears, therefore, that the dopaminergic mechanism plays an inhibitory role in RGH secretion. Intravenous injection of 5–HTP had no significant effect on plasma RGH in our experiments (3). However, *Collu et al.* (5) reported a rise in plasma RGH after the intraventricular injection of serotonin. The reason for this discrepancy must be studied further.

Intravenous injection of phentolamine or isoproterenol was capable of raising plasma RPr in urethane-anesthetized male rats (20). This suggests the involvement of the adrenergic mechanism in regulating RPr secretion.

Plasma RPr rose significantly in response to varying doses of 5-HTP (20). Since 5-HTP is known to be converted to serotonin in the brain, it seems likely that serotonin enhances RPr secretion. The rise in plasma RPr elicited by 5-HTP was inhibited completely by the combined administration of L-dopa, a precursor of dopamine. Dopamine was reported to lower plasma RPr, with an increase in blood levels of prolactin-inhibiting factor in pituitary portal vessels (12).

The results of our experiments in rats was summarized in table II. Both RGH and RPr secretions seem to be inhibited by either α-adrenergic or dopaminergic mechanism and enhanced by β-adrenergic mechanism. Serotonin appears to be capable of enhancing RPr secretion, although the effect of the amine on RGH secretion remains to be studied further. These results are similar to those observed in HPr secretion, but opposite to those observed in HGH secretion. The reason for the difference in receptor mechanisms between HGH and

RGH secretions remains unclear. One possible explanation is that HGH secretion is regulated by growth hormone-releasing factor, whereas RGH secretion is under the control of growth hormone-inhibiting factor. Further studies are required to solve the problem.

Possible Sites of Action of Monoamines

Wurtman (23) suggested six possible loci in the central or peripheral nervous system and the pituitary, at which monoamines might participate in the control of secretion of the anterior pituitary hormones. In an attempt to study this problem further, the effect of pretreatment with reserpine or α-methyl-*p*-tyrosine (α-MT), on plasma RGH response to chlorpromazine, was studied in rats (4). Both reserpine, a depletor of catecholamines and indolamines, and α-MT, a specific central catecholamine depletor, significantly suppressed plasma RGH response to chlorpromazine. This suggests that the site of action of monoamines is located in the central nervous system.

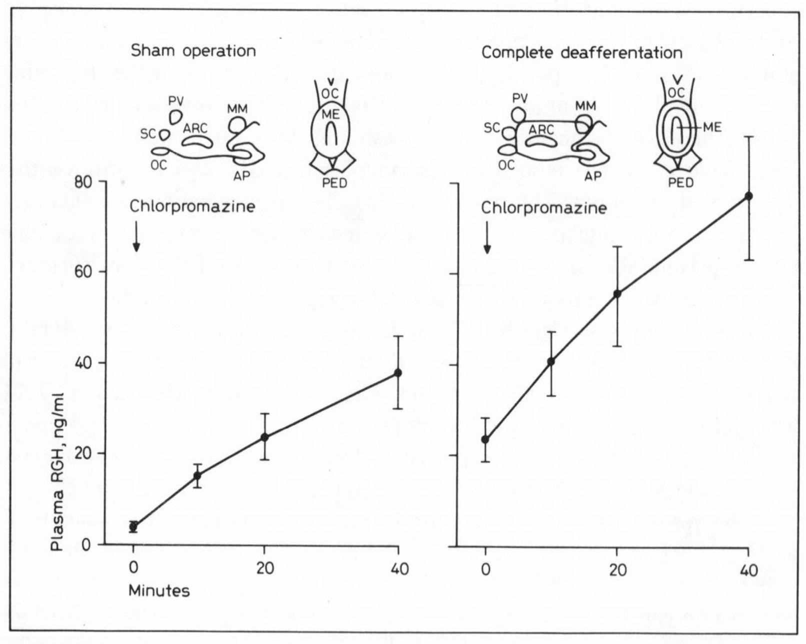

Fig. 3. Effect of the intravenous infusion of chlorpromazine (200 μg/100 g body weight) on plasma RGH levels in rats with the completely deafferentated hypothalamus and in sham-operated rats. Means ± SEM are shown.

We studied further the effect of chlorpromazine on plasma RGH in rats with the completely deafferentated hypothalamus. As shown in figure 3, basal RGH levels rose significantly 1 week after the complete hypothalamic deafferentation, although their response to chlorpromazine was maintained (21). Plasma RGH response to isoproterenol was also observed in rats after the complete hypothalamic deafferentation. These results lead to the conclusion that the site of action of these agents is in the hypothalamo-hypophyseal system. Although the direct action on the pituitary is not ruled out from our experiments, *Kamberi et al.* (12) reported that dopamine injected into the pituitary portal vessels was ineffective in inducing inhibition of prolactin release. It is most likely, therefore, that monoamines act at the level of the hypothalamus.

Conclusions

The role of the adrenergic, dopaminergic and serotonergic mechanisms in regulating secretions of HPr, RGH and RPr seems very similar, whereas HGH secretion is regulated in an opposite way by the adrenergic and dopaminergic mechanisms. The reason for this discrepancy remains to be solved. The stimulation of RGH secretion by chlorpromazine or isoproterenol was observed in rats with the completely deafferentated hypothalamus, but not in rats pretreated with α-MT or reserpine.

References

1 *Blackard, W.G. and Heidingsfelder, S.A.:* Adrenergic receptor control mechanism for growth hormone secretion. J. clin. Invest. *47:* 1407–1414 (1968).

2 *Boyd, A.E., III; Levobitz, H.E., and Pfeiffer, J.B.:* Stimulation of human growth hormone secretion by L-dopa. New Engl. J. Med. *283:* 1425–1429 (1970).

3 *Chihara, K.; Ohgo, S.; Kato, Y., and Imura, H.:* Studies on the regulation of growth hormone secretion. Folia endocrin. jap. *49:* 498 (1973) (abstr.).

4 *Chihara, K.; Ohgo, S.; Kato, Y., and Imura, H.:* Unpublished observation.

5 *Collu, R.; Frashini, F.; Visconti, P., and Martini, L.:* Adrenergic and serotonergic control of growth hormone secretion in adult male rats. Endocrinology, Springfield *90:* 1231–1237 (1972).

6 *Friesen, H.; Guyda, H.; Hwang, P.; Tyson, J.E., and Barbeau, A.:* Functional evaluation of prolactin secretion: a guide to therapy. J. clin. Invest. *51:* 706–709 (1972).

7 *Imura, H.; Kato, Y.; Ikeda, M.; Morimoto, M.; Yawata, M., and Fukase, M.:* Increased plasma levels of growth hormone during the infusion of propranolol. J. clin. Endocrin. *28:* 1079–1081 (1968).

8 *Imura, H.; Kato, Y.; Ikeda, M.; Morimoto, M., and Yawata, M.:* Effect of adrenergic-blocking or -stimulating agents on plasma growth hormone, immunoreactive insulin, and blood-free fatty-acid levels in man. J. clin. Invest. *50:* 1069–1079 (1971).

9 *Imura, H.; Nakai, Y.; Matsukura, S., and Matsuyama, H.:* Effect of intravenous infusion of L-dopa on plasma growth hormone levels in man. Horm. Metab. Res. *5:* 41–45 (1973).

10 *Imura, H.; Nakai, Y., and Yoshimi, T.:* Effect of 5-hydroxytryptophan (5-HTP) on growth hormone and ACTH release in man. J. clin. Endocrin. *36:* 204–206 (1973).

11 *Imura, H.; Nakai, Y.; Kato, Y.; Yoshimoto, Y., and Moridera, K.:* Effect of adrenergic agents on growth hormone and ACTH secretion; in Proc. 4th Int. Congr. Endocrinol. (Excerpta Medica, Amsterdam 1973).

12 *Kamberi, I.A.; Mical, R.D., and Porter, J.C.:* Effect of anterior pituitary perfusion and intraventricular injection of catecholamines on prolactin release. Endocrinology, Springfield *88:* 1012–1020 (1971).

13 *Kansal, P.C.; Buse, J.; Talbert, O.R., and Buse, M.G.:* The effect of L-dopa on plasma growth hormone, insulin, and thyroxine. J. clin. Endocrin. *34:* 99–105 (1972).

14a *Kato, Y.; Dupre, J., and Beck, J.C.:* Plasma growth hormone in the anesthetized rat: effects of dibutyryl cyclic AMP, prostaglandin E_1, adrenergic agents, vasopressin, chlorpromazine and L-dopa. Endocrinology, Springfield *93:* 135–146 (1973).

14b *Kato, Y.; Nakai, Y.; Imura, H.; Chihara, K., and Ohgo, S.:* Effect of 5-hydroxytryptophan (5-HTP) on plasma prolactin levels in man. J. clin. Endocrin. *38:* 381–383 (1974).

15 *Lucke, C. and Glick, S.M.:* Experimental modification of the sleep-induced peak of growth hormone secretion. J. clin. Endocrin. *32:* 729–736 (1971).

16 *Massara, F. and Camanni, F.:* Effect of various adrenergic receptor-stimulating and blocking agents on human growth hormone secretion. J. Endocrin. *54:* 195–206 (1972).

17 *Nakai, Y. and Imura, H.:* Effect of adrenergic-blocking agents on plasma growth hormone response to *L*-5-hydroxytryptophan (5-HTP) in man (in press, 1974).

18 *Nakai, Y.; Imura, H.; Sakurai, H., and Kurahachi, H.:* Effect of cyproheptadine on human growth hormone secretion J. clin. Endocrin. *38:* 100–103 (1974).

19 *Nakano, Y.; Imura, H.; Yawata, M.; Shinpo, S.; Ikeda, M.; Morimoto, M.; Manabe, S.; Kato, Y., and Fukase, M.:* Plasma immunoreactive insulin (IRI), growth hormone (HGH) and urinary catecholamines in patients with pheochromocytoma. Excerpta med. Int. Congr. Ser. *157:* 63 (1968) (abstr.).

20 *Ohgo, S.; Chihara, K.; Kato, Y., and Imura, H.:* Effect of amine precursors on prolactin secretion in rats. Folia endocrin. jap. *49:* 354 (1973) (abstr.).

21 *Takahashi, K.; Tsushima, T., and Irie, M.:* Effect of catecholamines on plasma growth hormone in dogs. Endocrin. jap. *20:* 323–330 (1973).

22 *Vinik, A.I. and Joubert, S.M.:* Growth hormone-adrenergic relationship. Life Sci. *9:* part I, 541–546 (1970).

23 *Wurtman, R.J.:* Brain catecholamines and the control of secretion from the anterior pituitary gland; in *Meites* Hypophysiotropic hormones of the hypothalamus: assay and chemistry, pp. 184–189 (Williams & Wilkins, Baltimore 1970).

Authors' address: Prof. *H. Imura* and Dr. *Y. Kato,* Third Division, Department of Medicine, Kobe University School of Medicine, 7 Kusunoki-cho, *Ikuta-ku, Kobe 650* (Japan)

Psychoneuroendocrinology. Workshop Conf. Int. Soc. Psychoneuroendocrinology, Mieken 1973, pp. 251–258 (Karger, Basel 1974)

Neurotransmission
A Proposed Mechanism of Steroid Hormones in the Regulation of Brain Function[1]

Antonia Vernadakis

Department of Psychiatry and Pharmacology, University of Colorado School of Medicine, Denver, Colo.

Introduction

Our early studies on the effects of hormones on brain development and function have shown that steroid hormones exert both excitatory and inhibitory effects depending on the stage of brain maturation at the time of hormone administration. For example, cortisol decreases brain excitability as assessed by electroshock seizure threshold when given to rats prior to 8 days postnatally and it increases brain excitability when given to rats after 8 days postnatally (*Vernadakis and Woodbury,* 1963). Cortisol treatment in the adult increases brain excitability (*Woodbury,* 1958). We interpret these results to mean that steroid hormones act on mechanisms of neurotransmission. Early work by *Woodbury* and associates (*Woodbury,* 1958) has shown that cortisol increases intracellular Na. Recent studies by *Vernadakis and Woodbury* (1971) have shown that cortisol increases intracellular K when given either for 6 h or 4 days to rats under 8 days of age, whereas it decreases intracellular K when given to 21-day-old rats. These changes in intracellular K concentration suggest changes in membrane potential of brain cells; an increase in intracellular K concentration, a hyperpolarized membrane and a decrease in intracellular K concentration, a lower membrane potential. In the same study *Vernadakis and Woodbury* (1971) found that cortisol decreases brain GABA levels when given to rats after 8 days postnatally. GABA has been implicated as an inhibitory neurotransmitter in the cerebral cortex (*Curtis,* 1969).

1 Supported by a Public Service Research Grant (NS 09199), University of Colorado General Research Support GRS 388, 416, 11, and a Research Scientist Development Award KO2 MH-42479 from the National Institute of Mental Health, NIH.

To further explore the hypothesis that cortisol may exert its effects on CNS activity by influencing neurotransmitter mechanisms, experiments were designed to investigate the effects of cortisol on the uptake of norepinephrine (NE) in the developing brain of the chick.

Uptake of Norepinephrine in Cerebral Hemispheres and Cerebellum

Prior to storage within noradrenergic nerves, exogenous NE is taken up across the neuronal membrane and subsequently across the membrane of the storage granule in which it is ultimately stored. In the adult brain it is possible to demonstrate neuronal uptake of NE *in vitro* in slices (*Dengler et al.*, 1962; *Rutledge and Jonason,* 1967; *Rutledge,* 1970), in isolated synaptosomes (*Davis et al.*, 1967) and in homogenates (*Snyder and Coyle,* 1969).

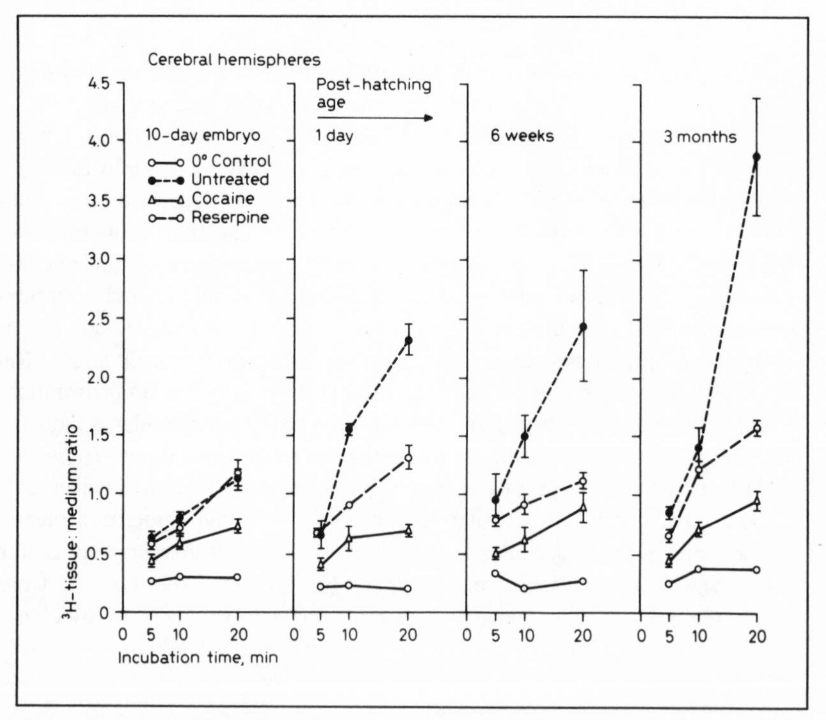

Fig. 1. Uptake of ^{3}H-NE into slices of cerebral hemispheres from 10-day chick embryos and from chicks at 1 day, 6 weeks and 3 months after hatching. The effects of cocaine (3×10^{-5} *M*) and reserpine (10^{-6} *M*) are illustrated. Points with vertical lines represent mean ± SE for 5–6 determinations.

In the present study, uptake of NE was studied *in vitro* using slices of the cerebral cortex and cerebellum from 10- and 20-day chick embryos, and at 1 day, 6 weeks, and 3 months after hatching. A modification of the procedure of *Snyder et al.* (1968) was followed and is described in detail elsewhere (*Kellogg et al.*, 1971).

There was a marked increase in the rate of accumulation of ^{3}H-NE in the cerebral hemispheres from chicks 1 day after hatching compared to cerebral hemispheres from 10-day chick embryos; also, there was a marked increase between 1 day after hatching and 3 months after hatching (fig. 1). Incubation of the tissue slices in the presence of cocaine (3×10^{-5} *M*), an agent known to inhibit neuronal uptake, resulted in a marked inhibition of the accumulation of ^{3}H-NE at both embryonic and post-hatching ages. When 10^{-6} *M* reserpine, an agent which inhibits uptake of biogenic amines into storage granules, was added to the incubation medium there was marked inhibition of ^{3}H-NE only in 20-day

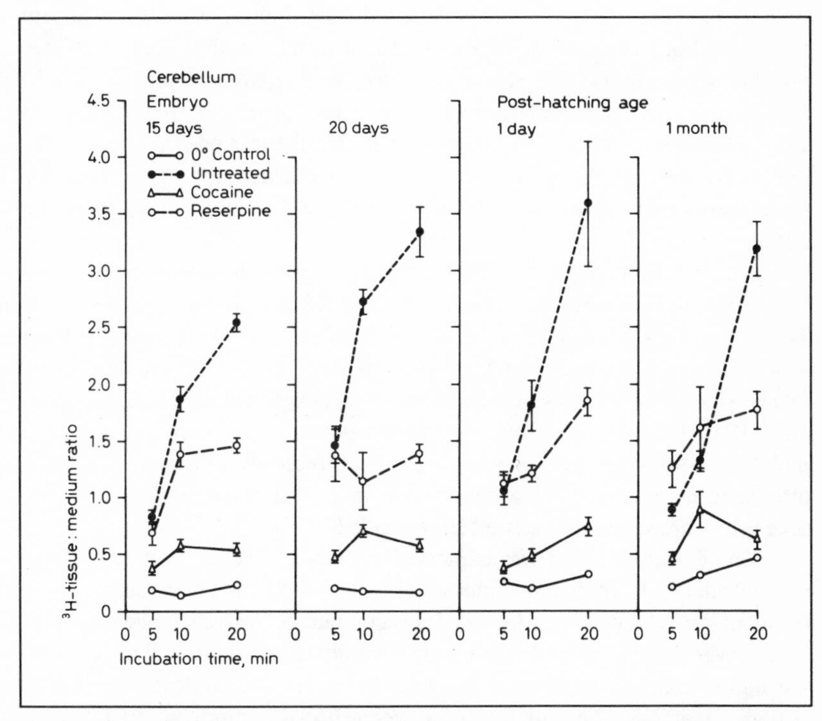

Fig. 2. Uptake of ^{3}H-NE into slices of cerebellum from 15- and 20-day chick embryos, and from chicks at 1 day and 1 month after hatching. The effects of cocaine (3×10^{-5} *M*) and reserpine (10^{-6} *M*) are illustrated. Points with vertical lines represent mean ± SE for 5–6 determinations.

embryos, not shown in figure 1 (*Kellogg et al.*, 1971) and in chicks after hatching. It is concluded from these data that uptake processes in the cerebral hemispheres develop earlier than the mechanisms for storage of NE.

Uptake and storage of ^{3}H-NE by cerebellar slices occurred in chicks at 15 days of embryonic age (fig. 2). No changes were observed in the uptake or storage of NE after 20 days of embryonic age. It appears from these data that adrenergic mechanisms reach maturation earlier in the cerebellum than in the cerebral hemispheres in chicks.

Extraneuronal Uptake of Norepinephrine

The foregoing finding shows that in both the cerebral hemispheres and cerebellum of chicks, ^{3}H-NE accumulation progressively increases during brain maturation, reaching maximum levels in the cerebral hemispheres at 3 months after hatching and in the cerebellum at 20 days of embryonic age. The continuous increase in ^{3}H-NE accumulation during brain maturation is interpreted to reflect both maturation of the neuronal uptake and storage processes and possible accumulation of ^{3}H-NE in cells other than neurons, i.e. glial cells. Evidence that NE accumulates in glial cells has recently been reported by *Henn and Hamberger* (1971) who found that glial cell-enriched brain fractions accumulate NE. *Iversen* (1971) has reported extraneuronal uptake (Uptake$_2$) of NE in the peripheral nervous system with different affinity properties than neuronal uptake (Uptake$_1$).

The present experiments were designed to investigate kinetic properties of NE uptake in cerebellar slices from chicks at 14 and 20 days of embryonic age. These ages were chosen because of the difference in the glial cell population; glial cells begin to proliferate around 20 days of embryonic age (*Hanaway*, 1967). Cerebellar slices were incubated at varying doses of ^{3}H-NE, ranging from 10^{-7} *M* to 10^{-6} *M*, for 5 min. This time interval was chosen because NE accumulation in brain slices in a 5-min incubation represents primarily uptake of NE into neurons, whereas NE accumulation in a 20-min incubation represents uptake into storage granules as well (*Iversen*, 1967).

Amine accumulation was expressed as *p*mol of ^{3}H per 5 min/g of tissue and was calculated from the specific activity of ^{3}H-NE in the medium. The reciprocals of ^{3}H-NE concentration and its accumulation into slices were analyzed by Lineweaver-Burk plots and Km values were obtained (table I). The Km value for the higher affinity component has been interpreted to reflect neuronal uptake (Uptake$_1$) of NE, the Km for the lower affinity component to reflect extraneuronal uptake (Uptake$_2$) of NE. However, before definite conclusions can be drawn from these data, experiments using inhibitors of Uptake$_1$ and Uptake$_2$ must be performed. In the peripheral nervous system, it has been shown that

Table I. Kinetic constants for the accumulation of ^{3}H-NE into cerebellar slices from chicks during embryonic age

Embryonic age, days	Km values[1]	
	Uptake$_1$	Uptake$_2$
14	5.0×10^{-6} *M*	–
16	2.0×10^{-6} *M*	–
20	0.6×10^{-6} *M*	5.0×10^{-6} *M*

1 For evaluation of kinetics of NE accumulation, the reciprocals of NE accumulation and NE concentration for 5 min incubation were analyzed by Lineweaver-Burk plots.

extraneuronal uptake (Uptake$_2$) occurs with high concentrations of epinephrine or norepinephrine and that Uptake$_2$ is a low affinity system (*Iversen,* 1971). Moreover, Uptake$_2$ is inhibited by different drugs than those found to be effective as Uptake$_1$ inhibitors (*Iversen,* 1971).

Effects of Hormones on the Uptake of Norepinephrine

Recently *Iversen and Salt* (1970) found that certain steroids are particularly powerful inhibitors of Uptake$_2$ and most of these substances are not active as inhibitors of Uptake$_1$. Based on this evidence, experiments were designed to study the effects of cortisol on the uptake of NE in cerebellar tissue of chick embryos.

In this study, cerebellar explants were removed from 16-day chick embryos and cultured for 4 h in the presence or absence of cortisol. A schematic representation of the set-up is illustrated in figure 3 (*Vernadakis,* 1971). Cerebellar explants were removed from 16-day chick embryos and were oriented on a triangular stainless steel organ culture grid. Platforms with explant were placed in organ culture dishes with a center well and an absorbent ring. The medium, 0.5 ml, Eagle's basal medium (1955) was added to the center well. When cortisol was added to the medium, the final concentrations were either 1×10^{-5} *M* or 2.76×10^{-5} *M.* Humidity was maintained by saturating the absorbent ring with distilled water. Explants were cultured at 35–36 °C for 4 h; the gas phase was 95 % O_2–5 % CO_2. After 4 h explants were weighed and removed from the grids and placed in vials containing Krebs-Henseleit incubation medium and processed for a 20-min ^{3}H-NE uptake, as described previously (*Kellogg et al.,* 1971). The results are summarized in table II. These preliminary findings show that the accumulation of ^{3}H-NE at 10^{-6} *M* was lower in explants cultured in the

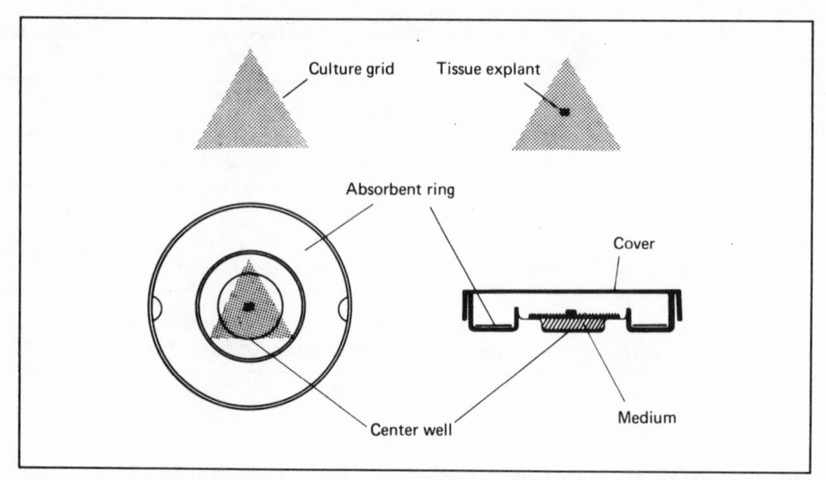

Fig. 3. A diagram for an organ culture set-up (from *Vernadakis and Woodbury*, 1971, fig. 1).

Table II. Tissue/medium ratios for 20 min. ^{3}H-NE accumulation in cerebellar explants removed from 16-day chick embryos and cultured in the presence of cortisol for 4 h

Treatment	^{3}H-NE concentration	
	10^{-7} *M*	10^{-6} *M*
Control	2.17 ± 0.24[1]	2.27 ± 0.34
Cortisol, 1 × 10^{-5} *M*	–	2.52 ± 0.41
Cortisol, 2.76 × 10^{-5} *M*	2.25 ± 0.30	1.61 ± 0.11 (< 0.02)[2]

1 Each value represents the mean ± SE of 6–8 samples from two experiments.
2 Represents P value for comparison to control.

presence of cortisol, 2.76 × 10^{-5} *M.* Since the inhibitory effect of cortisol was observed with the high concentration of NE (10^{-6} *M*) it is suggested that extraneuronal (glial) uptake may be primarily affected. This proposal is consistent with the findings of *Iversen* (1971) that some steroid hormones inhibit extraneuronal uptake ($Uptake_2$) of NE in the peripheral nervous system where high concentrations of NE are used. Before definite conclusions can be drawn, experiments using glial cells and neurons alone will provide more concrete information regarding the kinetic properties of uptake of NE by glial cells and neurons. For instance, the fractionation studies by *Henn and Hamberger* (1971) mentioned earlier have shown that glial cell-enriched brain fractions accumulate NE.

Conclusions

The importance of extraneuronal uptake of NE in the CNS can only be speculative at the present time. If the role of glial cells in neurotransmission is to provide a safety valve, i.e. to limit the possible build up of neurotransmitter substances extracellularly, then inhibition of glial cell uptake could lead to an intracellular-extracellular imbalance, and result in deleterious cellular effects. For example, excessive amounts of NE in the synaptic cleft would make more NE available to stimulate the CNS and would result in CNS hyperexcitability known to occur with cortisol treatment (*Woodbury,* 1958). Moreover, we suggest that the function of glial cells in neurotransmission may be age dependent and may vary among the CNS areas. For example, during early brain development, glial cells may be involved in neuronal growth, i.e. myelination. After maturation is reached, glial cells would begin to act as regulators of the extraneuronal environment including neurotransmission. During aging, glial cells may take over some neuronal functions. For example, we have found that with aging the uptake of NE markedly declines in the cerebral hemispheres but not in the cerebellum (*Vernadakis,* 1973b). Since glial cells continue to proliferate with age in both the cerebral hemispheres and cerebellum (*Vernadakis,* 1973a), then one can assume that the function of the glial cells in the cerebral hemispheres differs with age from that in the cerebellum.

References

Curtis, D.R.: Central synaptic transmitters; in *Jasper, Ward and Pope* Basic mechanisms of the epilepsies, pp. 105–129 (Little, Brown, Boston 1969).

Davis, J.M.; Goodwin, F.K.; Bunney, W.E.; Murphy, D.L., and Colburn, R.W.: Effects of ions in uptake of norepinephrine by synaptosomes. Pharmacologist *9:* 184 (1967).

Dengler, G.J.; Michaelson, I.A.; Spiegel, H.E., and Titus, E.: The uptake of labeled norepinephrine by isolated brain and other tissues of the cat. Int. J. Neuropharmacol. *1:* 23–28 (1962).

Eagle, H.: Nutrition needs of mammalian cells in tissue culture. Science *122:* 501–504 (1955).

Hanaway, J.: Formation and differentiation of the external granular layer of the chick cerebellum. J. comp. Neurol. *131:* 1–14 (1967).

Henn, F.A. and Hamberger, A.: Glial cell function: uptake of transmitter substances. Proc. nat. Acad. Sci., Wash. *68:* 2686–2690 (1971).

Iversen, L.L.: The uptake and storage of noradrenaline in sympathetic nerves (Cambridge University Press, Cambridge 1967).

Iversen, L.L.: Role of transmitter uptake mechanisms in synaptic neurotransmission. Brit. J. Pharmacol. *41:* 571–591 (1971).

Iversen, L.L. and Salt, P.J.: Inhibition of catecholamine $uptake_2$ by steroids in the isolated heart. Brit. J. Pharmacol. *40:* 528–530 (1970).

Kellogg, C.; Vernadakis, A., and Rutledge, C.D.: Uptake and metabolism of ^{3}H-norepinephrine in the cerebral hemispheres of chick embryos. J. Neurochem. *18:* 1931–1938 (1971).

Rutledge, C.D. and Jonason, J.: Metabolic pathways of dopamine and norepinephrine in rabbit brain *in vitro.* J. Pharmacol. exp. Ther. *157:* 493–502 (1967).
Rutledge, C.D.: The mechanisms by which amphetamine inhibits oxidative deamination of norepinephrine in brain. J. Pharmacol. exp. Ther. *171:* 188–195 (1970).
Snyder, S.H. and Coyle, J.T.: Regional differences in ^{3}H-norepinephrine and ^{3}H-dopamine uptake into rat brain homogenates. J. Pharmacol. exp. Ther. *165:* 78–86 (1969).
Snyder, S.H.; Green, A.I., and Hendley, E.D.: Kinetics of ^{3}H-norepinephrine accumulation into slices from different regions of rat brain. J. Pharmacol. exp. Ther. *164:* 90–102 (1968).
Vernadakis, A.: Hormonal factors in the proliferation of glial cells in culture; in *Ford* Influence of hormones on the nervous system, pp. 42–55 (Karger, Basel 1971).
Vernadakis, A.: Changes in nucleic acid content and butyrylcholinesterase activity in CNS structures during the life span of the chicken. J. Geront. *28:* 281–286 (1973a).
Vernadakis, A.: Uptake of ^{3}H-norepinephrine in the cerebral hemispheres and cerebellum of the chicken throughout the life span. Mech. Aging Develop. *2:* 371–379 (1973b).
Vernadakis, A. and Woodbury, D.M.: Effect of cortisol on the electroshock seizure threshold in developing rats. J. Pharmacol. exp. Ther. *139:* 110–113 (1963).
Vernadakis, A. and Woodbury, D.M.: Effects of cortisol on maturation of the central nervous system; in *Ford* Influence of hormones on the nervous system, pp. 85–97 (Karger, Basel 1971).
Woodbury, D.M.: Relation between the adrenal cortex and the central nervous system. Pharmacol. Rev. *10:* 275–357 (1958).

Author's address: Dr. *Antonia Vernadakis,* Departments of Psychiatry and Pharmacology, University of Colorado School of Medicine, *Denver, CO 80220* (USA)

Psychoneuroendocrinology. Workshop Conf. Int. Soc. Psychoneuroendocrinology, Mieken 1973, pp. 259–266 (Karger, Basel 1974)

Possible Involvement of Cyclic Nucleotides in the Stimulation of Pituitary Function Elicited by Reserpine

A. Guidotti, B. Zivkovic and E. Costa

Laboratory of Preclinical Pharmacology, National Institute of Mental Health, Saint Elizabeth Hospital, Washington, D.C.

Adrenal steroid hormones participate in the regulation of neuronal function in several ways: one of them, which is particularly pertinent to noradrenergic neurons, appears to be linked with the induction of regulatory enzymes. When drugs or environmental stimuli determine a persisting increase of the function of noradrenergic neurons, these adapt to this increase of function with a delayed and long-term increase of the formation of tyrosine hydroxylase (TH) and dopamine-β-hydroxylase (DBH) (*Axelrod,* 1971). Steroid hormones appear to exert a permissive action for the induction of TH and DBH (*Muller et al.*, 1970; *Gewirtz et al.,* 1971) and brain tryptophan hydroxylase activities (*Millard et al.*, 1972; *Zivkovic et al.,* 1974). The synthesis of enzymes regulating monoamine formation may be altered in affective disorders as suggested by *Wise and Stein* (1973) who showed that DBH activity in brain of patients affected by schizophrenia is lower than that found in the brain of non-schizophrenic. The decrease in schizophrenia of enzyme activities which are regulated by steroid hormones, prompts the pharmacologist to study the molecular nature of pituitary function regulation because the output of pituitary hormones may be a final control of the mechanisms whereby noradrenergic and serotonergic neurons adapt to the environment. These studies may establish new trends in the development of drugs to be used in the treatment of neuropsychiatric disorders. In various mammalian tissues, cyclic 3′,5′-adenosine monophosphate (cAMP) and cyclic 3′,5′-guanosine monophosphate (cGMP) function as intracellular second messengers and transduce stimuli acting on the cell membrane into a cascade of intracellular biochemical events. In neurons, cAMP and cGMP possess short-term and long-term effects. The former include phosphorylation of membrane proteins to facilitate membrane depolarization and participation in the release of transmitters from storage sites, and it appears that these short-term effects involve an activation of protein kinase (*Greengard et al.*, 1972). Among the long-term effects there is the regulation of synthesis of rate-limiting enzyme that makes possible long-term adaptation of neuronal activity to an increase of functional output (*Guidotti and Costa*, 1973). This report concerns the measurement of cAMP and cGMP concentrations in hypothalamus, pituitary, and adrenal gland of rats receiving reserpine. This drug produces a decrease of monoamines in brain (*Brodie et al.*, 1955; *Holzbauer and Vogt*, 1956), and elicits long-term increases of tyrosine hydroxylase in brain dopaminergic tracts (*Segal et al.*, 1971) and of tryptophan hydroxylase in brain serotonergic axons (*Zivkovic et al.*, 1974).

Methods

Sprague Dawley male rats (150–180 g) were purchased from Zivic Miller (Allison Park, Pa.) and housed for at least 5 days at 23 °C in rooms illuminated from 6 a.m. to 6 p.m. Each experiment was terminated by exposing the rat head to microwave radiation (1–2 sec) produced by a 2 kw generator at 2.5 GHz. This procedure causes a complete and instantaneous inactivation of brain enzyme, thus minimizing the changes of cAMP and cGMP concentration in tissues resulting from postmortem events. The concentrations of cAMP and cGMP were measured by a procedure reported elsewhere (*Mao and Guidotti*, 1974). The plasma corticosteroids were measured with the method reported by *Zenker and Bernstein* (1958).

Results

a) Concentration of cAMP and cGMP in Pituitary

We have analyzed the concentrations of cAMP and cGMP in anterior and posterior pituitary in normal rats. The data reported in table I show that the concentrations of cAMP are higher in the posterior than in the anterior pituitary, whereas the concentrations of cGMP are lower in the posterior pituitary than in the anterior. The cAMP:cGMP ratio in these two tissues is therefore different, being 5-fold greater in the posterior than in the anterior pituitary.

Table I. cAMP and cGMP concentrations in anterior and posterior pituitary of rat. Each value is the mean ± SE of 4–5 determinations

Tissue	pmol/mg protein	
	cAMP	cGMP
Anterior pituitary	6.2 ± 1	0.75 ± 0.1
Posterior pituitary	14.0 ± 2	0.33 ± 0.02

b) Effect of Reserpine Injections on Cyclic Nucleotide Concentrations in Pituitary and Other Brain Areas

Reserpine injected i.p. in dose of 16 μmol/kg produces a rapid increase of cyclic AMP in anterior pituitary. The zenith of the response occurs within 15 min when the concentration of cAMP is about 12 times greater than that of untreated rats (fig. 1). The cAMP concentration after 2 h is still five times the concentration measured in the anterior pituitary of saline-treated animals (fig. 1). Reserpine does not affect the concentration of cGMP of anterior pituitary at any time (fig. 1). Reserpine fails to change the cGMP or cAMP concentrations in hypothalamus or striatum (fig. 1). Reserpine also produces an 8-fold increase of cAMP concentration in posterior pituitary with a time constant similar to that described in the anterior pituitary. In our dissection the posterior pituitary contains also the infundibulum and at this time we cannot decide whether the change observed occurs in this tissue.

In table II are reported the concentrations of cAMP in adrenal cortical tissue at various times after reserpine injection. As can be seen, reserpine increases the cAMP concentrations of this organ. Moreover, the changes of cAMP concentrations elicited by reserpine are completely abolished by hypophysectomy. This dependence strongly suggests that the activation of the adenylate cyclase system of adrenal cortex is dependent from pituitary activation and presumably involves the release of ACTH from pituitary.

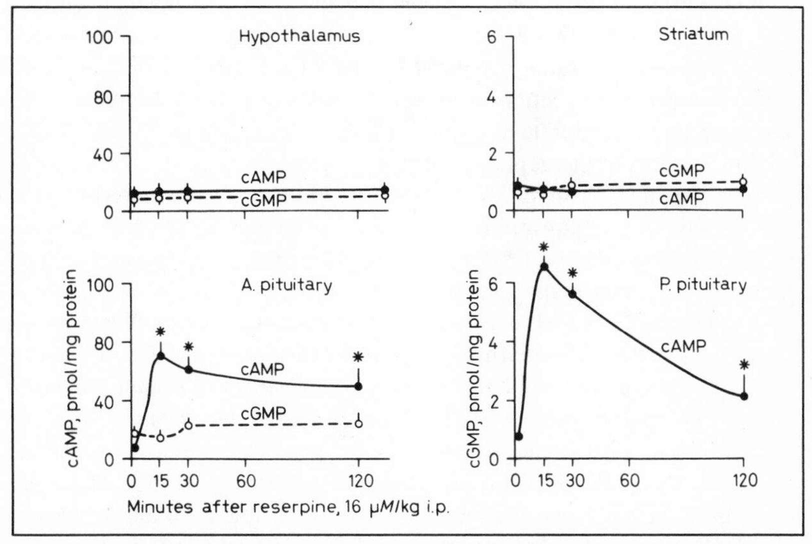

Fig. 1. cAMP and cGMP concentrations in different brain areas of rat receiving reserpine. Each value is the mean of five determinations. Vertical bars represent SE of the mean. $*p < 0.01$.

Table II. cAMP concentrations in adrenal cortex of rats receiving reserpine 16 μmol/kg, i.p. Each value is the mean of at least 4 experiments

Minutes after reserpine	cAMP pmol/mg protein, mean value ± SE
0	8.8 ± 1.2
15	110.0 ± 15.0[1]
30	86.0 ± 23.0[1]
60	26.0 ± 6.0

1 $p < 0.01$.

c) Role of Monoaminergic Neurons in the Regulation of Pituitary cAMP

The activity of catecholaminergic terminals impinging upon hypothalamus regulates ACTH secretion, probably by reducing the secretion of corticotropic releasing factor (*Van Loon,* 1973). For example, reserpine, a drug which impairs the function of catecholaminergic neurons increases ACTH excretion (*Martel et al.,* 1962). Serotonin, which is also depleted by reserpine treatment, may concur to the regulation of ACTH release (*Kreiger and Rizzo,* 1969). To elucidate whether the change of cAMP in anterior pituitary and the ensuing activation of adrenal cortex were related to monoaminergic neuronal function, we measured pituitary cAMP concentration and corticosteroid concentrations in plasma after administration of drugs which change the function of monoaminergic neurons. The data reported in figure 2 show that the increase of plasma corticosteroid and cAMP in anterior pituitary induced by reserpine is blocked by pargyline injection. However, the injection of deprenyl, another monoamine oxidase inhibitor which selectively inhibits dopamine metabolism (*Yang and Neff,* 1973) and inhibits selectively the depletion of brain dopamine caused by reserpine, did not prevent the increase of cAMP in anterior pituitary and the increase of plasma corticosteroids concentrations elicited by reserpine (fig. 2). Administration of parachlorophenylalanine (PCPA), a drug which blocks serotonin synthesis and therefore decreases brain serotonin concentrations (*Koe and Weissman,* 1966), fails to change the cAMP concentrations in anterior pituitary. When reserpine is injected to PCPA-pretreated rats, this alkaloid increases cAMP concentrations in pituitary and corticosteroid concentrations in plasma in a fashion similar to that observed in normal animals (fig. 3). These results are in agreement with the view that an impairment of the function of noradrenergic neurons may be the most important factor involved in the activation of adrenal cortex elicited by reserpine. In addition, it may be suggested that adrenergic function deficiency may be responsible for the increase of pituitary cAMP concentrations and that the balance between serotonergic and noradrenergic function does not seem operative in this regulation. Moreover, cAMP may be viewed as the second messenger,

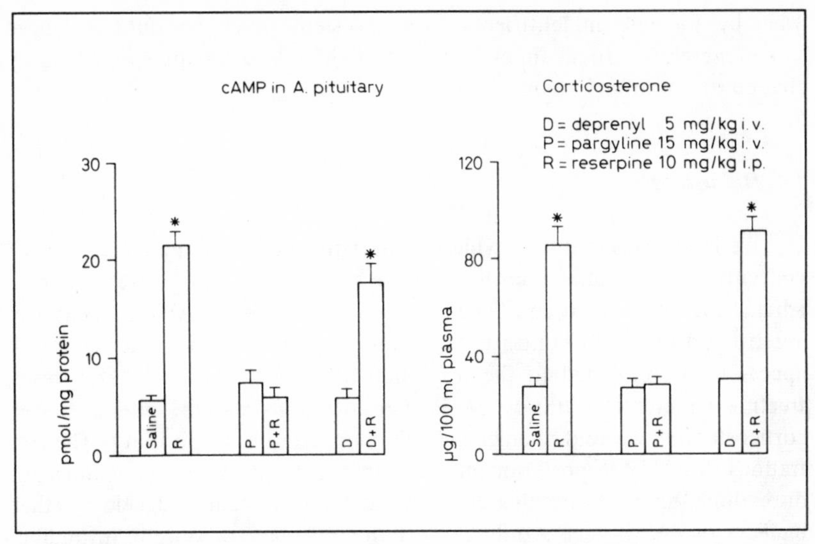

Fig. 2. Effect of pargyline and deprenyl on reserpine-elicited increase of pituitary cAMP and plasma corticosterone. Deprenyl or pargyline were injected 5 min before reserpine. cAMP and corticosteroids were assayed 2 h after treatment with drugs. Each value represents the mean ± SE of 5–6 animals. $*p < 0.01$.

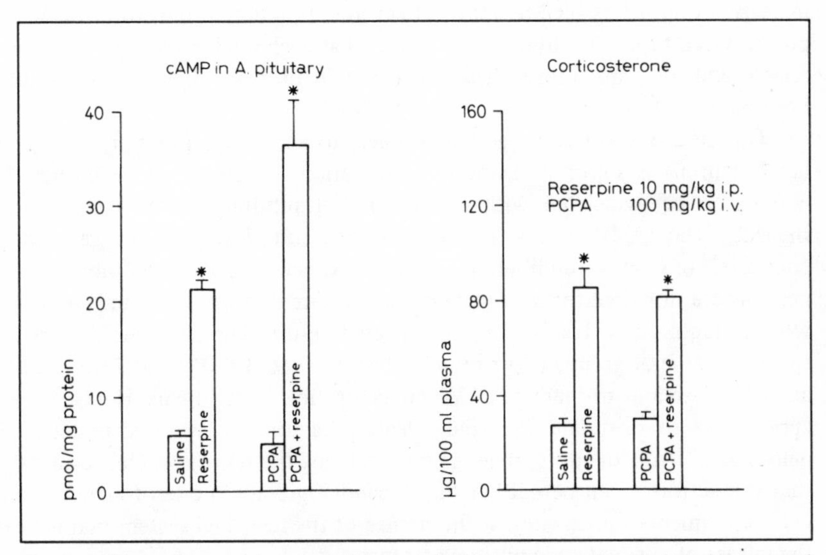

Fig. 3. Effect of parachlorophenylalanine (PCPA) on reserpine-elicited increase of pituitary cAMP and plasma corticosterone. PCPA methyl ester was injected 48 h before reserpine. cAMP and corticosterone were assayed 2 h after treatment with reserpine.

whereby the yet unidentified neuronal system which is counterbalanced by noradrenergic neuronal function regulates the release of the specific hormones that control pituitary function.

Discussion

We have presented data which support the view that the increase of cAMP concentrations in adrenal cortex and the secretion rate of corticosteroids from adrenal are interrelated and depend upon an increase of cAMP concentration in anterior pituitary. In contrast, the anterior pituitary concentrations of cGMP appear to be unrelated to the regulation of adrenal cortex. In fact, reserpine treatment increases pituitary cAMP concentrations and the plasma content of corticosteroids without changing cGMP concentration in pituitary. The concentrations of cAMP in posterior pituitary are increased by reserpine. Since, in our dissection, this part contains also infundibulum we cannot decide whether the increase of cAMP occurs only in posterior pituitary or only in infundibulum. Both cAMP and cGMP are probably involved in the processes of hormone release from storage sites in pituitary (*Peake,* 1973). However, the specific role of cAMP or cGMP in the release of specific anterior pituitary hormones has not been clearly defined. We can advance the hypothesis that the increase of pituitary cAMP may reflect the activation of a particular pituitary cell type, whereas the increase of pituitary cGMP may reflect the functional activation of another pituitary cell type. This hypothesis implies that each nucleotide may regulate the release and, perhaps, biosynthesis of a certain pituitary hormone or group of hormones.

Our results seem to suggest that when noradrenergic function of hypothalamic neurons is suppressed, then the activation of adrenal cortex ensues. This appears to be associated with an increase of pituitary cAMP concentration. Indeed, if the cAMP increase in the posterior pituitary reflects changes of cAMP concentrations in infundibulum, it is unexpected that when noradrenergic neurons are impaired the adenylate cyclase in target cells is activated. Additional work is required to resolve this unexpected result. The releasing factors were shown by several groups (*Zorr et al.,* 1970; *Steiner,* 1970) to activate pituitary adenylcyclase and to increase cAMP concentration in this tissue. Hence, it seems appropriate to postulate that the releasing factors are secreted under a dual neuronal control, the noradrenergic system being the repressor. Our results show that the activator can be neither the dopaminergic nor the serotonergic system. Our experiments fail to suggest the nature of the neuronal system that activates the release of corticotropic releasing factors.

If we want to follow the hypothesis, that changes in the function of monoaminergic neurons at the hypothalamic level may be of importance in the regulation of the releasing factors, then we could expect changes of cyclic nucleotides in the hypothalamus. As shown in figure 1, reserpine changes cAMP concentration in A. pituitary without changing cAMP or cGMP concentration in hypothalamus. Therefore, we can hypothesize that the hypothalamic peptides are released independently from changes of cyclic nucleotides or that cyclic nucleotides may change in very defined hypothalamic areas such as the infundibulum. However, we also can support the possibility suggested by some authors, that amines may alter the functional state of hypothalamic-hypophyseal portal blood flow and by such mechanisms influence indirectly the arrival of hypothalamic releasing factors to the anterior hypophysis.

In closing, we would like to summarize the concepts emerging from the present report: using reserpine as a pharmacological tool to produce a decrease in catecholaminergic nerve function, we can establish that this process is paralleled by changes of cAMP in pituitary adrenal axis and increase of corticosteroid in plasma. At the hypothalamic pituitary level the regulation of cyclic AMP appears suppressed by noradrenergic nerves while changes of serotonin or dopamine concentrations are not involved. We are now studying if an interaction between norepinephrine, cAMP, and corticosteroids, similar to that observed after reserpine administration, applies also to other drugs and stimuli that produce delayed, long-lasting changes in specific neuronal enzymes.

Summary

Reserpine (16 μmol/kg i.p.) drastically increases (about 10-fold) adenosine 3′,5′-monophosphate (cAMP) concentrations in rat anterior pituitary without changing guanosine 3′,5′-monophosphate (cGMP) concentrations. Reserpine also increases cAMP concentrations in posterior pituitary and adrenal cortex but does not change cAMP and cGMP concentrations in hypothalamus or caudate. Reserpine (16 μmol/kg i.p.) produces an increase of plasmatic corticosteroids.

To investigate whether the activation of pituitary adrenal axis by reserpine was related to monoaminergic neuronal function, we measured pituitary cAMP and plasma corticosteroids after administration of drugs which change the function of monoaminergic neurons.

Pargyline, 15 mg/kg i.v. prevents the reserpine-induced changes of pituitary cAMP and plasma corticosteroids. Deprenyl (5 mg/kg i.v., a monoamine oxidase inhibitor specific for dopamine) and parachlorophenylalanine (PCPA) do not prevent the changes induced by reserpine. These results support the concept that norepinephrine and not dopamine or serotonin express a negative control on the hypothalamus pituitary adrenal function.

References

Axelrod, J.: Noradrenaline. Fate and control of its biosynthesis. Science *173:* 589–606 (1971).

Brodie, B.B.; Pletscher, A., and Shore, P.A.: Evidence that serotonin has a role in brain function. Science *122:* 968 (1955).

Gewirtz, G.P.; Kvetnansky, R.; Weise, V.K., and Kopin, I.J.: Effect of hypophysectomy on adrenal dopamine-β-hydroxylase activity in rat. Molec. Pharmacol. *7:* 163–168 (1971).
Greengard, P.; McAfee, D.A., and Kebabian, J.W.: in *Greengard, Paoletti and Robison* Advances in cyclic nucleotides research, vol. 1, pp. 337–355 (Raven Press, New York 1972).
Guidotti, A. and Costa, E.: Involvement of adenosine 3',5'-monophosphate in the activation of tyrosine hychoxylase elicited by drugs. Science *179:* 902–904 (1973).
Holzbauer, M. and Vogt, M.: Depression by reserpine of noradrenaline concentration in the hypothalamus of cat. J. Neurochem. *1:* 8–11 (1956).
Koe, B.K. and Weissman, A.: *p*-Chlorophenylalanine. A specific depletion of brain serotonin. J. Pharmacol. exp. Ther. *154:* 499–516 (1966).
Kreiger, D.T. and Rizzo, F.: Serotonin mediation of circadian periodicity of plasma 17-hydroxycorticosteroids. Amer. J. Physiol. *217:* 1703–1707 (1969).
Mao, C.C. and Guidotti, A.: Simultaneous isolation of adenosine 3',5'-cyclic monophosphate (cAMP) and guanosine 3',5'-cyclic monophosphate (cGMP) in small tissue samples. Analyt. Biochem. (in press, 1973).
Martel, R.R.; Westermann, E.O., and Maickel, R.P.: Dissociation of reserpine-induced sedation and ACTH hypersecretion. Life Sci. *4:* 151–155 (1962).
Millard, S.A.; Costa, E., and Gal, E.M.: On the control of brain serotonin turnover rate by end-product inhibition. Brain Res. *40:* 545–551 (1972).
Muller, R.A.; Thoenen, H., and Axelrod, J.: Effect of pituitary and ACTH on the maintenance of basal tyrosine hydroxylase activity in the rat adrenal gland. Endocrinology, Springfield *86:* 751–755 (1970).
Peake, G.T.: The role of cyclic nucleotides in the secretion of pituitary growth hormone; in *Ganong and Martini* Frontiers in neuroendocrinology, pp. 173–208 (Oxford University Press, 1973).
Segal, D.S.; Sullivan, J.L.; Kuczenski, R.T., and Mandell, A.J.: Effect of long-term reserpine treatment on brain tyrosine hydroxylase and behavioral activity. Science *173:* 847–849 (1971).
Steiner, A.L.; Peake, G.T.; Utiger, R.; Karl, I.E., and Kipnis, D.M.: Hypothalamic stimulation of growth hormone and thyrotropin release *in vitro* and pituitary 3',5'-adenosine monophosphate. Endocrinology, Springfield *86:* 1354–1356 (1970).
Van Loon, G.R.: Brain catecholamines and ACTH secretion; in *Ganong and Martini* Frontiers in neuroendocrinology, pp. 209–248 (Oxford University Press, 1973).
Wise, C.D. and Stein, L.: Dopamine-β-hydroxylase deficits in the brain of schizophrenic patients. Science *181:* 444 (1973).
Yang, H.-Y.T. and Neff, N.H.: β-Phenylethylamine. A specific substrate for type B monoamine oxidase in brain. J. Pharmacol. exp. Ther. (in press, 1973).
Zenker, H. and Bernstein, D.E.: The estimation of small amounts of corticosterone in rat plasma. J. biol. Chem. *231:* 695–701 (1958).
Zivkovic, B.; Guidotti, A., and Costa, E.: On the regulation of tryptophan hydroxylase in brain. Adv. biochem. Psychopharm. *11:* 19–30 (1974).
Zorr, U.; Kaneko, T.; Schneider, H.P.G.; McCann, S.H., and Field, J.B.: Further studies of stimulation of anterior pituitary cyclic 3',5'-monophosphate formation by hypothalamic extract and prostaglandin. J. biol. Chem. *245:* 2883–2888 (1970).

Authors' address: Dr. *A. Guidotti*, Dr. *B. Zivkovic* and Dr. *E. Costa*, Laboratory of Preclinical Pharmacology, National Institute of Mental Health, Saint Elizabeth Hospital, *Washington, DC 20032* (USA)

Psychoneuroendocrinology. Workshop Conf. Int. Soc. Psychoneuroendocrinology, Mieken 1973, pp. 267–275 (Karger, Basel 1974)

Role of Cyclic AMP (cAMP) in the Mechanism of Action of Hypothalamic Hypophysiotropic Hormones

T. Kaneko, H. Oka, M. Munemura, S. Saito and N. Yanaihara

First Department of Internal Medicine, Faculty of Medicine, University of Tokyo, Tokyo; Department of Obstetrics and Gynecology, Kumamoto University, Kumamoto; Department of Clinical Laboratory, Tokushima University School of Medicine, Tokushima, and Department of Bioorganic Chemistry, Shizuoka College of Pharmacy, Shizuoka

Recent works by *Burgus et al.* on isolation, purification and structure determination of thyroid stimulating hormone-releasing hormone (TRH) (3, 4), luteinizing hormone-releasing hormone (LH-RH), *Matsuo et al.* (12), and growth hormone release-inhibiting factor (GIF) *Brazeau et al.* (2) created a new era to investigate the mechanism of action of these three hormones using synthetic pure preparations.

The present studies were undertaken to investigate the effect of synthetic analogs of LH-RH on cyclic AMP (cAMP) formation and LH release in rat anterior pituitary, and the effects of ions and phospholipase A on LH-RH stimulation of cAMP production in the gland, and also the effect of synthetic GIF on cAMP production and GH release in rat anterior pituitary to evaluate the role of cAMP in the action of hypothalamic releasing or inhibiting hormone.

Materials and Methods

The rat anterior pituitary glands were obtained as described previously (22). Whole intact anterior pituitary was preincubated at 37 °C in Krebs-Ringer bicarbonate buffer containing 1 mg/ml of glucose. The pituitary gland was then transferred to the same fresh buffer containing 1 mg/ml of glucose, 1 mg/ml of bovine serum albumin, 10^{-2} *M* theophylline and appropriate substance to be tested. The gas phase was 95 % O_2–5 % CO_2. Following the final incubation, the tissue was extracted with 0.2 ml of hot 0.1 *N* HCl. After centrifugation and neutralization with 0.1 *N* NaOH, an aliquot of the supernatant was assayed for cAMP concentration. cAMP was determined by the competitive protein binding assay (6) using purified rat liver protein kinase (9) and tracer amount of (^{3}H)-cAMP. For determination of cyclic GMP (cGMP), the neutralized supernatant was applied to Dowex I-X2 column (200–400 mesh) and was isolated cGMP with stepwise elution of formic acid (13). Fractions containing cGMP was lyophilized and dissolved in small amounts of 0.1 *M*

Tris buffer, pH 7.5 and assayed using purified cGMP binding protein obtained from pupal fat body and tracer amount of (^{3}H)-cGMP. The recovery of cGMP from the column was over 80 %.

LH and GH levels in the incubation medium were determined by radioimmunoassay, using kits for rat LH and GH, respectively, provided by the National Institute of Health, Bethesda, Md.

Synthetic LH-RH and its analogs, TRH and GIF were prepared by *Yanaihara et al.* (19, 20). Rat hypothalamic extract (NIAMD-Rat HE-RP-1) was generously supplied by the National Institute of Arthritis, Metabolic and Digestive Disease, Bethesda, Md. (^{3}H)-cAMP was purchased from New England Nuclear and (^{3}H)-cGMP was obtained from Radiochemical Center. The specific activity was 24 and 13.5 Ci/m*M*, respectively.

Results

The data in figure 1 demonstrated that cAMP production and LH release in the rat anterior pituitary were both stimulated by synthetic LH-RH. Synthetic LH-RH in a dose of 10 ng/ml significantly increased cAMP concentration and LH release, and a concentration of 100 ng/ml of LH-RH gave the maximum stimulation.

Synthetic TRH at a concentration of 10 ng/ml also increased cAMP levels and 100 ng/ml of TRH showed the maximum stimulation of cAMP production in the gland. However, the increase of cAMP which was stimulated by 100 ng/ml of LH-RH or TRH was greater than that due to 10 μg/ml of each hormone.

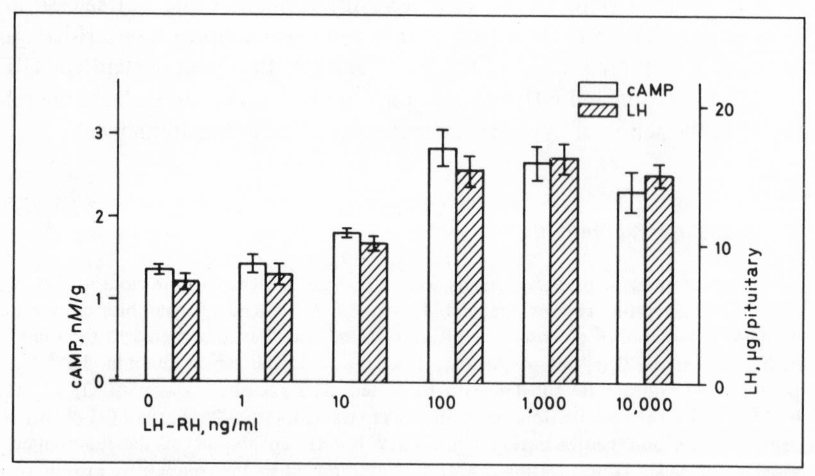

Fig. 1. The effects of synthetic LH-RH on cAMP levels and LH release in the rat anterior pituitary *in vitro.* The tissue was incubated for 15 min in Krebs-Ringer bicarbonate buffer containing 1 mg/ml of glucose, 1 mg/ml of albumin, 10^{-2} *M* theophylline and the hormone. The bars represent the mean value and the bracket represents SEM (n = 5).

Table I. The list of LH-RH analogs examined in the experiments

(Glu^{1})-LH-RH	(β-Ala^{6})-LH-RH
(Orotic Acid^{1})-LH-RH	(Sar^{6})-LH-RH
(Leu^{2})-LH-RH	(D-Ala^{6})-LH-RH
(Gln^{2})-LH-RH	(Gln^{8})-LH-RH
(Gly^{2})-LH-RH	(Leu^{8})-LH-RH
(Trp^{2}), (His^{3})-LH-RH	(Pro^{8}), (Arg^{9})-LH-RH
(Leu^{3})-LH-RH	
(Phe^{3})-LH-RH	
(Ala^{4}), (Phe^{5})-LH-RH	
des pGlu^{1}-LH-RH	
des pGlu^{1}, His^{2}-LH-RH	
des His^{2}-LH-RH	
des His^{2}, (Phe^{3})-LH-RH	
des His^{2}, Trp^{3}-LH-RH	
des His^{2}, (Phe^{5})-LH-RH	
des Trp^{3}-LH-RH	
pGlu-His-Pro-Trp-Ser-Tyr-Gly-Leu-Arg-Pro-Gly-NH$_2$	
pGlu-His-Trp-Ser-Tyr-Gly-Leu-Arg-Pro-Gly-OH	
Z-His-Trp-Ser-Tyr-Gly-Leu-Arg-Pro-Gly-NH$_2$	
pGlu-His-Trp-Ser-Tyr-Gly-Leu-OH	
pGlu-His-Trp-Ser-Tyr-Ala-NH$_2$	
pGlu-His-Trp-Ser-Tyr-Sar-NH$_2$	
H-Pro-His-Trp-Ser-Tyr-Gly-Leu-Arg-Pro-Gly-NH$_2$	

Stimulation of LH release by synthetic LH-RH was significant after incubation for 15 min, while cAMP formation was increased within 1 min incubation.

Although both LH-RH and TRH increased cAMP concentration, they did not stimulate cGMP formation in the anterior pituitary glands. However, ovine hypothalamic extract (4 mg/ml) showed marginal stimulation of cGMP formation in the gland, suggesting that some of the releasing hormones might act through guanylate cyclase-cGMP system.

Various analogs of LH-RH were synthetized and investigated with respect to their effects on cAMP formation and LH release in rat anterior pituitary (table I). Among LH-RH analogs substituted at position 1, 2, 3, 4 or 5 (Phe^{3})-LH-RH and (Ala^{4}), (Phe^{5})-LH-RH showed a marginal stimulation of cAMP formation, and LH release in the anterior pituitary previously reported (20).

All of three LH-RH analogs substituted at position 6 showed a significant stimulating effect on cAMP production at a concentration of 50 μg/ml, and the cAMP increasing activities of these three analogs were parallel to their activities of stimulation of LH release; i.e. (D-Ala^{6})-LH-RH showed the highest activity in both cAMP production and LH release, and (Sar^{6})-LH-RH had the lowest activity of cAMP formation and LH release.

(Gln^8)-LH-RH which possesses LH-releasing activity of approximately 5 % of natural pure LH-RH (19) showed cAMP-increasing activity, while the other two compounds, (Leu^8)-LH-RH and (Pro^8), (Arg^9)-LH-RH, did not exhibit any significant effects on these two parameters.

An analog in which a proline residue was inserted between position 2 and 3 of LH-RH increased cAMP concentration in the pituitary, and stimulated LH and FSH release into the incubation medium.

Among the other analogs listed in table I, pGlu-His-Trp-Ser-Tyr-Sar-NH_2 had a marginal stimulating effect on cAMP production, but others had no effect on cAMP formation and LH release.

Although high K^+ in the medium was reported to increase LH release and to show additive effect with LH-RH (14, 16, 17), high level (60 m*M*) of K^+ in the incubation medium did not give any effect on cAMP concentration in both control and LH-RH stimulated glands. Ca^{++}-free media neither affected the basal cAMP concentration nor modified the stimulating effect of LH-RH on cAMP production, though increasing hormone release caused by hypothalamic extract and LH-RH was reported to be markedly reduced in the absence of Ca^{++} in the medium (7, 14, 16, 17, 21).

Phospholipase A had little effect on basal cAMP level in rat anterior pituitary gland. Phospholipase A at a concentration of 1 U/ml showed a slight inhibition on cAMP production in the gland caused by LH-RH, but a higher dose of

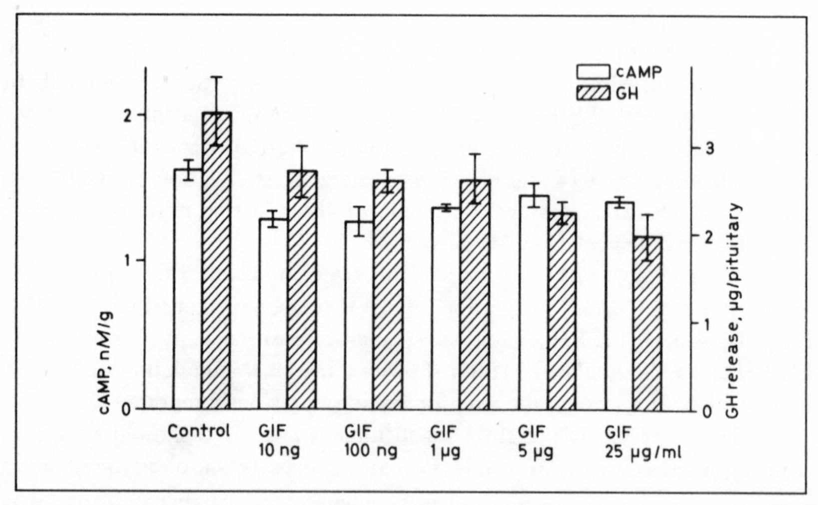

Fig. 2. The effect of synthetic GIF on cAMP levels and GH release in rat anterior pituitary. The tissue was incubated for 120 min in Krebs-Ringer bicarbonate buffer containing 1 mg/ml of glucose, 1 mg/ml of albumin, 10^{-2} *M* theophylline and synthetic GIF. The bars represent the mean value and the brackets represent SEM (n = 5).

phospholipase A abolished the effect of LH-RH on cAMP production in the anterior pituitary.

The effect of synthetic GIF, corresponding to the tetradecapeptide sequence proposed by *Brazeau et al.* (2), on cAMP concentration in rat anterior pituitary and GH release into the incubation medium is shown in figure 2. Synthetic GIF of 10 ng/ml concentration decreased significantly the basal level of cAMP in rat anterior pituitary and also decreased GH release into the incubation medium. Although synthetic GIF in a dose between 10 ng/ml and 25 μg/ml decreased basal cAMP concentration in the gland during 120-min incubation, decrease of GH release into the medium was more remarkable in the case of a higher dose. After 3-min incubation of the pituitary glands with rat hypothalamic extract (0.8 mg/ml), synthetic GIF (5 μg/ml) was added to the medium, or the glands were transferred to the new test tube containing 5 μg/ml of synthetic GIF. At 3-min incubation, cAMP concentration in the gland was markedly increased by hypothalamic extract, but in the following 7-min incubation, cAMP concentration was decreased in the case of the glands incubated with GIF or GIF and hypothalamic extract.

Discussion

Although stimulating effects of LH-RH or TRH on cAMP production in rat anterior pituitary were confirmed using synthetic pure materials (1, 8, 11), there still remain many problems to be solved in order to clarify the mechanism of action of hypothalamic-releasing or -inhibiting hormones.

In this study, various kinds of analogs of LH-RH were examined with regard to their effects on cAMP formation in the gland and LH release into the incubation medium. Among these compounds, (Phe^3)-LH-RH, (Ala^4), (Phe^5)-LH-RH, (Gln^8)-LH-RH, (Sar^6)-LH-RH, (β-Ala^6)-LH-RH, (D-Ala^6)-LH-RH and pGlu-His-Pro-Trp-Ser-Tyr-Gly-Leu-Arg-Pro-Gly-NH_2 showed significant stimulating activities of cAMP production and of LH release into the medium. On the other hand, the analogs of LH-RH which lack the LH-releasing activity did not show any stimulating effect of cAMP formation in the anterior pituitary gland. These data indicate that LH-releasing activity is closely related to the activity of increasing cAMP concentration in the pituitary gland and support strongly that cAMP plays an important role in the hormone release by LH-RH. However, the mechanism of the anterior pituitary hormone release seems more complicated. *Samli and Geschwind* (14) and *Wakabayashi et al.* (17) have reported the stimulating effect of a high level of K^+ in the medium on LH release. Similarly, release of some pituitary hormones are also stimulated by high K^+ in the medium, while a high level of K^+ did not produce any increase of cAMP formation in the anterior pituitary gland. The stimulation of anterior pituitary hormone release by high K^+ in the medium might be non-specific in contrast to the exquisite specificity shown by releasing hormones. The effect of high K^+ was additive with the stimulating effect of releasing hormone (14), suggesting both the mechanism of hormone release induced by releasing hormone and high K^+ are different. High K^+ in the medium might induce a depolarization and a permeability change in the membrane of the cells and, as a consequence, increase the hormone release in some manner.

Ca^{++} is thought to be the most important ion concerning the secretion of hormones in many endocrine tissues. In the absence of Ca^{++}, increasing LH release induced by hypothalamic extract and LH-RH were markedly reduced (7). However, the stimulating action of LH-RH and hypothalamic extract (15, 23) for cAMP production was not affected in the medium without Ca^{++}.

These studies provide additional information concerning the relationship between the adenylate cyclase-cAMP system and release of anterior pituitary hormones induced by hypothalamic-releasing hormones. However, they do not provide unequivocal evidence that hormone release is controlled by cAMP.

The importance of phospholipids in the membrane system is suggested by the observations that treatment of thyroid slices with phospholipase C inhibited the activity of adenylate cyclase-cAMP system induced by TSH (18) and, on the other hand, digestion of fat cell membrane with phospholipase A or C enhanced insulin binding to the membranes (5). The initial action of releasing hormone is probably the binding of the hormone to the receptor on the cell membrane. In the case of anterior pituitary gland treated with over 2 U/ml of phospholipase A, any significant change of cAMP concentration was not produced even in the gland incubated with LH-RH. These results are compatible with the phospholipids' involvement at the receptor site of LH secretory cells, and alterations of phospholipids in the membrane may be associated with either increased or decreased binding of hormone to the receptors.

Recently, the structure of growth hormone release-inhibiting factor was presented by *Brazeau et al.* (2). It is very interesting to see whether or not the hypothalamic inhibiting factor shows the inhibitory action on adenylate cyclase-cAMP system in the pituitary gland.

According to our experiments, synthetic GIF decreased the basal level of cAMP and GH release in rat anterior pituitary as shown in figure 2, suggesting that both hypothalamic-releasing and -inhibiting hormones might show their effect through adenylate cyclase-cAMP system by their binding to the membrane receptor. However, synthetic GIF inhibited the rat hypothalamic extract stimulation of cAMP production in rat anterior pituitary, but, in this condition, GH release into the medium was also inhibited. These results suggest that GH release from the pituitary gland in rat may be regulated mainly by GIF, not by GRF, and that synthetic GIF may affect the action of the other releasing hormone.

cGMP might be the second messenger of some peptide hormones, but in the case of LH-RH and TRH, guanylate cyclase-cGMP systems are not involved in their action on hormone release.

Recently, more than five different types of hormone-producing cells have been suggested to exist in the anterior pituitary (10). Thus, it should be very interesting to study various pituitary cells in isolation, details of the mechanism of hormone release, including the property of protein kinase activated by increased cAMP in the pituitary cells, the hormone receptor of the membrane, and also the mechanism of biosynthesis of pituitary hormones which might be controlled by releasing or inhibiting hormones.

Summary

Synthetic LH-RH and TRH increased cAMP concentration and stimulated the respected hormone release in rat anterior pituitary. The present studies showed that among the synthetic LH-RH analogs tested, only the compounds that increased cAMP production in the anterior pituitary gland stimulated the hormone release.

LH-RH and TRH did not stimulate cGMP formation in rat anterior pituitary.

A high K^+ and omission of Ca^{++} in the medium did not produce any effect on increasing cAMP formation caused by synthetic LH-RH.

The anterior pituitary treated with a high dose of phospholipase A was not stimulated by LH-RH in the cAMP concentration.

Synthetic GIF decreased the basal level of cAMP in the anterior pituitary and GH release into the incubation medium.

These results strongly support that the adenylate cyclase-cAMP system plays an important role in hormone release of the anterior pituitary gland and suggest that both hypothalamic-releasing and -inhibiting hormones may act through the cAMP system, initiated by the binding to the membrane receptor of the respective cells.

Acknowledgements

We wish to acknowledge the National Pituitary Agency, Endocrinology Study Section, and National Institute of Arthritis and Metabolic Diseases, USA, for the generous supplies of the radioimmunoassay kits for rat LH, FSH and GH, and of rat hypothalamic extracts.

References

1 *Borgeat, P.; Chavancy, G.; Dupont, A.; Labrie, F.; Arimura, A., and Schally, A.V.:* Stimulation of adenosine 3′,5′-cyclic monophosphate accumulation in anterior pituitary gland *in vitro* by synthetic luteinizing hormone-releasing hormone. Proc. nat. Acad. Sci., Wash. *69:* 2677–2681 (1972).

2 *Brazeau, P.; Vale, W.; Burgus, R.; Ling, N.; Butcher, M.; Rivier, J., and Guillemin, R.:* Hypothalamic polypeptide that inhibits the secretion of immunoreactive pituitary growth hormone. Science *179:* 77–79 (1973).

3 *Burgus, R.; Dunn, T.F.; Desiderio, D.M.; Vale, W. et Guillemin, R.:* Structure moléculaire du facteur hypothalamique hypophysiotrope TRF d'origine ovine. Mise en évidence par spectrométrie de masse de la séquence PCA-His-Pro-NH_2. C.R. Acad. Sci. *269:* 1870–1873 (1969).
4 *Burgus, R.; Dunn, T.F.; Desiderio, D.M.; Ward, D.N.; Vale, W., and Guillemin, R.:* Biological activity of synthetic polypeptide derivatives related to the structure of hypothalamic TRF. Endocrinology, Springfield *86:* 573–582 (1970).
5 *Cuatrecasas, P.:* Unmasking of insulin receptors in fat cells and fat cell membranes. Perturbation of membrane lipids. J. biol. Chem. *246:* 6532–6542 (1971).
6 *Gilman, A.G.:* A protein binding assay for adenosine 3',5'-cyclic monophosphate. Proc. nat. Acad. Sci., Wash. *67:* 305–312 (1970).
7 *Jutisz, M. and De la Llosa, M.P.:* Requirement of Ca^{++} and Mg^{++} ions for the *in vitro* release of follicle-stimulating hormone from rat pituitary glands and in its subsequent biosynthesis. Endocrinology, Springfield *86:* 761–768 (1970).
8 *Kaneko, T.; Saito, S.; Oka, H.; Oda, T., and Yanaihara, N.:* Effects of synthetic LH-RH and its analogs on rat anterior pituitary cyclic AMP, and LH and FSH release. Metabolism *22:* 77–80 (1973).
9 *Kumon, A.; Yamamura, H., and Nishizuka, Y.:* Mode of action of adenosine 3',5'-cyclic phosphate on protein kinase from rat liver. Biochem. biophys. Res. Commun. *41:* 1290–1297 (1970).
10 *Kurosumi, K.:* Fine structure of cells and tissues, vol. 5, pp. 2–47 (Igaku Shoin, Tokyo 1968).
11 *Makino, T.:* Study of the intracellular mechanism of LH release in the anterior pituitary. Amer. J. Obstet. Gynec. *115:* 606–614 (1973).
12 *Matsuo, H.; Baba, Y,; Nair, R.M.G.; Arimura, A., and Schally, A.V.:* Structure of the porcine LH- and FSH-releasing hormones. I. The proposed amino acid sequence. Biochem. biophys. Res. Commun. *43:* 1334–1339 (1971).
13 *Murad, F.; Manganiello, V., and Vaughan, M.:* A simple, sensitive protein binding assay for guanosine 3',5'-monophosphate. Proc. nat. Acad. Sci., Wash. *68:* 736–739 (1971).
14 *Samli, M.H. and Geschwind, I.I.:* Some effects of energy-transfer inhibitors and of Ca^{++}-free or K^{+}-enhanced media on the release of luteinizing hormone (LH) from the rat pituitary gland *in vitro*. Endocrinology, Springfield *82:* 225–231 (1968).
15 *Steiner, A.L.; Peak, G.T.; Utiger, R.D.; Karl, I.E., and Kipnis, D.M.:* Hypothalamic stimulation of growth hormone and thyrotropin release *in vitro* and pituitary 3',5'-adenosine cyclic monophosphate. Endocrinology, Springfield *86:* 1354–1360 (1970).
16 *Vale, W. and Guillemin, R.:* Potassium-induced stimulation of thyrotropin release *in vitro*. Requirement for presence of calcium and inhibition by thyroxin. Experientia *23:* 855–857 (1967).
17 *Wakabayashi, K.; Kamberi, I.A., and McCann, S.M.: In vitro* response of the rat pituitary to gonadotrophin-releasing factors and to ions. Endocrinology, Springfield *85:* 1046–1056 (1969).
18 *Yamashita, K.; Bloom, G.; Rainard, B.; Zor, U., and Field, J.B.:* Effects of chlorpromazine, propranolol and phospholipase C on thyrotropin and prostaglandin stimulation of adenyl cyclase-cyclic AMP system in dog thyroid slices. Metabolism *19:* 1109–1118 (1970).
19 *Yanaihara, N.; Yanaihara, C.; Hashimoto, T.; Kenmochi, Y.; Kaneko, T.; Oka,H.; Saito, S.; Schally, A.V., and Arimura, A.:* Syntheses and LH- and FSH-RH activities of LH-RH analogs substituted at position 8. Biochem. biophys. Res. Commun. *49:* 1280–1291 (1972).

20 *Yanaihara, N.; Hashimoto, T.; Yanaihara, C.; Tsuji, K.; Kenmochi, Y.; Ashizawa, F.; Kaneko, T.; Oka, H.; Saito, S.; Arimura, A., and Schally, A.V.:* Syntheses and biological evaluation of analogs of luteinizing hormone-releasing hormone (LH-RH) modified in position 2, 3, 4 or 5. Biochem. biophys. Res. Commun. *52:* 64–73 (1973).
21 *Zimmerman, G. and Fleischer, N.:* Role of calcium ions in the release of ACTH from rat pituitary tissue *in vitro.* Endocrinology, Springfield *87:* 426–429 (1970).
22 *Zor, U.; Kaneko, T.; Schneider, H.P.G.; McCann, S.M.; Lowe, I.P.; Bloom, G.; Borland, B., and Field, J.B.:* Stimulation of anterior pituitary adenyl cyclase activity and adenosine 3',5'-cyclic phosphate by hypothalamic extract and prostaglandin E_1. Proc. nat. Acad. Sci., Wash. *63:* 918–925 (1969).
23 *Zor, U.; Kaneko, T.; Schneider, H.P.G.; McCann, S.M., and Field, J.B.:* Further studies of stimulation of anterior pituitary cyclic adenosine 3',5'-monophosphate formation by hypothalamic extract and prostaglandins. J. biol. Chem. *245:* 2883–2888 (1970).

Author's address: Dr. *T. Kaneko,* First Department of Internal Medicine, Faculty of Medicine, University of Tokyo, 7-chome, Hongo, Bunkyo-ku, *Tokyo* (Japan)

Psychoneuroendocrinology. Workshop Conf. Int. Soc. Psychoneuroendocrinology, Mieken 1973, pp. 276–284 (Karger, Basel 1974)

Novel Enzymes Involved in the Inactivation of Hypothalamo-Hypophyseal Hormones

N. Marks and F. Stern

New York State Research Institute for Neurochemistry and Drug Addiction, Ward's Island, New York, N.Y.

The polypeptide nature of known hypothalamic regulatory hormones poses some important questions concerning the mechanisms of endocrine control. These relate to the enzymatic mechanisms involved in the formation of the peptide sequence and their subsequent inactivation, the anatomical and subcellular sites involved in their turnover, and finally the mechanisms of transport. With respect to the formation of hypothalamic factors two distinct possibilities exist: (1) biosynthesis *de novo* from precursor amino acids, and (2) formation by breakdown of pre-existing hormones or other precursor materials. The termination of biological activity is mediated by proteolytic enzymes resulting in the release of inactive peptides or of free amino acids. Currently, very little is known about the mechanisms responsible for regulating protein and peptide turnover in cells. In brain itself, most proteins in the adult animal are in steady-state with processes of synthesis matching those of breakdown (14, 15). Despite the many brilliant advances in our knowledge of synthetic processes very little information is available on the pathways involved in degradation which are essential to maintain turnover. This paucity of knowledge extends to the mechanisms involved in protein and polypeptide turnover in the specific neurosecretory regions. It is the purpose of this account to focus attention on some of the enzymes present in nerve tissue capable of degrading biologically active peptides. In several cases the structures of peptidyl hormones suggest that they are readily susceptible to classical exo- and endopeptidases; in other cases a unique structural feature of the peptide may imply a specific degrading enzyme system. The structures for five of the hypothalamic regulatory hormones have been proposed and these contain from 3–14 amino acids. Three of the factors TRH, MIF, and LRF have *N*- or and *C*-terminal protected groups in the form of pyroglutamyl and amide moieties suggesting that enzymes with novel specificities are involved in their metabolism (27, 29).

Formation of Biologically Active Peptides by Biosynthesis

There are reports that fragments of rat hypothalamus incubated in presence of labelled Pro, Glu, His for periods up to 1 h *in vitro* are capable of forming a tripeptide indistinguishable from TRH (2, 23, 24). *Reichlin and Mitnick* (23) have termed this system 'TRF synthetase' which is dependent on Mg^{2+} and ATP; synthetic activity appears to be mediated by a non-ribosomal system similar to that first proposed by *Lipmann* (13). Formation from precursor amino acids implies that there are enzymatic systems capable of cyclizing glutamate to form the pyroglutamate (lactam) *N*-terminal group, and also capable of amidating the *C*-terminal Pro to form Pro.NH_2. It is of interest in this connection that formation of α-MSH from ACTH in the *pars intermedia* by a tryptic-like cleavage of ACTH implies three supplementary reactions: carboxypeptidase action, amidation and acetylation. In subsequent studies *Reichlin and Mitnick* (23, 24) have observed formation of factors with LRF, PRF, and GH-RF-like activities on incubation in the presence of the appropriate amino acid mixtures. In the case of TRF-synthetase the level of synthetic activity appears to be under regulation by the levels of circulating T_3 and T_4. As yet, detailed confirmatory studies of the biosynthetic pathways are not available.

Few studies exist on the cellular sites for the various 'synthetase' activity. This is of importance with respect to the structures involved and modes of transport and storage. According to *Reichlin and Mitnick* (23), TRF synthetase is a completely soluble system devoid of ribosomes. It is of interest that *Edwardson et al.* (8) reported nerve-endings, isolated from sheep hypothalamus, contain CRF, PIF and perhaps all of the hypophysiotrophic factors which control the secretion of hormones by the anterior pituitary gland.

Formation of Hormones by Processes of Breakdown

Formation of a new species by a process of breakdown is of general occurrence, and other examples include activation of proteolytic enzymes from zymogens (trypsin, chymotrypsin, pepsin, etc.), formation of insulin (from proinsulin) glucagon, parathyroid hormone, gastrin, etc. In the case of hypothalamic and pituitary peptides the mechanisms are not completely elucidated although there is some evidence that vasopressin, growth hormone, α-MSH, MIF, MRF are formed from precursor materials (25, 30). As noted, there is evidence for the formation of α-MSH and an ACTH-like peptide by breakdown of corticotrophin involving a tryptic-like enzyme in the *pars intermedia* (26). Another example is the formation of the factor inhibiting the release of MSH. Incubation of hypothalamic fragments with oxytocin led to the production of Pro-Leu-Gly.NH_2 the terminal tripeptide of oxytocin (6, 20, 30). The origin of MRF is less clear but it might be noted that it is identical to the pentapeptide sequence Cys-Tyr-Ile-Gln-Asn.OH of oxytocin (7). It must be stressed that the validity of several hypothalamic factors have been questioned (2, 10). This largely is related to the assay procedures employed and the criteria used for defining a hormonal-releasing factor. Without entering this controversy it might be noted that it was only in the 1920s and 1930s that we arrived at a satisfactory definition of a neurotransmitter substance. One of the features of a neurotransmitter is its destruction at the site of activity. By analogy, releasing factors, which form a new category of chemical messengers are presumably inactivated at target sites by proteolytic enzymes. Therefore, some distinction must be made between mechanisms of breakdown which result in the formation of new biologically active species, and those involved in the termination of hormonal activity as described below.

Inactivation of Biologically Active Peptides in Brain

Brain, like other tissues, contains a variety of exo- and endopeptidases capable of hydrolyzing a wide variety of proteins and polypeptides (table I).

Table I. Proteinase and peptidase content in brain and neurosecretory areas

Substrate	Cortex	Tissue enzymes, μM/g tissue/h			
		hypothalamus		pituitary	
		anterior	posterior	anterior	posterior
1 Hemoglobin (pH 7.6)	9.3	7.3	11	12	11
2 Hemoglobin (pH 3.2)	15	14	20	44	46
3 Leu-Gly-Gly	360	350	490	700	700
4 Arg-Arg-βNA	27	16	20	29	29
5[1] Pro-Leu-Gly.NH_2	3.2	2.6	3.0	4.3	3.5
6[2] pGlu-His-Pro NH_2	2.2	4.0	5.3	3.9	tr
7[3] pGlu .... Gly.NH_2	0.35	0.57	0.62	1.2	0.70

Neutral proteinase, aminopeptidase and arylamidase assayed according to the procedures described elsewhere (14, 15). Hypophysiotrophic factors assayed with 50 μM of substrate, for 4 h at 37° with 0.25–1 mg of homogenate protein in 2.5 mM Tris-HCl buffer containing 0.5 mM Cleland's reagent and fixed with 3-percent sulfosalicylic acid. Breakdown products determined by a modified analysis system illustrated in figure 1. Results in all cases are the mean of 3–6 determinations agreeing within 10%.

1 MSH-I-RF (MIF).
2 TRH.
3 LRF, pGlu-His-Trp-Ser-Tyr-Gly-Leu-Arg-Pro-Gly.NH_2.

Inactivation of Oxytocin and Vasopressin

Breakdown of oxytocin has attracted considerable interest for several reasons, chief amongst which are: (1) the reported hormonal properties of its breakdown products, and (2) its unique structure (a cyclic peptide with an alicyclic chain containing a *C*-terminal amide group). Studies on oxytocin inactivation in various tissues reveal interesting species and tissue differences suggesting the presence of more than one mechanism. In blood it is reported that oxytocin is rapidly inactivated by an *N*-terminal breakdown with a dramatic increase in enzyme activity ('oxytocinase' see ref. 14) during pregnancy. In contrast to blood, oxytocin is not degraded to any marked extent in terms of *N*-terminal products in whole brain homogenates (18, 28). Incubation with brain extracts however, led to the rapid release of the dipeptide Leu-Gly.NH_2 and glycinamide (17). These results would suggest that in brain the disulfide bridge prevents attack by the typical tissue aminopeptidases, but permits *C*-terminal cleavage. In more detailed studies it was shown that the dipeptide is cleaved prior to the formation of glycinamide. Since the dipeptide is an excellent substrate for arylamidases, more than one enzyme may be involved in the inactivation process in crude extracts.

The ability to cleave the *C*-terminal moieties is dependent on the tissue and the species; thus enzyme from rat kidney has been reported to remove glycinamide as compared to enzyme from human uterus removing only the *C*-terminal glycinamide (11, 30). 'Phaseolain' obtained from French-bean leaves was reported to cleave glycinamide from oxytocin, but not other *C*-terminal amides present in secretin (..Val.NH_2) α-MSH (..Val.NH_2), eledoisin (..Met.NH_2) (5). The question of tissue specificity or anatomical localization is especially relevant to breakdown of oxytocin in brain. Our recent studies suggest a higher concentration of some arylamidases and aminopeptidases in the neurosecretory regions compared to cerebral cortex (19). More recently *Walter et al.* (31) reported the presence of enzyme(s) in the hypothalamus capable of slowly splitting oxytocin by *N*-terminal cleavage to form MIF. In the latter case the enzyme was not purified and the possibility exists that scission of the disulfide ring preceded normal aminopeptidase action since linear analogs of oxytocin and vasopressin including the MIF itself, are susceptible readily to hydrolysis by brain extracts (17, 18).

The presence of enzymes specific for *C*-terminal amides is of particular interest since biologically active peptides with such groupings abound in nature. In the case of brain, the primary split appears to occur at the -Pro^7-Leu^8- of oxytocin. Except for the human uterus, enzymes specific for *C*-terminal peptidyl amides have not been described or well characterized. There are a few reports of novel dipeptidyl carboxylases in bacteria (12) and in swine kidney (32, 33) but these generally are inactive on model substrates containing a *C*-terminal amide:

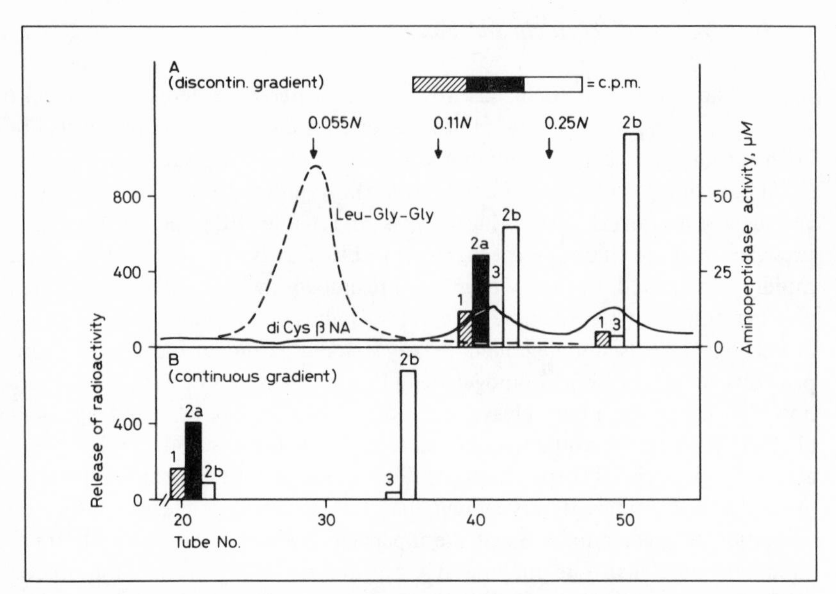

Fig. 1. For purification of enzymes inactivating oxytocin (□), and vasopressin (▨■), 10 g of rat brain was homogenized in 10 volumes of 0.32 *M* sucrose (A) or 30 m*M* phosphate buffer (B) and centrifuged at 100,000 *g* to remove debris. 50 ml supernatant was passed through a DEAE-cellulose column (20 × 1.5 cm) and eluted with a stepwise gradient of NaCl (A), or a continuous gradient 0.05–0.5 *M* (B). Fractions were incubated with 2 μg of labelled peptide for periods up to 24 h at 37° and then analyzed. The products formed were as follows: 1: Arg (^{14}C)-Gly.NH_2; 2a and 2b: ^{14}C-Gly.NH_2; 3: Leu (^{14}C)-Gly.NH_2. In method A, the first peak liberated relatively high concentrations of the *C*-terminal dipeptides. In both methods the second peak gave largely Gly.NH_2 as the major radioactive product. The distribution of two peptidases assayed with Leu-Gly.Gly (aminopeptidase) and dicystine-β-naphthylamide (cystine aminopeptidase) is appended for comparison. These were assayed by methods previously described (1).

in the case of the kidney the enzyme appears to be specific for substrates with Pro as the penultimate group (33).

Attempts to purify the brain enzymes liberating *C*-terminal products from oxytocin have met only with limited success. The enzyme is present largely in the soluble supernatant with only trace activities in subcellular organelles. Extraction with 0.32 *M* sucrose as compared to hyptonic buffers appeared to give preparations that led to a greater release of glycinamide versus the dipeptide (18). More than 90 % of activity was lost on passage through DEAE-cellulose, yielding two peaks of activity on elution with a step-wise or salt gradient (fig. 1). The first peak gave a mixture of dipeptide and glycinamide in the case of oxytocin and vasopressin, and the second peak gave largely glycinamide (*Marks and Pirotta*, unpublished findings).

Angiotensin and Kinins

Our studies have shown that brain extracts have the capacity to degrade these vasoactive peptides, but by different enzymatic mechanisms. A highly purified sample of arylamidase – an *N*-terminal peptidase purified on the basis of an arylamide substrate such as Arg-βNA, or dicystine βNA (fig. 1) (14) – was shown capable of cleaving Ile^5-angiotensin-11 with rapid release of the first five *N*-terminal amino acids, and a much slower breakdown of the terminal tripeptide Hist-Pro-Phe (1). The specificity of this degradation is demonstrated by the fact that another purified amino-peptidase – purified with Leu-Gly-Gly as substrate (fig. 1) – failed to inactivate angiotensin. It appears that cells contain batteries of peptide hydrolases displaying specificities towards biologically active peptides (9, 15).

Another example of specificity of brain enzyme was provided by breakdown of bradykinin. No breakdown was observed with purified aminopeptidases but inactivation rapidly occurred in crude supernatant extracts. Purification of the supernatant led to the unexpected finding that breakdown was attributable to a neutral endopeptidase with initial cleavage at the -Phe^5-Ser^6- bond followed by secondary cleavage of the split peptides by other enzymes (16). These findings were subsequently confirmed by *Camargo et al.* (4) who reported the presence of at least two neutral proteinases, one leading to the formation of only Arg-Pro-Pro-Gly-Phe, and Ser-Pro-Phe-Arg and separate from contaminating peptide hydrolases. Considering the fact that both angiotensin-11 and bradykinin are linear peptides with free *N*-terminal and *C*-terminal groups the specificity of the brain enzymes is most impressive. In the case of bradykinin the presence of three prolines may impose conformational limitations that reflect on the specificity of breakdown.

Hypothalamic Regulatory Factors

MIF. As observed in previous studies cerebral cortex contains a spectrum of proteinases and peptide hydrolases (14, 15). We have now extended these studies to the major neurosecretory regions as summarized in table I. It will be noted that in general, activity with the model proteins and peptide substrates was higher than with the hypothalamic peptides. Nevertheless, the activity displayed with such factors was sufficient to completely inactivate the level of hormone incubated within a period of 1–4 h. An important question arises as to whether an enzyme with a K_m in the range of 10^{-4}–10^{-6} *M* is operative in the presence of a substrate at concentrations in the range of 10^{-9} *M* or even lower. This leaves aside the question of accessibility of enzyme to substrate and localized concentration of the hormone in relation to other components. Brain is exceedingly rich in aminopeptidase cleaving model peptides such as Leu-Gly-Gly, Leu-Gly, dicystine-βNA, Arg-Arg-βNA, Arg-βNA, etc. (14). It is also noteworthy that when activity of these enzymes is expressed on a fresh weight basis the activity in the neurosecretory regions is markedly higher than in the cerebral cortex (table I). This is also true in the case of MIF with activity highest in the anterior pituitary.

The mechanisms of breakdown of MIF are of interest in terms of the enzymes involved. It would appear from our studies that inactivation is by removal of the *N*-terminal amino acids. The appearance of Pro was always accompanied by equal levels of Leu and Gly.NH_2. Even in short-term incubation studies, no detectable levels of other intermediates were observed such as Leu-Gly.NH_2. This may be understandable since the breakdown of dipeptides with amides on the *C*-terminal is particularly rapid in tissues (see table I). Leu-Gly.NH_2 is a good substrate for brain arylamidases and is also cleaved by sera of a variety of species. Presumably, inactivation of MIF is a two-step reaction with removal of *N*-terminal Pro as the limiting step. Since the factors are transported by blood their stability is of considerable importance. In general, hypothalamic factors and other hormones have relatively short half-lives following injection (2, 22). Surprisingly, however, we noted a marked species difference in the case of MIF with considerably slower rates in human blood compared to that of rodents. Our results are in contrast to those reported by *Nair et al.* (21) using MIF labeled in the Pro and Leu positions since they reported the major end-products as Pro and Leu-Gly.NH_2. The finding of the dipeptide as the major cleavage products must be reconciled with the known lability of this component in blood and tissue (17,18).

TRH. Several studies have demonstrated rapid inactivation of TRH by rat and human plasma with a half-life of circulating hormone in the range of 4 min (2, 22). In brain, the levels of enzyme activity are lower than with MIF; activity was higher in the case of the neurosecretory regions except for the posterior pituitary (table I). The major products detected on the amino acid analyzer were His and Pro.NH_2, indicating the presence of a pyroglutamyl peptidase.

LRF. We have now conducted extensive studies with this decapeptide both in terms of its distribution and mechanisms of degradation using different analogs. Activity was lower than with the other two hypothalamic peptides tested. In general the range of activity in crude brain homogenate was between 0.3–1.2 μM of fresh weight compared to 3–4.3 for MIF, and 350–700 for Leu-Gly-Gly (table I). Nevertheless, the activity in pituitary, especially in the adenohypophysis, was more than double that of the hypothalamus and cortex. In all regions the yield of His and Gly.NH_2 was significantly lower than other amino acids. This would suggest an internal cleavage followed by other peptide hydrolases. The mechanisms would be similar to that envisaged for breakdown of bradykinin by brain neutral proteinase. It would appear from the breakdown of pyro-Glu-His-Trp that this bond is the least susceptible when in tripeptide linkage since pyro-Glu-His and des-pyro-Glu-LRF are split readily in terms of the *N*-terminal residues.

References

1 *Abrash, L.; Walter, R., and Marks, N.:* Inactivation studies of angiotensin-11 by purified brain enzymes. Experientia *27:* 1325–1353 (1971).

2 *Blackwell, R.E. and Guillemin, R.:* Hypothalamic control of adenohypophyseal secretions. Ann. Rev. Physiol. *35:* 357–390 (1973).

3 *Burgus, R.; Ling, N.; Butcher, M., and Guillemin, R.:* Primary structure of somatostatin, or hypothalamic peptide that inhibits the secretion of pituitary growth hormone. Proc. nat. Acad. Sci., Wash. *70:* 684–688 (1973).

4 *Camargo, A.C.M.; Shapanka, R., and Greene, L.J.:* Preparation, assay, and partial characterization of a neutral endopeptidase from rabbit brain. Biochemistry *12:* 1838–1844 (1973).

5 *Carey, W.F. and Wells, R.E.:* Phaseolain. J. biol. Chem. *247:* 5573–5579 (1972).

6 *Celis, M.E.; Taleisnik, S., and Walter, R.:* Regulation of formation and proposed structure of the factor inhibiting the release of melanocyte-stimulating hormone. Proc. nat. Acad. Sci., Wash. *68:* 1428–1435 (1971).

7 *Celis, M.E.; Taleisnik, S., and Walter, R.:* Release of pituitary melanocyte stimulating hormone by the oxytocin fragment H-Cys-Tyr-Ile-Glu-Asn.OH. Biochem. biophys. Res. Commun. *45:* 564–569 (1971).

8 *Edwardson, J.A.; Bennett, G.W., and Bradford, H.F.:* Release of amino acids and neurosecretory substances after stimulation of nerve endings (synaptosomes) isolated from the hypothalamus. Nature, Lond. *240:* 554–556 (1972).

9 *Ganten, D.; Minnich, J.L.; Granger, P.; Hayduk, K.; Brecht, H.M.; Barbeau, A.; Boucher, R., and Genest, J.:* Angiotensin-forming enzyme in brain tissue. Science *173:* 64–65 (1971).

10 *Grant, N.H.; Clark, D.E., and Rosanoff, E.I.:* Evidence that Pro-Leu-Gly.NH_2, tocinoic acid and des-cys-tocinoic acid do not affect section of MSH. Biochem. biophys. Res. Comm. *51:* 100–106 (1973).

11 *Koida, M.; Glass, J.D.; Schwartz, I.L., and Walter, R.:* Mechanisms of inactivation of oxytocin by rat kidney enzyme. Endocrinology, Springfield *88:* 633–643 (1971).

12 *Lee, H.J.; LaRue, J.N., and Wilson, I.B.:* Dipeptidyl carboxypeptidase from *coryne bacterium equi.* Biochim. biophys. Acta *250:* 608–613 (1971).

13 *Lipmann, F.:* Attempts to map a process evolution of peptide biosynthesis. Science *173:* 875–884 (1971).

14 *Marks, N.:* Exopeptidase of the nervous system. Int. Rev. Neurobiol. *11:* 57–97 (1968).

15 *Marks, N. and Lajtha, A.:* Protein and polypeptide breakdown; in *Lajtha* Handbook of neurochemistry, vol. 5 A, pp. 49–139 (Plenum Press, New York 1971).

16 *Marks, N. and Pirotta, M.:* Breakdown of bradykinin and its analogs by rat brain proteinase. Brain Res. *33:* 565–567 (1971).

17 *Marks, N. and Walter, R.:* MSH-release-inhibiting factor: inactivation by proteolytic enzymes. Proc. Soc. exp. Biol. Med. *140:* 673–676 (1972).

18 *Marks, N.; Abrash, L., and Walter, R.:* Degradation of neurohypophyseal hormones of brain extracts and purified brain enzymes. Proc. Soc. exp. Biol. Med. *142:* 455–460 (1973).

19 *Marks, N.; Galoyan, A.; Grynbaum, A., and Lajtha, A.:* Protein and peptide hydrolases of the rat hypothalamus and pituitary. J. Neurochem. (in press, 1974).

20 *Nair, R.M.G.; Kastin, A.J., and Schally, A.V.:* Isolation and structure of hypothalamic MSH-release-inhibiting factor. Biochem. biophys. Res. Commun. *43:* 1376–1381 (1971).

21 *Nair, R.M.G.; Redding, T.W.; Kastin, A.J., and Schally, A.V.:* Site of inactivation of MSH-R-IF by human plasma. Biochem. Pharmacol. *22:* 1915–1919 (1973).

22 *Redding, T.W. and Schally, A.V.:* On the half-life of TRH in rats. Neuroendocrinology *9:* 250–256 (1972).

23 *Reichlin, S. and Mitnick, M.:* Enzymatic synthesis of GH-RF by rat incubates and by extracts of rat and porcine hypothalamic tissue. Proc. Soc. exp. Biol. Med. *142:* 497–501 (1972). With *C. Valverde-R.:* Evidence for 'PRF-synthetase'. Proc. Soc. exp. Biol. Med. *143:* 418–421 (1973).

24 *Reichlin, S. and Mitnick, M.:* Biosynthesis of hypothalamic hypophysiotropic factors; in *Ganong and Martini* Frontiers in neuroendocrinology, pp. 61–88 (Oxford University Press, New York 1973).

25 *Sachs, H.:* Neurosecretion; in *Lajtha* Handbook of neurochemistry, vol. 4, pp. 373–428 (Plenum Press, New York 1970).

26 *Scott, A.P.; Ratcliffe, J.G.; Rees, L.H.; Landon, J.; Bennett, H.P.J.; Lowry, P.J., and McMartin, C.:* Pituitary peptide. Nature, Lond. *244:* 65–67 (1973).

27 *Schally, A.V.; Arimura, A., and Kastin, A.J.:* Hypothalamic regulatory hormones. Science *179:* 341–350 (1973).

28 *Simmons, W.H. and Brecher, A.S.:* Inactivation of MSH-I-RF by a Mn^{2+} stimulated bovine brain amino peptidase. J. biol. Chem. *248:* 5780–5784 (1973).

29 *Vale, W.; Grant, G., and Guillemen, R.:* Chemistry of the hypothalamic-releasing factors – studies on structure-function relationships. Frontiers of neuroendocrinology, pp. 375–411 (Oxford Univ. Press, New York 1973).

30 *Walter, R.:* in Peptides 1972. Proc. 12th Europ. Peptide Symp., pp. 363–378 (North Holland, Amsterdam 1973).

31 *Walter, R.; Griffith, E.C., and Hooper, K.C.:* Production of MSH-release-inhibiting hormone by a particulate preparation of hypothalami; mechanisms of oxytocin inactivation. Brain Res. *60:* 449–457 (1973).

32 *Yang, H.Y.T.; Erdos, E.G., and Levin, Y.A.:* A dipeptidyl carboxypeptidase that converts angiotensin I and inactivates bradykinin. Biochim. biophys. Acta *214:* 374–376 (1970).

33 *Yang, H.Y.T. and Erdos, E.G.:* Prolylcarboxypeptidase: a recently described lysosomal enzyme. Excerpta Medica Inter. Congr. Series Immunopathol. Inflammat *229:* 146–148 (1970).

Authors' address: Dr. *N. Marks* and *F. Stern,* New York State Research Institute for Neurochemistry and Drug Addiction, Ward's Island, *New York, NY 10035* (USA)

Psychoneuroendocrinology. Workshop Conf. Int. Soc. Psychoneuroendocrinology, Mieken 1973, pp. 285–294 (Karger, Basel 1974)

Oxytocin and Other Peptide Hormones as Prohormones[1]

Roderich Walter

Department of Physiology and Biophysics, Mount Sinai School of Medicine of the City University of New York, New York, N.Y.

The enzymatic formation of H-Pro-Leu-Gly-NH_2 from oxytocin as proposed by *Celis et al.* (8) constitutes the first example which showed on a molecular level that a peptide hormone with well-characterized physiological properties can serve as a precursor molecule (prohormone second order)[2] for a smaller peptide with hormonal activities completely different from those of the parent principle. The tripeptide H-Pro-Leu-Gly-NH_2 was proposed to be the natural factor inhibiting the release of melanocyte-stimulating hormone (MSH) from the pituitary; oxytocin and its analogs and intermediates tested, with the exception of H-Pro-Leu-Gly-NH_2, were inactive *per se.* Since our report in 1971, a picture emerges which indicates that the above example with oxytocin is by no means an isolated case. Synthetic peptides, comprised of partial sequences of neurohypophyseal hormones, have been found to act as hormone-releasing and hormone-release-inhibiting factors (see table I). Furthermore, certain neurohypophyseal hormone fragments act in the central nervous system to alter behavior and memory (table I). This author has no doubt that additional biological activities will be discovered for active fragments already reported and, furthermore, that other neurohypophyseal hormone fragments will be found to possess distinct biological properties.

1 This review was aided by US Public Health Service Grant No. AM-13567.

2 Two types of ribosomally-formed hormone precursors do exist, and it is suggested that they be distinguished by referring to them as 'prohormone first order' and 'prohormone second order'. The prohormone first order has no known physiological activity, or possesses only weak activities characteristic for the hormone released enzymatically from the precursor. This group would include proinsulin(s) (37, 38) and precursors for such hormones as gastrin (52), parathyroid hormone (1, 34), ACTH (53, 54), growth hormone (2, 36), glucagon (24, 25, 39, 40), etc. When the higher-molecular-weight precursor molecule has well-characterized physiological properties of its own, as with oxytocin, vasopressin or ACTH, etc., it may be referred to as prohormone second order, in relation to new hormones enzymatically generated.

Table I. Oxytocin and vasopressin fragments possessing activity on the pituitary and central nervous system[1,2]

Peptide	Activity
(1) H-Pro-Leu-Gly-NH_2 [3]	MSH-release-inhibiting factor[4] Antidepressant and antiparkinsonian[5]
(2) H-Cys-Tyr-Ile-Gln-Asn-Cys-OH	MSH-release-inhibiting activity[6]
(3) H-Cys-Tyr-Ile-Gln-Asn-OH	MSH-releasing activity[7]
(4) Cys-Tyr-Phe-Gln-Asn-Cys-Pro-Lys-OH[8]	Facilitation of avoidance behavior[9] Modification of memory consolidation[10]
(5) Cys-Tyr-Phe-Gln-Asn-Cys-OH	ACTH-releasing activity[11]

1 With the exception of H-Pro-Leu-Gly-NH_2 there is no published evidence to date that the factors are formed enzymatically from the neurohypophyseal hormones.
2 With certain fragments difficulties have been encountered in reproducing the activities consistently or at all, e.g. with peptide *(1)* see refs. 12 and 16; peptide *(2)* see refs. 7, 12 and 22; peptide *(3)* see ref. 12; and peptide *(5)* see refs. 15 and 30.
3 Found in bovine hypothalamic extracts, see ref. 21.
4 See ref. 8.
5 See ref. 26.
6 See ref. 5.
7 See ref. 9.
8 Found in porcine pituitary extracts, see ref. 19.
9 See refs. 19 and 51.
10 See ref. 18.
11 See refs. 14 and 31.

Other peptide hormones should likewise be able to serve as precursors for new hormonal principles. In line with this contention it has very recently been proposed (33) that ACTH is in part degraded in the pars intermedia of rat and pig pituitaries to yield, after a series of enzymatic steps, α-MSH and a 'corticotrophin-like intermediate lobe peptide' (CLIP):

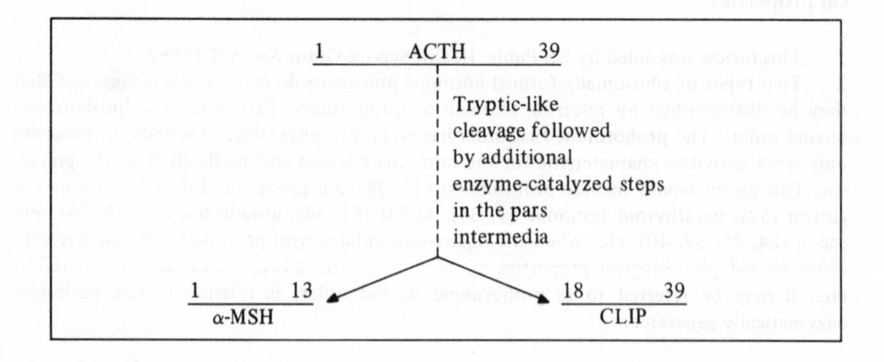

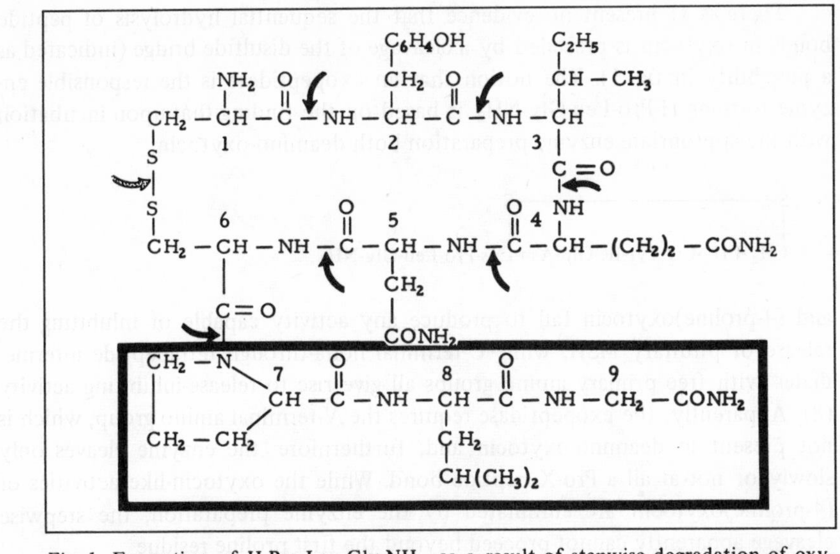

Fig. 1. Formation of H-Pro-Leu-Gly-NH_2 as a result of stepwise degradation of oxytocin by a hypothalamic membrane-bound exopeptidase. Solid arrows indicate sites of stepwise hydrolytic cleavage of peptide bonds by the exopeptidase. Dashed arrow indicates possible, albeit unlikely, cleavage of disulfide bridge preceding the exopeptidase reactions.

Returning to H-Pro-Leu-Gly-NH_2, this factor results from controlled, stepwise degradation of oxytocin, beginning at the N-terminal cysteine residue in position 1 of the hormone (8), by a membrane-bound, hypothalamic enzyme (47) (fig. 1). It has been considered (44) that the enzyme is present in specific oxytocinergic neurons which invade the pars intermedia (3, 11) and possess a different enzymic activity profile (*expression* of enzyme activity is different) from those which are involved in transport of oxytocin through the hypothalamo-neurohypophyseal tract to the posterior pituitary (13, 32). In general, as judged from research on pituitary cells (20, 35) and insulin (17) the peptide secretory process appears to involve synthesis of the precursor molecule in the rough endoplasmic reticulum of a given cell, transport to the peripheral elements of the Golgi apparatus, and packing of the protein into vacuoles which eventually become secretory granules. These granules subsequently fuse with the outer cell membrane (plasma membrane) to release, by exocytosis, the peptide hormone and inactive fragments of the precursor. In the case of neurohypophyseal hormones, which themselves seem to be synthesized as inactive precursors (for vasopressin, see ref. 29), the enzymes responsible for release of oxytocin from its precursor and of the C-terminal tripeptide from oxytocin are possibly located in the perikaryon and are packed along with the prohormone molecules into the secretory granules.

There is at present no evidence that the sequential hydrolysis of peptide bonds in oxytocin is preceded by a cleavage of the disulfide bridge (indicated as a possibility in fig. 1). The notion that an exopeptidase is the responsible enzyme forming H-Pro-Leu-Gly-NH_2 is based on the finding that upon incubation with the appropriate enzyme preparation both deamino-oxytocin,

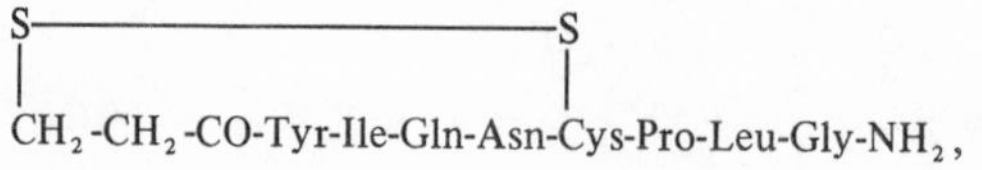

S————————————S
| |
CH_2-CH_2-CO-Tyr-Ile-Gln-Asn-Cys-Pro-Leu-Gly-NH_2,

and (4-proline)oxytocin fail to produce any activity capable of inhibiting the release of pituitary MSH, while *C*-terminal nona-through tetrapeptide intermediates with free primary amino groups all give rise to release-inhibiting activity (8). Apparently, the exopeptidase requires the *N*-terminal amino group, which is not present in deamino-oxytocin and, furthermore, the enzyme cleaves only slowly or not at all a Pro-X peptide bond. While the oxytocin-like activities of (4-proline)oxytocin are eliminated by the enzyme preparation, the stepwise cleavage apparently cannot proceed beyond the first proline residue

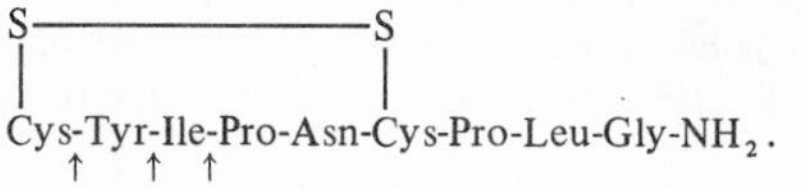

S——————————S
| |
Cys-Tyr-Ile-Pro-Asn-Cys-Pro-Leu-Gly-NH_2.
↑ ↑ ↑

The inability of the enzyme to cleave the Pro-X bond also explains the accumulation of H-Pro-Leu-Gly-NH_2 in hypothalamic extracts (47). Incubation with the microsomal enzyme preparation of analogs possessing a 20-membered ring identical to that of oxytocin, but having different peptide sequences in the linear portion, and assay for the MSH-release-inhibiting activity of the enzyme digest, is a somewhat indirect approach to gain information as to what degree the structure of H-Pro-Leu-Gly-NH_2 can be modified with retention of activity (fig. 2). In addition to the results shown in figure 2, enzymatic digests of lysine- and arginine-vasopressin inhibited the release of MSH; the factors responsible are thought to be produced as follows:

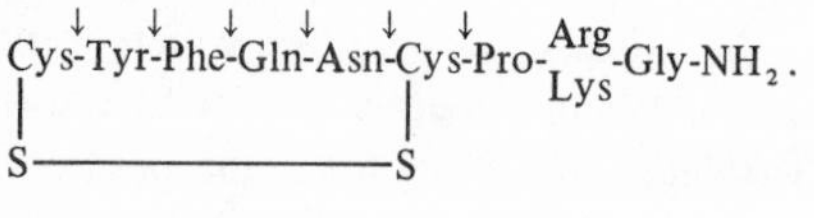

↓ ↓ ↓ ↓ ↓ ↓
Cys-Tyr-Phe-Gln-Asn-Cys-Pro-$^{Arg}_{Lys}$-Gly-NH_2.
| |
S——————————S

Some of the active tripeptides such as H-Pro-Leu(or Lys or Arg)-Gly-NH_2, presumed to be formed in the enzymatic digest, were synthesized. All three peptides inhibited the release of pituitary MSH *in vitro* and *in vivo* (7, 8). These limited structure-activity studies (for some additional results, see refs. 7 and 8) seem to suggest that H-Pro-Leu-Gly-NH_2 is most sensitive in terms of its MSH-

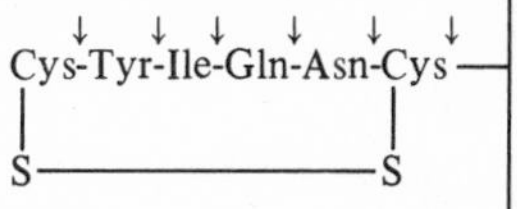

	Peptide	Inhibition of release of MSH[1]
	Pro-Leu-Sar-NH_2	–
	Pro-Leu-Gly-OH	–
↓ ↓ ↓ ↓ ↓ ↓	Pro-Leu-Gly-$N(CH_3)_2$	–
Cys-Tyr-Ile-Gln-Asn-Cys —	Pro-Ser-Gly-NH_2	+
| |	Pro-Phe-Gly-NH_2	+
S ——— S	Pro-Ala-Gly-NH_2	+
	Gly-Leu-Gly-NH_2	–

1 Active C-terminal tripeptides presumed to be released *in situ* which possess MSH-release-inhibiting activity are denoted +, while inactive peptides are denoted –.

Fig. 2. MSH-release-inhibiting activity after incubation of hypothalamic microsomal preparations of male rats with oxytocin analogs.

release-inhibiting activity to modification or substitution of the *N*-terminal Pro and the carboxamide of Gly.

These results in conjunction with spectroscopic data from proton nmr and with preliminary crystallographic parameters led *Walter et al.* (46) to propose a preferred three-dimensional structure for H-Pro-Leu-Gly-NH_2 (fig. 3). The conformation consists of a 10-membered β-turn closed by a $4 \rightarrow 1$ hydrogen bond between the *trans* carboxamide proton of the glycinamide moiety and the carbonyl oxygen of the proline moiety. Determinations of spin-lattice relaxation times (T_1) by ^{13}C-nmr indicate that H-Pro-Leu-Gly-NH_2 has not only in dimethylsulfoxide, but also in D_2O a relatively compact conformation (10). Semiempirical energy calculations by *Ralston et al.* (27) for H-Pro-Leu-Gly-NH_2 in water, pH 7, also suggest that this tripeptide is folded but flexible with a fair number of interconvertible conformations. After structure refinement, H-Pro-Leu-Gly-NH_2 possesses at least three preferred compact conformations. Two of these conformations occupy rather broad and flat energy troughs, while the third occupies a narrow and deep potential energy well. This third structure of H-Pro-Leu-Gly-NH_2 (fig. 3) is the very same which was proposed for MSH-R-IF in dimethylsulfoxide (46) and which was subsequently also found to exist in the crystalline state as determined by X-ray analysis (28).

In view of its simple structure, H-Pro-Leu-Gly-NH_2 lends itself well to a conformation-activity analysis as attempted previously with neurohypophyseal hormones (43, 50). Oxytocin, vasopressin and other naturally occurring neurohypophyseal peptides are comprised of ten-membered β-turns (41, 43) like H-Pro-Leu-Gly-NH_2, although it should be noted that different moieties in the two classes of hormones are involved in forming the loops – e.g. compare figure 1b in *Urry and Walter* (41) and figure 3 in this report. From a correlation

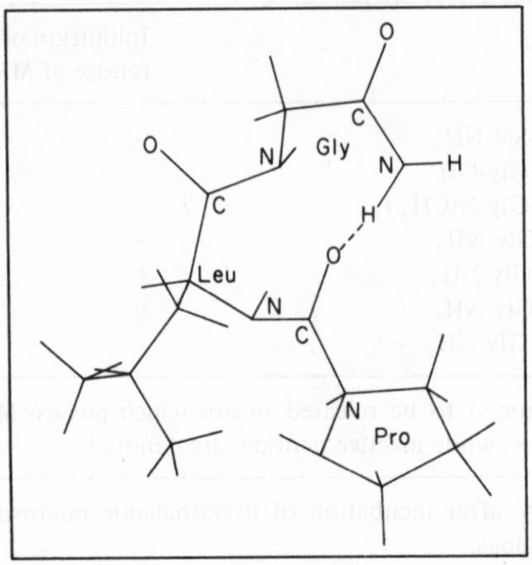

Fig. 3. Preferred three-dimensional structure of H-Pro-Leu-Gly-NH_2 in dimethylsulfoxide.

of structure-activity studies of neurohypophyseal hormones with their proposed conformation it emerged that substitutions of amino acid residues in the corners of the β-turns – residue i + 1 and i + 2 according to *Venkatachalam* (42) – give rise to selective changes in the activity profile, i.e. enhancement of certain biological activities and reduction or elimination of different ones, while changes in positions other than i + 1 or i + 2 result either in inhibitors or in a general alteration of activities, i.e. all activities are enhanced or reduced. In H-Pro-Leu-Gly-NH_2 the Leu and Gly residues occupy, respectively, positions i + 1 and i + 2, while Pro would be considered residue i and the amide moiety of the glycinamide corresponds to 'residue' i + 3. Hence the fact that H-Pro-Arg(or Lys)-Gly-NH_2 (7) and other tripeptides with substitutions of residue i + 1 (8) possessed some of the activity of H-Pro-Leu-Gly-NH_2 itself, is in accord with conformational considerations. Moreover, it should be possible to find active analogs in which the Gly residue is replaced, and it would not be surprising if analogs in which the Leu or Gly residues are substituted would show biological activities different from those of H-Pro-Leu-Gly-NH_2 (see table I). As a note of caution it should be stressed that the above analysis does not imply that the biological activity exhibited by H-Pro-Leu-Gly-NH_2 resides in residue i + 1 or i + 2 *per se;* it means that the β-turn conformation places the residues which make up the molecule into such a conformational arrangement as to enable the receptor to recognize the topography of the hormonal peptide as a whole. Thus, the average preferred conformation places the critical moieties of the hormonal messenger into optimal spatial arrangement for multiple-point interaction with the receptor. Without such conformational specificity, even for small acyclic

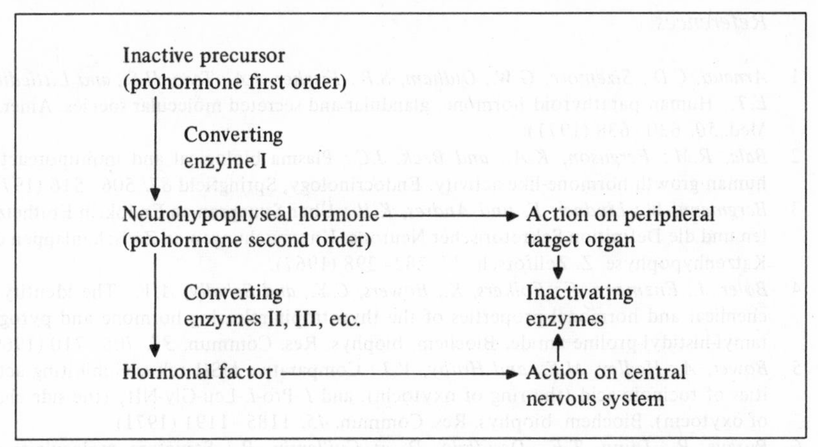

Fig. 4. Proposed route of formation and inactivation of peptide hormones as illustrated with neurohypophyseal hormones.

peptides like H-Pro-Leu-Gly-NH_2 or TRH (4, 6), regulatory processes induced by hormones could not exist.

The enzymatic breakdown of H-Pro-Leu-Gly-NH_2 has been studied in some detail. Nevertheless, experimental results from different laboratories vary considerably. *Nair et al.* (23) suggested that incubation of H-Pro-Leu-Gly-NH_2 with plasma causes cleavage of the Pro-Leu bond with formation of Pro and H-Leu-Gly-NH_2 as the main products. Although we concur, on the basis of experiments involving a different approach from that taken by *Nair et al.*, that the rate-limiting step in the inactivation of H-Pro-Leu-Gly-NH_2 is the hydrolysis of the Pro-Leu bond, we are unable to recover significant amounts of H-Leu-Gly-NH_2 due to its rapid inactivation (49). Moreover, in contrast to *Nair et al.* we found a major species difference in the inactivation of H-Pro-Leu-Gly-NH_2 by plasma or serum. It appears from our results that serum of a species with a well-defined pars intermedia (rat) rapidly inactivates MSH-R-IF, while the serum of a species in which the pars intermedia fails to develop (bird) or regresses during the individual's lifetime (man) does not inactivate this hypothalamic regulatory factor (48). The reader is referred to ref. 45 for a recent review of this area of research.

In summary, neurohypophyseal hormones are probably derived enzymatically from high-molecular-weight precursors (prohormones first order), but can serve themselves as precursors (prohormones second order) for new hormonal factors (fig. 4). This general scheme of sequential release of biologically-active peptides from large inactive precursors formed by ribosomal synthesis should, in principle, apply to other peptide hormones as well.

References

1 *Arnaud, C.D.; Sizemore, G.W.; Oldham, S.B.; Fischer, J.A.; Tsao, H.S., and Littledike, E.T.:* Human parathyroid hormone: glandular and secreted molecular species. Amer. J. Med. *50:* 630–638 (1971).

2 *Bala, R.M.; Ferguson, K.A., and Beck, J.C.:* Plasma biological and immunoreactive human growth hormone-like activity. Endocrinology, Springfield *87:* 506–516 (1970).

3 *Bargmann, W.; Lindner, E. und Andres, K.H.:* Über Synapsen an Endokrin Epithelzellen und die Definition Sekretorischer Neurone. Untersuchungen an Zwischenlappen der Katzenhypophyse. Z. Zellforsch. *77:* 282–298 (1967).

4 *Bøler, J.; Enzmann, F.; Folkers, K.; Bowers, C.Y., and Schally, A.V.:* The identity of chemical and hormonal properties of the thyrotropin-releasing hormone and pyroglutamyl-histidyl-proline amide. Biochem. biophys. Res. Commun. *37:* 705–710 (1969).

5 *Bower, A.; Hadley, M.E., and Hruby, V.J.:* Comparative MSH release-inhibiting activities of tocinoic acid (the ring of oxytocin), and *L*-Pro-*L*-Leu-Gly-NH_2 (the side chain of oxytocin). Biochem. biophys. Res. Commun. *45:* 1185–1191 (1971).

6 *Burgus, R.; Dunn, T.F.; Desiderio, D. et Guillemin, R.:* Structure moleculaire du facteur hypophysiotrope TRF d'origine ovine: mise en evidence par spectrometrie de masse de la sequence PCA-His-Pro-NH_2. C.R. Acad. Sci. (D) (Paris) *269:* 1870–1873 (1969).

7 *Celis, M.E.; Hase, S., and Walter, R.:* Structure-activity studies of MSH-release-inhibiting hormone. FEBS Letters *27:* 327–330 (1972).

8 *Celis, M.E.; Taleisnik, S., and Walter, R.:* Regulation of formation and proposed structure of the factor inhibiting the release of melanocyte-stimulating hormone. Proc. nat. Acad. Sci., Wash. *68:* 1428–1433 (1971).

9 *Celis, M.E.; Taleisnik, S., and Walter, R.:* Release of pituitary melanocyte-stimulating hormone by the oxytocin fragment, H-Cys-Tyr-Ile-Gln-Asn-OH. Biochem. biophys. Res. Commun. *45:* 564–569 (1971).

10 *Deslauriers, R.; Walter, R., and Smith, I.C.P.:* Intramolecular motion in peptides determined by ^{13}C NMR: a spin-lattice relaxation time study on MSH-release-inhibiting factor. FEBS Letters *37:* 27–32 (1973).

11 *Etkin, W.:* Relation of the pars intermedia to the hypothalamus; in *Martini and Ganong* Neuroendocrinology, vol. 2, pp. 261–282 (Academic Press, New York 1967).

12 *Grant, N.H.; Clark, D.E., and Rosanoff, E.F.:* Evidence that pro-leu-gly-NH_2, tocinoic acid, and des-cys-tocinoic acid do not affect secretion of melanocyte-stimulating hormone. Biochem. biophys. Res. Commun. *51:* 100–106 (1973).

13 *Heller, H.:* Occurrence, storage and metabolism of oxytocin; in *Caldeyro-Barcia and Heller* Oxytocin, pp. 3–23 (Pergamon Press, London 1960).

14 *Hiroshige, T.:* Attempts to physiologically validate the circadian rhythm of corticotropin-releasing activity in the rat hypothalamus. Fourth Intl. Mtg. of the Intl. Soc. for Neurochem., Tokyo, Japan, p. 147, Abstr. No. R-6-6 (1973).

15 *Hiroshige, T.:* Personal communication.

16 *Hruby, V.J.; Smith, C.W.; Bower, A., and Hadley, M.E.:* Melanophore stimulating hormone: release inhibition by ring structures of neurohypophyseal hormones. Science *176:* 1331–1332 (1972).

17 *Kemmler, W.; Peterson, J.D.; Rubenstein, A.H., and Steiner, D.F.:* On the biosynthesis, intracellular transport and mechanism of conversion of proinsulin to insulin and C-peptide. Diabetes, N.Y. *21:* 572–581 (1972).

18 *Lande, S.; Flexner, J.B., and Flexner, L.B.:* Effect of corticotropin and desglycinamide[9]-lysine vasopressin on suppression of memory by puromycin. Proc. nat. Acad. Sci., Wash. *69:* 558–560 (1972).

19 *Lande, S.; Witter, A., and De Wied, D.:* Pituitary peptides: an octapeptide that stimulates conditioned avoidance acquisition in hypophysectomized rats. J. biol. Chem. *246:* 2058–2062 (1971).
20 *McShan, W.H. and Hartley, M.W.:* Production, storage, and release of anterior pituitary hormones. Ergebn. Physiol. *56:* 264–296 (1965).
21 *Nair, R.M.G.; Kastin, A.J., and Schally, A.V.:* Isolation and structure of hypothalamic MSH release-inhibiting-hormone. Biochem. biophys. Res. Commun. *43:* 1376–1381 (1971).
22 *Nair, R.M.G.; Kastin, A.J., and Schally, A.V.:* Isolation and structure of another hypothalamic peptide possessing MSH-release-inhibiting activity. Biochem. biophys. Res. Commun. *47:* 1420–1425 (1972).
23 *Nair, R.M.G.; Redding, T.W.; Kastin, A.J., and Schally, A.V.:* Site of inactivation of melanocyte-stimulating hormone-release-inhibiting hormone by human plasma. Biochem. Pharmacol. *22:* 1915–1919 (1973).
24 *Noe, B.D. and Bauer, G.E.:* Evidence for glucagon biosynthesis involving a protein intermediate in islets of the angler fish *(lophius americanus).* Endocrinology, Springfield *89:* 642–651 (1971).
25 *Noe, B.D. and Bauer, G.E.:* Further characterization of a glucagon precursor from angler fish islet tissue. Proc. Soc. exp. Biol. Med. *142:* 210–213 (1973).
26 *Plotnikoff, N.P.; Kastin, A.J.; Anderson, M.S., and Schally, A.V.:* Deserpidine antagonism by a tripeptide, *L*-prolyl-*L*-leuclyglycinamide. Neuroendocrinology *11:* 67–71 (1973).
27 *Ralston, E.; De Coen, J.-L., and Walter, R.:* Tertiary structure of MSH-release-inhibiting factor derived by conformational energy calculations. Proc. nat. Acad. Sci., Wash. (in press, 1974).
28 *Reed. L.L. and Johnson, P.L.:* Solid state conformation of the C-terminal tripeptide of oxytocin, *L*-pro-*L*-leu-gly-$NH_2 \cdot {}^1/_2 H_2O$. J. amer. chem. Soc. *95:* 7523–7524 (1973).
29 *Sachs, H.; Fawcett, P.; Takatabake, Y., and Portanova, R.:* Biosynthesis and release of vasopressin and neurophysin. Recent. Progr. Hormone Res. *25:* 447–491 (1969).
30 *Saffran, M.:* Personal communication.
31 *Saffran, M.; Pearlmutter, A.F.; Rapino, E., and Upton, G.V.:* Pressinoic acid: a peptide with potent corticotrophin-releasing activity. Biochem. biophys. Res. Commun. *48:* 748–751 (1972).
32 *Scharrer, E. und Scharrer, B.:* Neurosekretion; in *Möllendorf und Bargmann* Handbuch der mikr. Anat. des Menschen, vol. 6, part V, pp. 953–1066 (Springer, Berlin 1954).
33 *Scott, A.P.; Ratcliffe, J.G.; Rees, L.H.; Landon, J.; Bennett, H.P.J.; Lowry, P.J., and McMartin, C.:* Pituitary peptide. Nature New Biol., Lond. *244:* 65–67 (1973).
34 *Sherwood, L.M.; Rodman, J.J., and Lundberg, W.B.:* Evidence for a precursor to circulating parathyroid hormone. Proc. nat. Acad. Sci., Wash. *67:* 1631–1638 (1970).
35 *Smith, R.E. and Farquhar, M.G.:* Lysosome function in the regulation of the secretory process in cells of the anterior pituitary gland. J. Cell Biol. *31:* 319–347 (1966).
36 *Stachura, M.E. and Frohman, L.A.:* Large growth hormone: evidence for the association of growth hormone with another protein moiety in the rat pituitary. Endocrinology, Springfield *92:* 1708–1713 (1973).
37 *Steiner, D.F.; Cunningham, D.; Spigelman, L., and Alten, B.:* Insulin biosynthesis: evidence for a precursor. Science *157:* 697–700 (1967).
38 *Steiner, D.F. and Oyer, P.E.:* The biosynthesis of insulin and a probable precursor of insulin by a human islet cell adenoma. Proc. nat. Acad. Sci., Wash. *57:* 473–480 (1967).

39 *Tager, H.S. and Steiner, D.F.:* Isolation of a glucagon-containing peptide: primary structure of a possible fragment of proglucagon. Proc. nat. Acad. Sci., Wash. *70:* 2321–2325 (1973).

40 *Tung, A.K. and Zerega, F.:* Biosynthesis of glucagon in isolated pigeon islets. Biochem. biophys. Res. Commun. *48:* 387–395 (1971).

41 *Urry, D.W. and Walter, R.:* Proposed conformation of oxytocin in solution. Proc. nat. Acad. Sci., Wash. *68:* 956–958 (1971).

42 *Venkatachalam, C.M.:* Stereochemical criteria for polypeptides and proteins. V. Conformation of a system of three linked peptide units. Biopolymers *6:* 1425–1436 (1968).

43 *Walter, R.:* Conformations of oxytocin and lysine-vasopressin and their relationships to the biology of neurohypophyseal hormones; in *Margoulies and Greenwood* Structure-activity relationships of protein and polypeptide hormones, part I, pp. 181–193 (Excerpta Medica, Amsterdam 1971).

44 *Walter, R.:* Discussion. Recent. Progr. Hormone Res. *28:* 278–279 (1972).

45 *Walter, R.:* The role of enzymes in the formation and inactivation of peptide hormones; in *Hanson and Jakubke* Peptides 1972, Proc. 12th intern. Peptide Symp., pp. 363–377 (Elsevier, Amsterdam 1973).

46 *Walter, R.; Bernal, I., and Johnson, L.F.:* Has the MSH-release-inhibiting hormone a preferred conformation?; in *Meienhofer* Chemistry and biology of peptides, pp. 131–140 (Ann Arbor Sci. Publ., Ann Arbor 1972).

47 *Walter, R.; Griffiths, E.C., and Hooper, K.C.:* Production of MSH-release-inhibiting hormone by a particulate preparation of hypothalami. Mechanisms of oxytocin inactivation. Brain Res. *60:* 449–457 (1973).

48 *Walter, R. and Marks, N.:* A major species difference in the inactivation of H-pro-leu-gly-NH_2 (MSH-R-IF). Endocrinology, Springfield *92:* 279 A (1973).

49 *Walter, R. and Marks, N.:* In preparation.

50 *Walter, R.; Schwartz, I.L.; Darnell, J.H., and Urry, D.W.:* Relation of the conformation of oxytocin to the biology of neurohypophyseal hormones. Proc. nat. Acad. Sci., Wash. *68:* 1355–1359 (1971).

51 *Wied, D. De; Greven, H.M.; Lande, S., and Witter, A.:* Dissociation of the behavioral and endocrine effects of lysine vasopressin by tryptic digestion. Brit. J. Pharmacol. *45:* 118–122 (1972).

52 *Yalow, R.S. and Berson, S.A.:* Size and charge distinctions between endogenous human plasma gastrin in peripheral blood and heptadecapeptide gastrins. Gastroenterology *58:* 609–615 (1970).

53 *Yalow, R.S. and Berson, S.A.:* Size heterogeneity of immunoreactive human ACTH in plasma and in extract of pituitary glands and ACTH-producing thymoma. Biochem. biophys. Res. Commun. *44:* 439–445 (1971).

54 *Yalow, R.S. and Berson, S.A.:* Characteristics of 'big ACTH' in human plasma and pituitary extracts. J. clin. Endocrin. Metab. *36:* 415–423 (1973).

Author's address: Dr. *R. Walter*, Department of Physiology and Biophysics, Mount Sinai School of Medicine of the City University of New York, *New York, NY 10029* (USA)

Psychoneuroendocrinology. Workshop Conf. Int. Soc. Psychoneuroendocrinology, Mieken 1973, pp. 295–303 (Karger, Basel 1974)

Experimental Electron Microscopic Studies on the Neurosecretory System of Rabbits

Y. Shiotani

Department of Anatomy, Osaka University Medical School, Osaka

Introduction

The regulatory mechanism of hormone secretion in the neurosecretory system is not yet fully understood. In our laboratory, light microscopic studies have been done on the hypothalamus-posterior pituitary system of rabbits under various experimental conditions including hypothalamic stimulation – *Shimazu et al.* (2); *Tanimura et al.* (4); *Shiotani et al.* (3). On the other hand, electron microscopic investigations on the neurosecretory system of rabbits were very scanty as compared with other experimental animals. In the present paper, we would like to describe the ultrastructure of the neurosecretory system of rabbits in normal and some experimental conditions.

Materials and Methods

Adult rabbits of both sexes were used in this study. After the experiments, rabbits were perfused with 4-percent solution of glutaraldehyde in 0.1 *M* phosphate buffer under Nembutal anesthesia. Posterior pituitary glands were removed carefully and cut into small pieces under dissecting microscope. Blocks were post-fixed with 2-percent solution of O_sO_4, dehydrated with graded series of ethanol and embedded in Epon 812. The neurosecretory nucleus of the hypothalamus was also prepared for electron microscopic study. Ultra-thin sections were cut on LKB Ultratome, stained with uranyl acetate and lead hydroxide, and observed with Hitachi electron microscope type 11B and 12.

Results

Ultrastructure of the Posterior Pituitary Gland of Normal Rabbits

In the neurosecretory nerve fibers of the posterior pituitary gland, two kinds of secretory granules were observed (fig. 1). One was a large granule (L) of low electron density, 200–250 mμ in diameter, surrounded by an irregular perigranular membrane, and the other was a small granule (S) of high electron density, 100–120 mμ in diameter. Different kinds of granules were not coexistent in an axon in general. Together with these granules, small vesicles (V) of 30–50 mμ in diameter, similar to so-called synaptic vesicles, were contained in the nerve endings. Axons with small dark granules were less numerous than those with large light granules, and tended to terminate close to the pericapillary space (fig. 2). Axons with large light granules, on the other hand, sometimes made a large swelling filled with many granules (fig. 3), and it was considered to correspond to the Herring's body in the light microscopic study.

The pituicytes of the rabbit posterior pituitary gland were roughly divided into two types (fig. 4). The first type (I) had a light nucleus and rich cytoplasm, while the second type (II) had a dark nucleus and scanty cytoplasm. The first type contained filamentous structures besides the ordinary organelle, and had the complicated contact with neurosecretory axons (fig. 5), occasionally engulfing those axons in it (arrow). Lipid droplets, reported in other species of animals, were scarcely observed in the pituicytes of the rabbit.

Fig. 1. A part of the posterior pituitary gland of a normal male rabbit. L = large light granules; S = small dark granules; V = vesicles. × 10,000.

Fig. 2. Pericapillary region of the posterior pituitary gland of a normal female rabbit. × 10,000.

Fig. 3. So-called Herring's body in the posterior pituitary gland of a normal male rabbit. × 10,000.

Fig. 4. Two types of pituicytes, I and II, in the posterior pituitary gland of a normal male rabbit. × 10,000.

Fig. 5. Neurosecretory axons in (arrow) and around the I type of pituicyte of a normal male rabbit. × 10,000.

Fig. 6. A part of the posterior pituitary gland of a reserpine-treated female rabbit. × 10,000.

Fig. 7. Pericapillary region of the posterior pituitary gland of a 4-day dehydrated male rabbit. Arrows indicate the nerve endings containing vesicles only. × 10,000.

Fig. 8. Flattened sac-like structures in the nerve endings (arrow) in the posterior pituitary gland of a 4-day dehydrated male rabbit. × 10,000.

Fig. 9. Amorphous substances (arrow) in the posterior pituitary gland of a rabbit just after the parturition. × 10,000.

Fig. 10. Pericapillary region of the posterior pituitary gland of a lactating rabbit. × 10,000.

Fig. 11. Amorphous substances (arrow) in the posterior pituitary gland of a lactating rabbit. × 10,000.

Fig. 12. A part of a supraoptic neurosecretory neuron of a normal female rabbit. Ly = lysosome. × 10,000.

Fig. 13. A part of a supraoptic neurosecretory neuron of a 4-day dehydrated female rabbit. G = Golgi apparatus; rER = rough surfaced endoplasmic reticulum. An arrow indicates a granule in the Golgi apparatus. × 10,000.

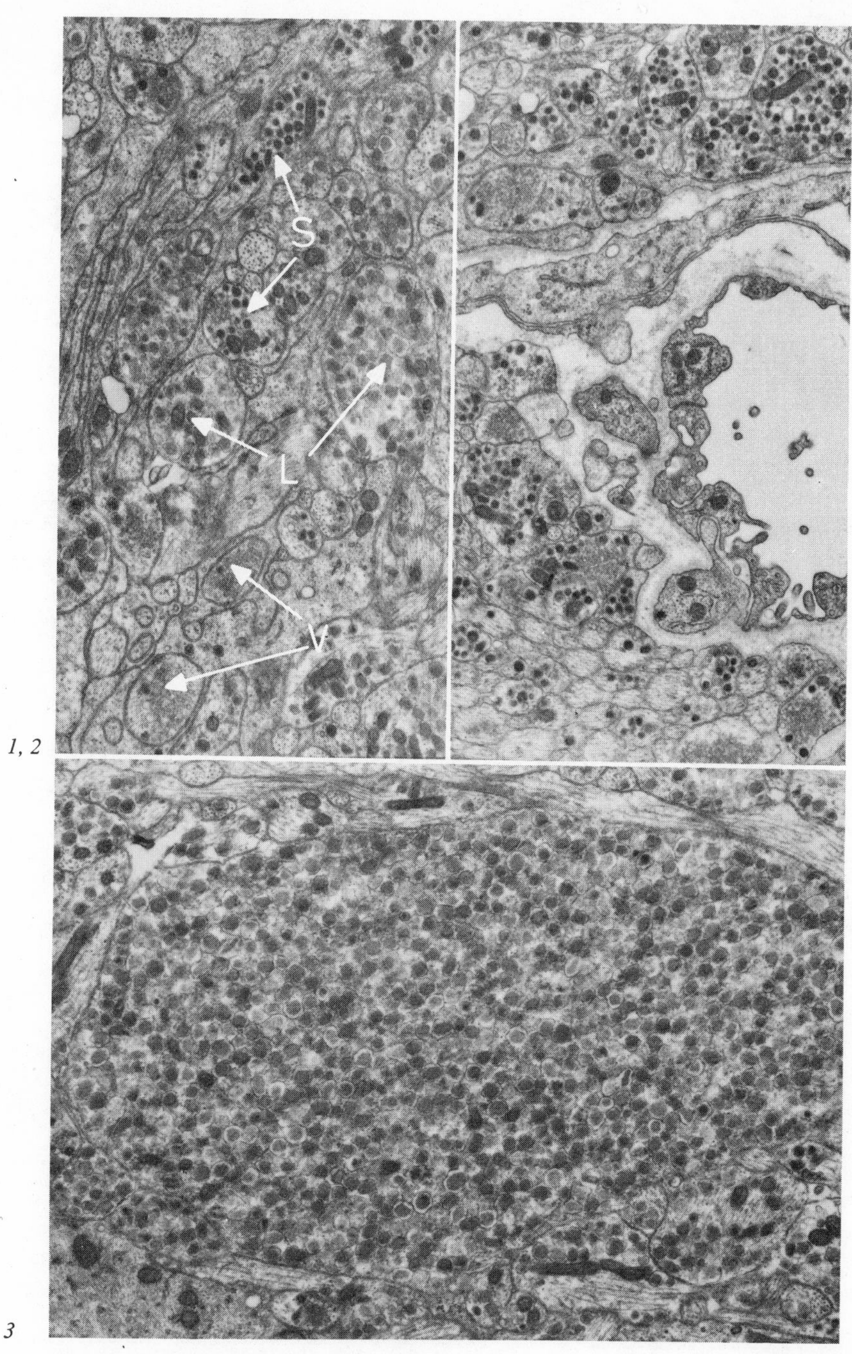

1, 2

3

(For legends see page 296)

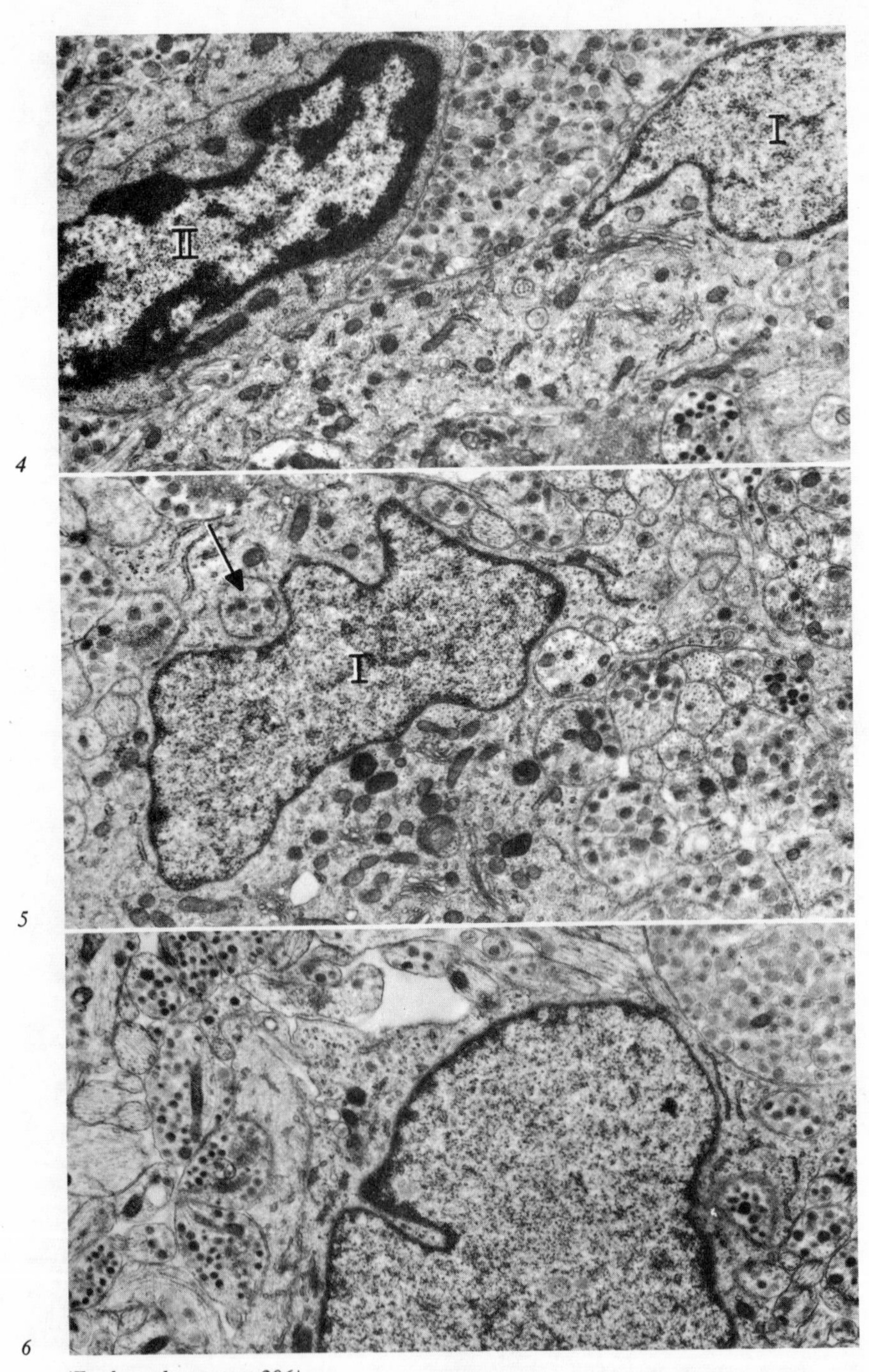

(For legends see page 296)

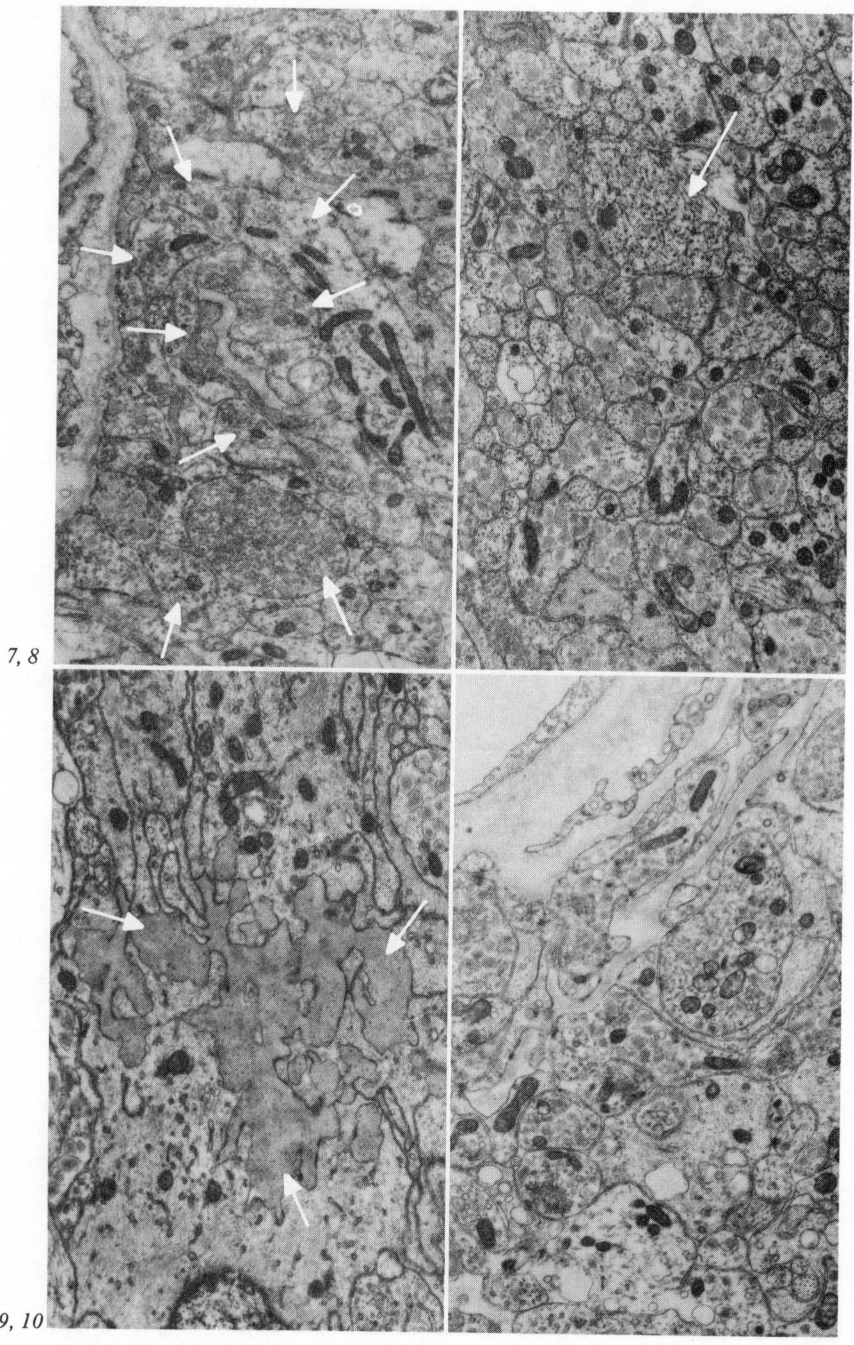

7, 8

9, 10

(For legends see page 296)

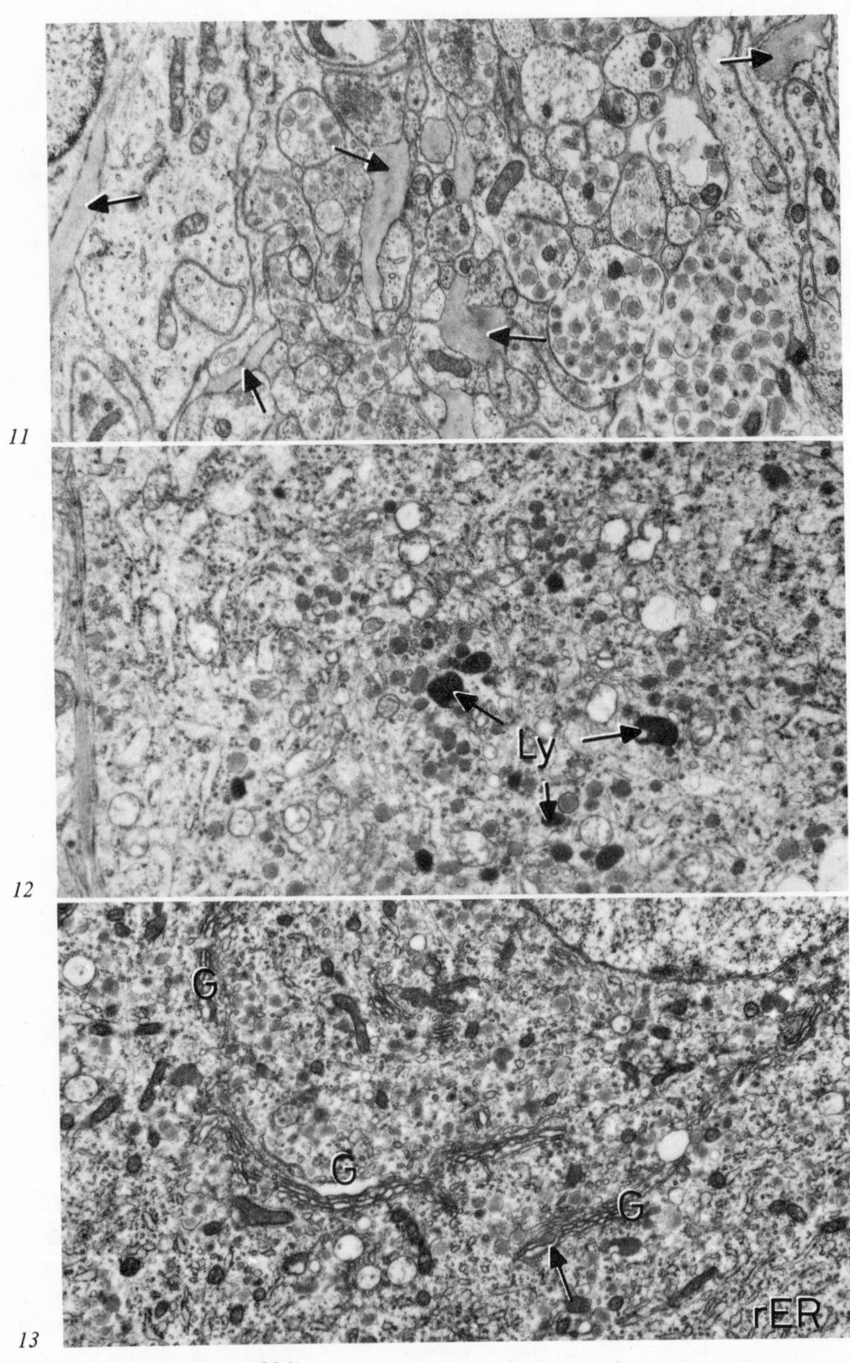

(For legends see page 296)

Ultrastructural Changes of the Posterior Pituitary Gland in Some Experimental Conditions

a) Reserpine Administration

In order to exclude the possibility that small granules might contain catecholamines, reserpine (1 mg/kg) was injected daily to a female rabbit for 3 days. Small granules, however, did not disappear from the posterior pituitary gland at all (fig. 6).

b) Dehydration

When the rabbits were deprived of water for 4 or 5 days, secretory granules in the nerve endings were decreased in number, especially around the pericapillary space (fig. 7), and some nerve endings contained only the vesicles (arrows). Sometimes, flattened sac-like structures, reported by *Kurosumi* (1) in the dehydrated rat, were observed in the nerve endings (fig. 8, arrow). Herring's bodies were, however, still observed at this stage.

c) Parturition and Lactation

In the posterior pituitary gland of a female rabbit just after the parturition, amorphous substances of medium electron density were observed in the dilated intercellular space (fig. 9, arrows). On the second day of lactation, neurosecretory granules were decreased in number, especially in the pericapillary region (fig. 10), and the amorphous substances were still observed in the intercellular spaces (fig. 11, arrows).

Ultrastructure of the Hypothalamic Neurosecretory Neurons

In the perikaryon of the neurosecretory neurons in the supraoptic nucleus, a number of large light granules were observed (fig. 12). Besides the ordinary organelle, lysosomes (Ly) were frequently seen. When the rabbit was deprived of water for 4 or 5 days (fig. 13), the rough-surfaced endoplasmic reticulum (rER) increased in the cytoplasm and the Golgi apparatus (G) developed markedly. Some forming granules were observed within the Golgi apparatus (arrow).

Discussion

As is well known, vasopressin and oxytocin are produced in the hypothalamic neurosecretory neurons and carried to the posterior pituitary gland through their axons. Electron microscopically, these hormones are considered to be contained in the neurosecretory granules, though the differences between vasopressin granules and oxytocin granules are quite obscure. Interestingly, in the rabbit posterior pituitary gland, unlike the other species of animals, two kinds of secretory granules, large light granules and small dark ones, were observed. When the secretion of the posterior pituitary hormone was accelerated, such as in dehydration, parturition or lactation, secretory granules were definitely decreased in number in the posterior pituitary gland, but it was rather difficult to determine which kind of granules were specifically released, because in the nerve endings only the vesicles were observed. However, activation of large granule-containing neurons in the supraoptic nucleus of dehydrated rabbits might suggest the existence of vasopressin in the large granules. Moreover, the posterior pituitary gland of rabbits was reported to contain more vasopressin than oxytocin, and this would be coincident with the majority of the large granules. On the other hand, CRF-activity was repeatedly demonstrated in the extract of the posterior pituitary gland, and those releasing factors also might exist in some granules in the axons. Furthermore, catecholamine nerve terminals were identified in the posterior pituitary gland of some animals, though the small granules in rabbits were found not to be catecholamines in this study. From all the data mentioned above, the following two hypotheses would be possible: (1) large granules contain vasopressin while small granules contain oxytocin; (2) large granules contain both vasopressin and oxytocin while small granules contain releasing factors.

In order to solve these problems, further studies would be necessary, especially on the hypothalamic neurosecretory neurons in various experimental conditions.

Summary

(1) In the neurosecretory nerve endings of the posterior pituitary gland of rabbits, (a) large light granules of 200–250 mμ in diameter, (b) small dark granules of 100–120 mμ and (c) vesicles of 30–50 mμ were observed.

(2) When the secretion of the posterior pituitary hormones was accelerated, those granules were decreased in number.

(3) In the supraoptic neurosecretory cells, large granules were observed. These neurons showed an enhanced secretory activity after the dehydration.

References

1 *Kurosumi, K.:* Ultrastructure of the rat posterior pituitary gland, with special reference to the release mechanism of the neurosecretory substance. Med. J. Osaka Univ. *21:* 53–73 (1971).

2 *Shimazu, K.; Okada, M.; Ban, T., and Kurotsu, T.:* Influence of stimulation of the hypothalamic nuclei upon the neurosecretory system in the hypothalamus and the neurohypophysis of rabbit. Med. J. Osaka Univ. *5:* 701–727 (1954).
3 *Shiotani, Y. and Ban, T.:* Effect of long-term electrical stimulations of the hypothalamus on pituitary-target gland systems in rabbits. Med. J. Osaka Univ. *20:* 119–155 (1969).
4 *Tanimura, H.; Momose, T.; Tsutsui, H., and Ban, T.:* Neurosecretion during pregnancy, parturition and lactation in rabbits. Med. J. Osaka Univ. *11:* 95–121 (1960).

Author's address: Dr. *Y. Shiotani,* Department of Anatomy, Osaka University Medical School, 33, Joancho, *Kitaku, Osaka* (Japan)

Psychoneuroendocrinology. Workshop Conf. Int. Soc. Psychoneuroendocrinology,
Mieken 1973, pp. 304–312 (Karger, Basel 1974)

Mass Fragmentography: A Method to Study Monoamine Neurotransmitters in Brain Nuclei

E. Costa, M. Bjegovic and S.H. Koslow

Laboratory of Preclinical Pharmacology, National Institute of Mental Health,
Saint Elizabeth Hospital, Washington, D.C.

Future neurochemical adventures in neuroendocrinology will require specific and sensitive methods to measure the concentrations of neurotransmitters present in nerve terminals impinging upon specific diencephalic nuclei and to assess the changing rates of their biosynthesis in conjunction with neuroendocrine regulation.

In this report, we will present the principles of the technique called mass fragmentography. This technique appears to be the only suitable way to approach the problems confronting the neuroendocrinologist who wishes to localize and study the neurochemical control of the pituitary function and the regulation of various endocrine tissues.

Technical Considerations

Mass fragmentography takes advantage of the possibility of simultaneously focusing specific ion fragments of the original molecule with the detection system of a mass spectrometer. The positively charged fragments are generated by electron impact under vacuum in the ionization chamber of the mass spectrometer. This chamber receives the effluent from a gas chromatography (GC) apparatus. Schematically, the instrumentation can be represented as follows:

$$A \rightarrow B \rightarrow C \rightarrow D \rightarrow E$$

where A is the GC. This instrument separates the derivatized molecule contained in the original sample from other tissue constituents. This separation depends on the particular partition coefficient of the molecule to be analyzed into the gas liquid phases of the GC column. The eluate from GC is purified from the gas carrier (usually He) in the molecular separator (B) and the molecules remaining in the gas stream pass into the ionization chamber (C). Here the molecules are bombarded by electron impact. At any given condition of vacuum, temperature and intensity of electron beam, each molecule exhibits a constant and characteristic cracking pattern. This is defined by the relative abundance of various ion species identified by their mass to charge (m/e) value. The device used for the separation of the various ion fragments according to their m/e is indicated in the scheme with D. In the scheme E is the detection device which measures the ion density of each ion characterized by its own m/e value. Presently, three separatory devices (Part D of above scheme) can be used: (a) time of flight; (b) quadrupole filter, and (c) magnetic deflection. The time of flight has not been used to any great extent for mass fragmentographic determinations. We have had direct experience with quadrupole and magnetic deflection devices; they are both equally suitable to mass fragmentography.

A. Mass Fragmentography Using Magnetic Deflection Devices (LKB 9000 GC-MS)

This instrument separates the various fragments by accelerating them with a constant potential and deflecting them within a flight tube passing through a magnetic field of variable strength. The modality, whereby various ions with different m/e values are related to the magnetic field strength (H), the accelerating voltage (V) and the curvature radius of the ion beam in the flight tube (r), is given by equation (1).

$$m/e = (H^2 r^2 / 2V) \quad (1)$$

Positive fragments of any m/e value within the mass range specification of the instrument can be focused by appropriately changing the values of H. In practice, to obtain a mass spectrum of a compound, the accelerating voltage is kept constant and H is increased at an exponential rate. A mass spectrum of a compound can be recorded when 10^{-6}–10^{-7} *M* of a compound are available. Equation (1) shows that since r is constant for each instrument, if we keep H and V constant, only a population of fragments characterized by a given m/e value will be detected. By keeping the H constant during the elution of a given compound from the GC the sensitivity of the detection can be increased by 10^7-fold; now only these ions with a given m/e are accelerated by H and they impinge on the ion collector at the end of the flight tube represented in our scheme with E. At this high level of sensitivity, absolute specificity of qualitative identification can be obtained by simultaneously monitoring two or more ion fragments of the compound (multiple ion detection).

Since the ion density ratio between two ion fragments generated by electron bombardment of a molecule is a characteristic and unique property of its mass spectrum, multiple ion detection gives valid structural information. Moreover, the total ionization current generated by a molecule is proportional to the concentration of this molecule in the ionization chamber until saturation is reached. Hence, multiple ion detection can yield quantitative information. In the LKB 9000 an accelerating voltage alternator allows for the simultaneous monitoring of the ion density of two or three positively charged fragments of the parent compound. This device is used in conjunction with a constant magnetic field strength and allows simultaneous monitoring of the ion density generated by each of three ions provided their m/e values are within a certain proportion from each other (10 % in the LKB instrument that we use). This type of analysis is explained with detail by *Hammar et al.* (1969). An actual limitation of the instruments with magnetic deflection is the limited range of m/e values that can be monitored simultaneously.

B. The Quadrupole Filter

This instrument obtains separation of ions according to their m/e by quadrupole mass filtering which uses a combination of DC and radio frequency current. A given species of ions characterized by m/e values is allowed to pass depending upon the DC and radio frequency. This instrument offers the advantage of being able to monitor multiple ion fragments within the total mass range limit of the instrument. Therefore, it has a greater flexibility in mass fragmentographic application than magnetic deflection instruments. We have experience with the Finnigan 3000 GC-MS and found this instrument eminently suitable to simultaneous multiple ion detection of up to eight fragments of the same molecule. This type of flexibility is very useful to perform mass fragmentography of several molecules and the necessary internal standard in one GC-MS run. Moreover, the cost of the MS with a quadrupole filter is inferior to that of instruments with magnetic deflection.

Mass Fragmentography of Putative Transmitters

A. Catecholamines

For exact details of the method, the reader is advised to consult the publication by *Cattabeni et al.* (1972a); *Koslow et al.* (1972); *Costa et al.* (1972), and *Koslow* (1973a). In brief, for GC separation, catecholamines are extracted from tissues (1–10 mg) by homogenization in 0.4 *N* formic acid, 50 m*M* ascorbic acid and centrifugation (10,000 *g* × 10 min at 4 °C). The supernatant is dried under nitrogen and reacted at 60 °C for 30 min with pentafluoropropionic anhydride in the presence of ethyl acetate to form acylated derivatives (PFP) with higher vapor pressure than the parent compounds. Before formation of the acylated derivatives, internal standards are added to the samples. We have used α-methylnorepinephrine (α-MNE) for norepinephrine (NE) and epinephrine (E), and α-methyldopamine (α-MDA) for dopamine (DA). However, these internal standards can be substituted by the analogous deuteriated compounds of DA, NE and E. Such a substitution offers the advantage of using the deuteriated compound as a carrier of the endogenous compound. This allows to perform the analysis below the limits of sensitivity imposed by adsorption of the material to be analyzed on the instrument components. In addition, the reactivity of the deuteriated compound with PFP has very similar rate constants for the forward and backward reaction to the endogenous non-deuteriated analogues.

After the acylated derivatives have been formed the excess of reagent and the by-products are evaporated. The residues are reconstituted with ethyl acetate for injection into the GC. Using a 12-ft glass column (i.d. 2 mm) packed with 3-percent OV-17 Gas Chrom Q and the other instrumental conditions specified in previous publications (*Cattabeni et al.*, 1972a; *Koslow et al.*, 1972), the concentrations of NE and DA have been assayed in cryostat sections of the rat sympathetic ganglion (10 μm thick). The sensitivity of the assay is in the order of 10×10^{-15} mol/sample (*Koslow*, 1973a). It was also possible to assay the NE and DA concentrations in nucleus coeruleus (*Cattabeni et al.*, 1972a).

The specificity of the assay depends on: (1) gas chromatographic retention time; (2) detection of m/e value of the characteristic base peak of the compound; (3) ratio between m/e values of pairs of ion fragments generated from the parent compound by electron impact. For details of the characteristic of the mass spectra of the compounds and the m/e values selected for mass fragmentography the reader should consult the original papers of *Cattabeni et al.* (1972a); *Costa et al.* (1972); *Koslow et al.* (1972) and *Koslow* (1973a).

B. Pentylethanolamine

Drs. *Willner and LeFevre* from our laboratory have identified and measured this compound in the hypothalamus of the rat. Its location in various brain nuclei is presently under investigation. The technique for this assay involves the use as internal standard of the deuteriated derivative of phenylethanolamine and the derivation of the compounds with PFP. Independently, phenylethanolamine has been found in rat brain also by *Saavedra and Axelrod* (1973).

C. Indolealkylamines

The compounds measured in the hypothalamus of the rat and in the pineal gland were serotonin (S), *N*-acetylserotonin (NAS), 5-methoxytryptamine (5MT) and melatonin (M). The procedures for the analysis of these compounds were reported previously (*Cattabeni et al.*, 1972b; *Koslow and Green,* 1973; *Koslow et al.*, 1973; *Koslow,* 1973b). These amines are extracted by homogenizing the tissue with 0.1 *M* $ZnSO_4$ followed by neutralization with 0.1 *M* $Ba(OH)_2$. An aliquot of the supernatant is dried in the presence of α-methylserotonin and *N*-acetyltryptamine used as internal standards. The residue is reacted for 3 h at 60 °C with PFP in the presence of ethyl acetate. The reaction mixture is dried, and the residue reconstituted with ethyl acetate is injected in the GC-MS.

Analysis of rat hypothalamic extracts verified the presence of S, M and 5MT in this tissue. The concentration of the three amines found is reported in table I (*Green et al.*, 1973).

Since it was believed that the compounds reported in table I were localized exclusively in pineal where, in fact, we have detected them (*Cattabeni et al.*, 1972b; *Koslow et al.*, 1973), we have reasoned that their presence in the hypothalamus could reflect an uptake of circulating indoles by this tissue. We have, therefore, performed analysis in pinealectomized rats and found that the concentration of these compounds is identical to that of normal rats (*Green et al.*, 1973; *Koslow,* 1974). Both the origin and the functional significance of M and 5MT in the hypothalamus remain to be established. The microspectrofluorometric studies of *Bjorklund et al.* (1970, 1971a and b) could suggest that 5MT may be another putative transmitter. Recently, *Banerjee and Snyder* (1973) have shown that using the methylation cofactor, 5-methyltetrahydrofolic acid (MTHF) proposed by *Laduron* (1972) and brain extract as a source of the enzyme S is predominantly methylated on the 5-hydroxyl group to form 5-methoxytryptamine. If it were to be confirmed that *o*-methylation of S by brain homogenates requires MTHF, then it could be assumed that the 5MT we

Table I. Indolylalkylamine in the rat hypothalamus

	*n*mol/g[1]
Serotonin	4.00 ± 0.23
N-acetylserotonin	not detectable
5-Methoxyserotonin	0.62 ± 0.02
Melatonin	1.50 ± 0.16

1 Each value is the mean ± SE of eight assays.

detected in brain could be either an S metabolite in a selective neuronal system or a new transmitter.

Comparing Methods to Detect Putative Neurotransmitters

A. Catecholamines

Various enzymatic methods have been proposed to measure tissue levels of catecholamines (*Coyle and Henry,* 1973). These methods take advantage of two enzymes: catechol-*o*-methyltransferase (COMT), and phenylethanolamine-*N*-methyltransferase (PNMT), which catalyze the transfer of a labeled methyl group into either amino or catechol group of the catecholamine. The sensitivity of these methods is very high (10^{-13} *M*) but the specificity depends always on partition coefficients of the radioactive amine formed between two solvent systems. Presently, when analyzing brain nuclei for catecholamines, we are called to differentiate in our assays, NE, E, DA and epinine (*Laduron,* 1972). It is impossible at the high level of sensitivity of the enzymatic method, to differentiate by radiometric assay E from NE and DA from epinine. Moreover, some cross contamination (about 10–20 %) is unavoidable. These considerations offer the mass fragmentography as the only way to analyze with no cross contamination and high sensitivity (10^{-14}–10^{-15} *M*) the four catecholamines mentioned above in the same sample. We have, in fact, found that by substituting the OV-17 column with SE-54 as suggested by *Karoum et al.* (1972), it is possible to resolve DA from epinine. In fact, such resolution is essential because epinine and DA have very similar mass spectra. We are presently attempting to detect epinine in rat tissues to confirm *Laduron*'s proposal (1972) that epinine is the normal

Table II. Increase of epinephrine elicited by dexamethasone in superior cervical ganglia of newborn rats[1]

	Superior cervical ganglion		Caudate nucleus	
	control	dexa-methasone	control	dexa-methasone
Epinephrine	4.4 ± 0.62	750 ± 125[2]	–	–
Norepinephrine	610.0 ± 74	930 ± 63[2]	15 ± 1.7	28 ± 8
Dopamine	71.0 ± 1.5	140 ± 17[2]	264 ± 26	239 ± 53

1 Values are the mean ± SE of three determinations expressed as *p*mol/mg protein.
2 $p < 0.05$. Newborn rats were treated for days 1–8 with dexamethasone sodium phosphate 0.1 μg/g, subcutaneously, and the above tissues analyzed on day 9.

precursor for E. Our attempts have been unsuccessful, but we have still to try some new ideas. We are confident that mass fragmentography is the only possible solution of this important problem in catecholamine research.

Recently, *Ciaranello et al.* (1973), by injecting dexamethasone to neonate rats have reported the appearance of PNMT activity in sympathetic ganglia. They attributed the presence of this enzyme in dexamethasone-treated rats to a hormonal stimulation of the enzyme present in the small intensely fluorescent cells of the sympathetic ganglia. We have found (table II) that as a result of dexamethasone injections, the concentration of E in sympathetic ganglia increases greatly. This increase has a time constant characteristic similar to the PNMT induction. Since E fails to increase in caudate and other peripheral tissue innervated by noradrenergic neurons, our results support the proposal that PNMT is localized in SIF cells. Moreover, with our experiments we have shown that the increase of PNMT activity has functional significance. These studies could not have been carried out with the enzymatic method proposed by *Coyle and Henry* (1973) for it does not differentiate E from NE.

B. Indolealkylamines

Using the most sensitive fluorescence methods to detect and measure indolealkylamines one has to cope with the low specificity of the fluorophore. In fact, the reagent used to form the fluorophore is *o*-phthalylaldehyde (OTP). This reagent yields fluorophores with identical spectral characteristics when it reacts with S, M, and 5MT and various other endogenous indoles. Moreover, the fluorescence yield of these three compounds per mole of reactant differs by several orders of magnitude. This different amount of fluorescence generated, amplifies the error due to possible cross contamination between 5MT and S deriving from organic extraction purification. With the radioimmunoassay procedures (*Pesker and Spector,* 1973) the specificity does not improve because 5MT is as active as S in inhibiting the ^{3}H-S binding to the specific (?) antibody.

Conclusions

Mass fragmentography is the only method for assaying putative monoamine transmitters and their metabolites in small brain nuclei as required for the progress of our present neurochemical knowledge that applies to neuroendocrine regulation. We have shown that this method is 10–100-fold more sensitive than any other method to measure monoamines including radioimmunoassay and radioenzymatic assays. In addition, mass fragmentography allows multiple amine measurement at very high sensitivity, avoiding the problem of cross contamination which plagues radioimmunoassay and radioenzymatic assays. Finally, mass fragmentography is the only method that protects the investigators from potential errors deriving from the presence in tissue of unsuspected chemical analogues of the putative neurotransmitter under study.

References

Banerjee, S.P. and Snyder, S.H.: Methyltetrafolic acid mediates *N*- and *O*-methylation of biogenic amines. Science *182:* 74–75 (1973).

Bjorklund, A.; Falck, B., and Stenevi, V.: On the possible existence of a new intraneuronal monoamine in the spinal cord of the rat. J. Pharmacol. exp. Ther. *175:* 525–532 (1970).

Bjorklund, A.; Falck, B., and Stenevi, V.: Classification of monoamine neurons in the rat mesencephalon. Distribution of a new monoamine neuronal system. Brain Res. *32:* 269–285 (1971a).

Bjorklund, A.; Falck, B., and Stenevi, V.: Microspectrofluorometric characterization of monoamines in the central nervous system. Evidence for a new neuronal monoamine-like compound; in Histochemistry of nervous transmission. Progr. Brain Res., vol. 34, pp. 63–73 (Elsevier, Amsterdam 1971b).

Costa, E.; Green, A.R.; Koslow, S.H.; LeFevre, H.F.; Revuelta, A., and Wang, C.: Dopamine and norepinephrine in noradrenergic axons. A study *in vivo* of their precursor product relationship by mass fragmentography and radiochemistry. Pharmacol. Rev. *24:* 167–190 (1972).

Cattabeni, F.; Koslow, S.H., and Costa, E.: Gas chromatography–mass fragmentography: a new approach to the estimation of amines and amine turnover; in Advances in biochemical psychopharmacology, vol. 6, pp. 37–59 (Raven Press, New York 1972a).

Cattabeni, F.; Koslow, S.H., and Costa, E.: Gas chromatographic–mass spectrometric assay of four indolealkylamines of rat pineal. Science *178:* 166–168 (1972b).

Ciaranello, R.D.; Jacobowitz, D., and Axelrod, J.: Effect of dexamethasone on phenylethanolamine *N*-methyltransferase in chromaffin tissue of the neonatal rat. J. Neurochem. *20:* 799–805 (1973).

Coyle, J.T. and Henry, D.: Catecholamines in fetal and newborn rat brain. J. Neurochem. *21:* 61–67 (1973).

Green, A.R.; Koslow, S.H., and Costa, E.: Identification and quantification of a new indolealkylamine in rat hypothalamus. Brain Res. *51:* 371–374 (1973).

Hammar, G.C.; Holmstedt, B.; Lindgren, J.-E., and Tham, R.: The combination of gas chromatography and mass spectrometry in the identification of drugs and metabolites. Adv. Pharmacol. *7:* 53–89 (1969).

Karoum, F.; Cattabeni, F., and Costa, E.: Gas chromatographic assay of picomole concentrations of biogenic amines. Analyt. Biochem. *47:* 550–561 (1972).

Koslow, S.H.: Application of mass fragmentography to the quantitation of endogenous catecholamines; in Frontiers in catecholamine research (Pergamon Press, London 1973a) (in press).

Koslow, S.H.: 5-Methoxytryptamine: a possible central nervous system transmitter; in Advances in biochemical psychopharmacology, vol. 11 (in press, 1974).

Koslow, S.H. and Green, A.R.: Analysis of pineal and brain indole alkylamines by gas chromatography–mass spectrometry; in Advances in biochemical psychopharmacology, vol. 7, pp. 33–43 (Raven Press, New York 1973).

Koslow, S.H.; Cattabeni, F., and Costa, E.: Quantitative mass fragmentography of some indolealkylamines of the rat pineal gland; in Biochemistry and physiology of the pineal gland (Spectrum, New York 1973) (in press).
Koslow, S.H.; Cattabeni, F., and Costa, E.: Norepinephrine and dopamine assay by mass fragmentography in the picomole range. Science *176:* 177–180 (1972).
Laduron, P.: *N*-methylation of dopamine to epinine in brain tissue using *N*-methyltetrahydrofolic acid as the methyl donor. Nature New Biol., Lond. *238:* 212–213 (1972).
Pesker, B. and Spector, S.: Serotonin. Radioimmunoassay. Science *179:* 1340–1341 (1973).
Saavedra, J.M. and Axelrod, J.: Demonstration and distribution of phenylethanolamine in brain and other tissues. Proc. nat. Acad. Sci., Wash. *70:* 769–772 (1973).

Authors' address: Dr. *E. Costa,* Dr. *M. Bjegovic* and Dr. *S.H. Koslow,* Laboratory of Preclinical Pharmacology, National Institute of Mental Health, Saint Elizabeth Hospital, *Washington, DC 20032* (USA)